D0848156

TOTALITY

The Great American Eclipses of 2017 and 2024

TOTALITY

The Great American Eclipses of 2017 and 2024

Mark Littmann and Fred Espenak

OXFORD
UNIVERSITY PRESS

Great Clarendon Street, Oxford, OX2 6DP,
United Kingdom

Oxford University Press is a department of the University of Oxford.
It furthers the University's objective of excellence in research, scholarship,
and education by publishing worldwide. Oxford is a registered trade mark of
Oxford University Press in the UK and in certain other countries

First Edition published in 2017

Impression: 1

Published in the United States of America by Oxford University Press
198 Madison Avenue, New York, NY 10016, United States of America

British Library Cataloguing in Publication Data
Data available

Library of Congress Control Number: 2016947984

ISBN 978–0–19–879569–8

Printed and bound by
CPI Litho (UK) Ltd, Croydon, CR0 4YY

To Peggy, to our children Beth and Owen, and to our grandchildren Liam and Adele, with love.

Mark Littmann

To my wife Pat, and granddaughters Valerie and Maggie who will see their first total eclipse in 2017.

Fred Espenak

Contents

Foreword

I have never seen a total solar eclipse.

That is a peculiar confession for an astronomer, especially one who prides himself on observing the sky both dark and sunlit. I've seen rocket launches and space debris re-entering, stars exploding, galaxies cosmically proximate and distant, planets in alignment, and meteors by the thousands.

So how can it be that in my life I've never witnessed this one phenomenon, universally (if I may be so bold) acknowledged as the most glorious sight the heavens provide?

Well, y'know, *life*. Timing, schedules, finances, weather—these mundanities can interfere. But to be honest, none of these has been the issue personally. The real root of my lack of experience is *circumstance*: Eclipses are particular in where they occur. For people of Europe, the great eclipse of 1999 was a godsend, but not so much for a newly minted American father who couldn't possibly travel that far to see it.

South America in 1994, Africa in 2001, Asia in 2009 . . . all these were too distant or too logistically difficult for me to attend.

I'm sure penguins enjoyed the 2003 eclipse in Antarctica. Me, and the rest of us, not so much.

But all this is about to change. In August of 2017 the shadow of the Moon will sweep across my home country, and will pass just a few hundred kilometers north of my home, an easy day's drive. When the Great American Eclipse occurs, I will be there, to bask in over two minutes of non-sunlight.

And when it happens I will be prepared, because I have read the book you now hold in your hands. After you read it, you will be too. The information you're about to receive is both critically important to prepare for the eclipse as it is delightful to read.

Let me be clear: This book will prepare you *logistically* for the eclipse. You will know where to be, when to be there, where to look, what to look for, how to look, and why it's all happening.

But the book also does a good job preparing you *emotionally* too. The stories by people who have witnessed eclipses for themselves, especially first-timers, are moving (I particularly liked the "Moments of Totality" chapter interstitials), and the photographs will give you a taste of what it will be like. Of course, no number of words, no matter how exquisitely crafted, will have you understand what it is like to witness such magnificence for yourself, to have your soul touched by the experience.

Or so I've been told. I'll find out in August, just as I hope you will.

After reading the book, I thought carefully about what that day, that hour, those just-more-than two minutes will be like. And I have decided not to take any pictures of the eclipse, at least not directly. I may take a few photos of the people around me staring in awe, or the shadows of the eclipse on the ground should circumstances fall that way, or maybe even set up a telescope and camera to shoot pictures automatically.

But when the moment comes, I want to stand there, with nothing between me and the Sun but a thin slip of atmosphere and the depths of space as the Moon joins us in alignment. I don't want to worry about terrestrial mundanities, I just want to enjoy the show brought to us care of celestial mechanics.

. . . at least, for a while. Note this: Due to the complex and subtle dance of gravity, the Moon is slowly receding from the Earth, drawing away from us at a rate of about 4 centimeters per year. In the past it was much closer, and in the future it will be more distant.

This means that in the remote past it appeared much larger in the sky, and would have blocked the Sun to a much greater degree. Back then, hundreds of millions of years ago, the glory of the Sun's atmospheric corona would scarcely have been visible; the Moon's larger size would have blocked most of it.

Conversely, in the very distant future the Moon will appear *smaller* in the sky, and won't be able to completely cover the surface of the Sun. In some far-flung day, hundreds of millions of years from now, there will be a very last total solar eclipse, and then every one after that will be annular, partial.

We live in the middle of a billion-year-long eclipse season. You'd better catch them while they last. Your great-great-great^{nth} descendants won't have this luxury.

But that is a concern for them, not us. For now, I'm content to look forward to August 21, 2017, to my very first total solar eclipse. And when that's over?

For those of us in the States (and Mexico and Canada!), there's always April 8, 2024.

Phil Plait, Ph.D.
Astronomer, author, and host of
Phil Plait's Bad Universe on the Discovery Channel
and
Crash Course Astronomy on YouTube

Acknowledgements

Enormous thanks to astronomers Charles Lindsey, Jay Pasachoff, and Larry Marschall for their generous advice and help through this and our earlier eclipse books. Thanks too to astronomers Joseph Hollweg and Alan Clark for sharing their expertise with us.

Totality: The Great American Eclipses of 2017 and 2024 is a richer work because noted authorities contributed vignettes (sidebars) for the book. We are very grateful to them:

Lucian V. Del Priore, M.D.
Stephen J. Edberg
Patricia Totten Espenak
Carl Littmann
Laurence A. Marschall
Luca Quaglia
Ken Willcox

There are 220 illustrations in this book. The magnificent eclipse photographs were graciously contributed by:

Greg Babcock
Fred Bruenjes
Ben Cooper
Terry Cuttle
Arne Danielsen
Arne Danielsen
Kris Delcourte
Miloslav Druckmüller
Alan Dyer
Patricia Totten Espenak
Charles Fulco
Blanchard Guillaume
Stephane Guisard
Stephan Heinsius
Stan Honda
Johnny Horne
Dave Kodama
Chris Kotsiopoulos
Thierry Legault
Chin Wei Loon
Sara Marwick
Stephen Mudge
Jay Pasachoff
Janne Pyykkö
Glenn Schneider
György Soponyai
Babak Tafresh
Tunç Tezel
Alex Tudorica
Jaime Manuel Pombo Vilinga
Alson Wong

The handsome and informative maps of the 2017 and 2024 eclipse paths are the work of cartographer Michael Zeiler. He created them for this book and we admire his achievement.

The excellent diagrams are primarily the work of solar astronomer Charles A. Lindsey, Will Fontanez and Tom Wallin of Cartographic Services at the University of Tennessee, and Fred Espenak. We are very grateful to them.

A number of eclipse veterans—both professional and amateur astronomers—graciously offered their experiences and advice for our chapters on Observing a Total Eclipse and The Strange Behavior of Man and Beast—Modern Times. We thank them for their insight and eloquence:

Jay Anderson
Paul and Julie (O'Neil) Andrews
Dave Balch
John Beattie
Richard Berry
Satyendra Bhandari
P. M. M. (Ellen) Bruijns
Joe Buchman
Kristian Buchman
Dennis di Cicco
Stephen J. Edberg
Alan Fiala
George Fleenor
Ruth S. Freitag
Thomas Hockey
Joseph V. Hollweg
Xavier Jubier
Jean Marc Larivière
Dawn Levy
Charles Lindsey
George Lovi
David Makepeace
Larry Marschall
Sarah Marwick
Frank Orrall
Jay M. Pasachoff
Patrick Poitevin
Luca Quaglia
Joe Rao
Leif J. Robinson
Michael Rogers
Gary and Barbara (Schleck) Ropski
Walter Roth
Glenn Schneider
Mike Simmons
Gary Spears
Roger W. Tuthill
Ken Willcox
Sheridan Williams
Michael Zeiler
Jack B. Zirker
Evan Zucker

Special additional thanks to John R. Beattie, who provided such a ringing moment-by-moment description of a total eclipse that it became the nucleus of our first chapter.

There are a host of other folks who helped us greatly in so many ways—astronomy, geology, archeoastronomy, history, mythology, translating, critiquing—in the course of earlier eclipse books and *Totality: The Great American Eclipses of 2017 and 2024*. Thanks so much to:

Julie Andsager
Anthony F. Aveni
Eric Becklin
Kenneth Brecher
John Carper
B. Ralph Chou
Kevin Dieke
Todd Dupuis
Patricia Totten Espenak
Andrew Fraknoi
Geoffrey K. Gay
Gerry Grimm

Robert S. Harrington
Karen Harvey
Don Hassler
Anne Hensley
Kevin Krisciunas
Carl Littmann
David and Esther Littmann
Jane Littmann
Owen and Machiko Littmann
Peggy Littmann
Eli Maor
Richard E. McCarron
Beth and James McGinnis
John McNair
Larry November
Bea and Tom Owens
David Pankenier
Ann and Paul Rappoport
Michael Rogers
Gary Rottman
Glenn Schneider
Sabatino Sofia
E. Myles Standish, Jr.
John Steele

Thanks to Phil Plait for providing the foreword to our book – and for his longstanding and ongoing distinguished service to public understanding of science.

We are grateful to Simon Mitton, Life Fellow, St. Edmund's College, University of Cambridge; Stuart Ryder, Head of International Telescopes Support, Australian Astronomical Observatory; and Mike Reynolds, Professor of Astronomy, Florida State College at Jacksonville, and Executive Director of the Association of Lunar and Planetary Observers, who read portions of this manuscript as Oxford University Press referees. They provided us with very helpful advice and encouragement.

Thanks to University of Tennessee science writing graduate students Christopher Samoray and Sean Simoneau who read drafts of the entire manuscript and offered wise suggestions.

The Chair of Excellence in Science Writing Endowment at the University of Tennessee, a gift of Tom Hill and Mary Frances Hill Holton, other benefactors, and the State of Tennessee, provided funds to help with expenses for this book, for which we are truly grateful.

Our deep gratitude to Dr. Sonke Adlung, Oxford University Press' Senior Editor, Physical Sciences, for his interest in and encouragement for *Totality: The Great American Eclipses of 2017 and 2024*. We greatly appreciate also Harriet Konishi, Assistant Commissioning Editor for Physical Sciences at Oxford University Press, for ever so graciously and effectively guiding this project.

Our thanks too to the Oxford University Press production group: Production Editor Shereen Karmali, Project Manager Marie Felina Francois and her team at Integra-Pondicherry, India, Copy Editor Kerry Boettcher, Marketing Director Sarah Russo, and all the others whose names we do not know.

To everyone who helped with this book, our profound thanks and our hope that you will enjoy many total eclipses of the Sun.

1

The Experience of Totality

In rating natural wonders, on a scale of 1 to 10, a total eclipse of the Sun is a million.

An observer who has seen 27 total eclipses[1]

First contact. A tiny nick appears on the western side of the Sun.[2] The eye detects no difference in the amount of sunlight. Nothing but that nick portends anything out of the ordinary. But as the nick becomes a gouge in the face of the Sun, a sense of anticipation begins. This will be no ordinary day.

Still, things proceed leisurely for the first half hour or so, until the Sun is more than half covered. Now, gradually at first, then faster and faster, extraordinary things begin to happen. The sky is still bright, but the blue is a little duller. On the ground around you the light is beginning to dim. Over the next 10 to 15 minutes, the landscape takes on a steely gray metallic cast.

As the minutes pass, the pace quickens. With about a quarter of an hour left until totality, the western sky is now darker than the east, regardless of where the Sun is in the sky. The shadow of the Moon is approaching. Even if you have never seen a total eclipse of the Sun before, you know that something amazing is going to happen, is happening now—and that it is beyond normal human experience.

Less than 15 minutes until totality. The Sun, a narrowing crescent, is still fiercely bright, but the blueness of the sky has deepened into blue-gray or violet. The darkness of the sky begins to close in around the Sun. The Sun does not fill the heavens with brightness anymore.

Five minutes to totality. The darkness in the west is very noticeable and gathering strength—a dark, amorphous form rising upward and spreading out along the western horizon. It builds like a massive storm, but in utter silence, with no rumble of distant thunder. And now the darkness begins to float up above the horizon, revealing a yellow or orange twilight beneath. You are already seeing through the Moon's narrow shadow to the resurgent sunlight beyond.

The acceleration of events intensifies. The crescent Sun is now a blazing white sliver, like a welder's torch. The darkening sky continues to close in around the Sun, faster, engulfing it.

Partial phases of the total solar eclipse of March 29, 2006, from Jalu, Libya. [Nikon D200 DSLR, Sigma 170–500 mm at 500 mm, f/11, 1/500 s, ISO 200, Thousand Oaks Type 3 solar filter. ©2006 Patricia Totten Espenak]

Minutes have become seconds. A ghostly round silhouette looms into view. It is the dark limb of the Moon, framed by a white opalescent glow that creates a halo around the darkened Sun. The corona, the most striking and unexpected of all the features of a total eclipse, is emerging. At one edge of the Moon the brilliant solar crescent remains. Together they appear as a celestial diamond ring.

Suddenly, the ends of the bare sliver of the Sun break into individual points of intense white light—Baily's beads—the last rays of sunlight passing through the deepest lunar valleys. The beads flicker, each lasting but an instant and vanishing as new ones form. And now there is just one left. It glows for a moment, then fades as if it were sucked into an abyss.

Totality.

Where the Sun once stood, there is a black disk in the sky, outlined by the soft, pearly white glow of the corona, about the brightness of a full moon. Small but vibrant reddish features stand at the eastern rim of the Moon's disk, contrasting vividly with the white of the corona and the black where the Sun is hidden. These are the prominences, giant clouds of hot gas in the Sun's lower atmosphere. They are always a surprise, each unique in shape and size, different yesterday and tomorrow from what they are at this special moment.

You are standing in the shadow of the Moon.

It is dark enough to see Venus and Mercury and whichever of the brightest planets and stars happen to be close to the Sun's position and above the horizon. But it is not the dark of night. Looking across the landscape at the horizon in all directions, you see beyond the shadow to where the eclipse is not total, an eerie twilight of orange and yellow. From this light beyond the darkness that envelops you comes an inexorable sense that time is limited.

Now, at the midpoint in totality, the corona stands out most clearly, its shape and extent never quite the same from one eclipse to another. And only the eye can do the corona justice, its special pattern of faint wisps and spikes on this day never seen before and never to be seen again.

Diamond ring effect at the total solar eclipse of July 11, 2010, from Easter Island. [Nikon D700, Borg 100ED & 2x teleconverter, fl = 1280 mm, ISO 800, f/12.8 at 1/4000 second. ©2010 Dave Kodama]

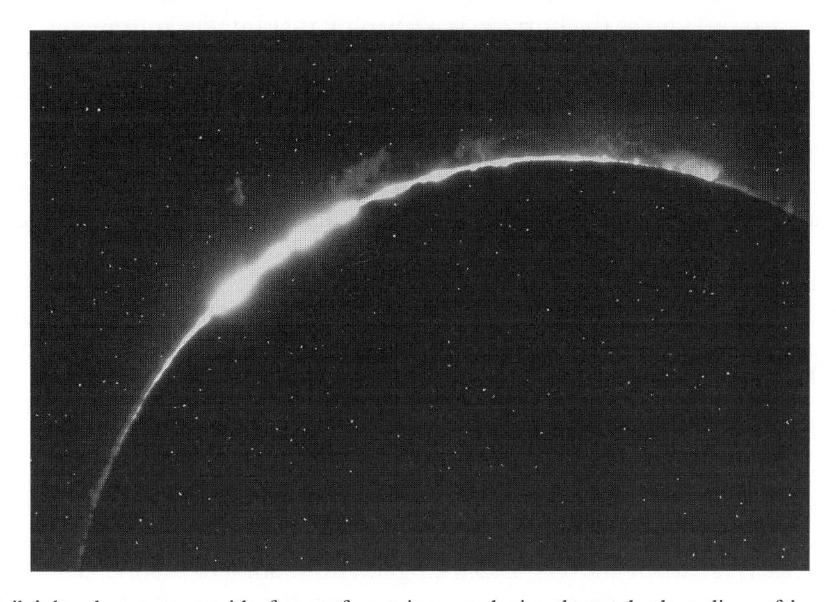

Baily's beads are seen amid a forest of prominences during the total solar eclipse of August 11, 1999, from Lake Hazar, Turkey. [Pentax SLR, 94 mm Brandon refractor, f/30, 1/125 s, Ektachrome V100 film pushed to ISO 200. ©1999 Greg Babcock]

The outer corona is revealed during the total eclipse of August 1, 2008, from Novosibirsk, Russia. [Canon 450D, 300 mm, ISO 100, f/8, 1/2 second. ©2008 Arne Danielsen]

Yet around you at the horizon is a warning that totality is drawing to an end. The west is brightening while in the east the darkness is deepening and descending toward the horizon. Above you, prominences appear at the western edge of the Moon. The edge brightens.

Suddenly totality is over. A point of sunlight appears. Quickly it is joined by several more jewels, which merge into a sliver of the crescent Sun once more. The dark shadow of the Moon silently slips past you and rushes off toward the east.

It is then you ask, "When is the next one?"[3]

NOTES AND REFERENCES

1. Epigraph: Fred Espenak, October 24, 2014.
2. In sky observations, the western side of the Sun or Moon refers to the edge of the Sun or Moon closer to the western horizon. For observers in mid-northern latitudes, the Sun is usually to the south. When facing south, east is to the left and west is to the right. This south-looking orientation can briefly confuse readers who are used to maps that are oriented north, so that east is to the right and west to the left.
3. Special thanks to John Beattie of New York City, upon whose experience and description this chapter is based.

○◑●◐

A MOMENT OF TOTALITY

Reaction to Totality

Eclipse veteran Sheridan Williams says that at the end of totality, the usual reaction of a first-time eclipse observer is "It was so short," "That was the most amazing thing I've ever seen," "I never realized it could be so beautiful," and "When is the next one?" Often they have tears in their eyes. There are not enough superlatives.

"After seeing a total eclipse," Sheridan says, "I have never, never heard anyone say, 'I don't see what all the fuss was about' or 'Why bother to see another one?'"[1]

[1] Sheridan Williams is a British rocket scientist (retired) and a Fellow of the Royal Astronomical Society.

2

The Great Celestial Cover-Up

If God had consulted me before embarking upon creation, I would have recommended something simpler.

Alfonso X, King of Castile (1252)[1]

A total eclipse of the Sun is exciting and even profoundly moving.

But what causes a total solar eclipse? The Moon blocks the Sun from view. And that is all you absolutely need to know to enjoy a solar eclipse. So you can now skip to the next chapter.

If however you are reading this paragraph, you are right: there is more to tell—about dark shadows and oblong orbits and tilts and danger zones and amazing coincidences. Yet before you venture further, promise yourself one thing. If for any reason your eyes begin to glaze over, stop reading this chapter immediately and go right on to the next. You must not let celestial mechanics, or this explanation of it, stand in the way of your enjoyment of the wild, wacky, and wonderful things people have thought and done about solar eclipses.

Moon Plucking

How big is the Moon in the sky? What is its angular size?

Extend your arm upward and as far from your body as possible. Using your index finger and thumb, imagine that you are trying to pluck the Moon out of the sky ever so carefully, squeezing down until you are just barely touching the top and bottom of the Moon, trapping it between your fingers. How big is it? The size of a grape? A plum? An orange?

It is the size of a pea. (You can win bets at cocktail parties with this question.) The Moon has an angular size of only half a degree.

Now, how large is the Sun in the sky? Your friends will almost all immediately guess that it is bigger. Before they damage their eyes by trying the Moon pinch on the Sun, just remind them that a total eclipse is caused by the Moon completely covering the Sun, so the Sun must appear no bigger

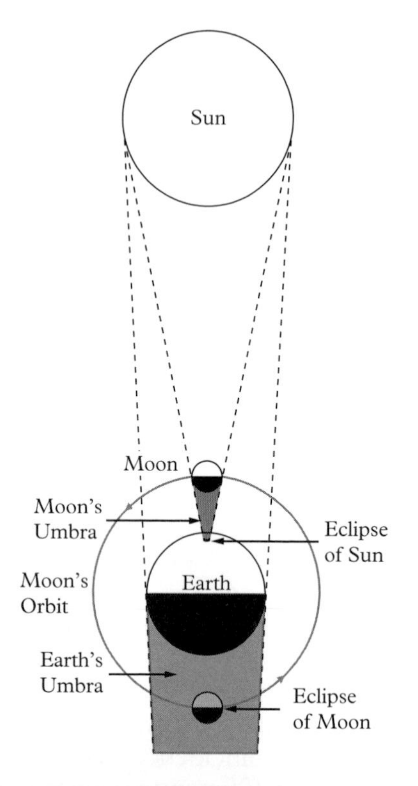

A total solar eclipse occurs when the Moon's umbra touches the Earth. A lunar eclipse occurs when the Moon passes into the Earth's shadow (umbra). The relative sizes and distances of the Sun, Moon, and Earth in this diagram are not to scale.

The solar corona during the total solar eclipse of June 21, 2001, from Zambia. [35 mm SLR, Vixen 102-ED/Great Polaris mount, f/6.4, 11 exposures: 1/1000 s to 1 s, Ektachrome 200 film, images combined in Photoshop. ©2001 Alson Wong]

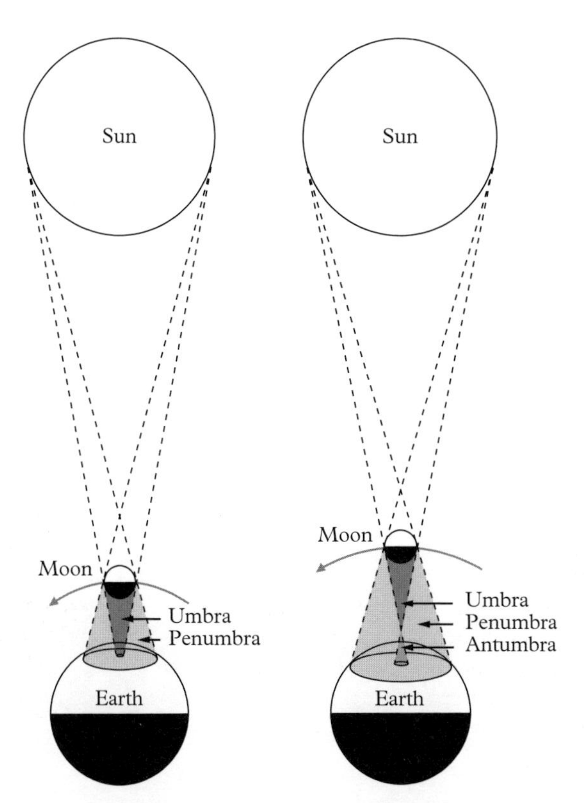

Configuration of a total solar eclipse *(left)* and an annular eclipse *(right)*. Within the Moon's umbra (dark converging cone), the entire surface of the Sun is blocked from view. In the penumbra (lighter diverging cone), a fraction of the sunlight is blocked, resulting in a partial eclipse. When the Moon's umbra ends in space *(right)*, a total eclipse does not occur. Projecting the cone through the tip of the umbra onto the Earth's surface defines the region in which an annular eclipse is seen.

than a pea in the sky as well. It is the brightness of the Moon and especially the Sun that deceives people into overestimating their angular size.

Now that you have collected on your bets and can lead a life of leisure, think about the remarkable coincidence that allows us to have total eclipses of the Sun. The Sun is 400 times the diameter of the Moon, yet it is about 400 times farther from the Earth, so the two appear almost exactly the same size in the sky. It is this geometry that provides us with the unique total eclipses seen on Earth when our Moon just barely covers the face of the Sun. If the Moon, 2,160 miles (3,476 km) in diameter, were 169 miles (273 km) smaller than it is, or if it were farther away so that it appeared smaller, people on Earth would never see a total eclipse.[2]

It is amazing that there are total eclipses of the Sun at all. As it is, total eclipses can just barely happen. The Sun is not always exactly the same angular size in the sky. The reason is that the Earth's orbit is not circular but elliptical, so the Earth's distance from the Sun varies. When the Earth is closest to the Sun (early January),[3] the Sun's disk is slightly larger in angular diameter, and it is harder for the Moon to cover the Sun to create a total eclipse.

An even more powerful factor is the Moon's elliptical orbit around the Earth. When the Moon is its average distance from the Earth or farther, its disk is too small to occult the Sun completely. In the midst of such an eclipse, a circle of brilliant sunlight surrounds the Moon, giving the event a ring-like appearance—hence the name *annular* eclipse (from the Latin *annulus*, meaning ring).

Because the angular diameter of the Moon is smaller than the angular diameter of the Sun on the average, annular eclipses are more frequent than total eclipses.

But the Moon does not just dangle motionless in front of the Sun. It is in orbit around the Earth. It catches up with and passes the Sun's position

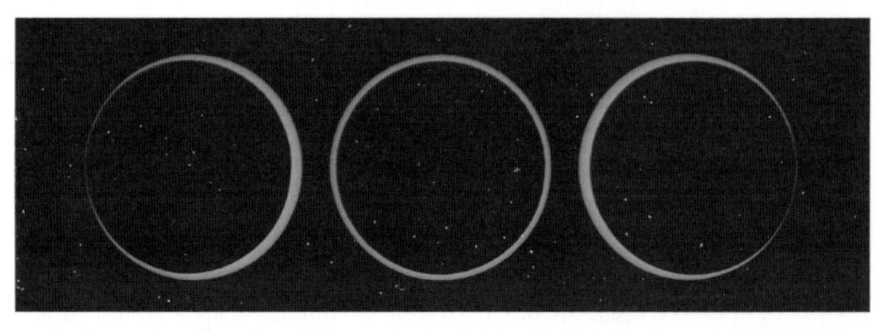

The beginning, middle, and end of annularity during the annular solar eclipse of October 3, 2005, from Spain. [Nikon D200 DSLR, Celestron 90 Maksutov, fl = 1000 mm, f/11, 1/125 s, ISO 200, Thousand Oaks Type II Filter, images combined in Photoshop. ©2005 Patricia Totten Espenak]

Apparent Size of Moon and Sun in Sky

	Diameter	Ratio—Moon to Sun
Moon	2,160 miles (3,476 km)	
Sun	864,989 miles (1,392,000 km)	1/400
	Mean Distance from Earth	
Moon	238,870 miles (384,400 km)	
Sun	92,960,200 miles (149,598,000 km)	1/389

Significance

The Sun has a diameter 400 times bigger than the Moon, but the Sun is about 400 times farther away than the Moon, so **the Moon and Sun appear to be nearly the *same size* as seen from Earth.**

The Shadow of the Moon

	Maximum	Minimum	Mean
Moon's distance from Earth (center to center)			
Miles	252,720	221,470	238,870
Kilometers	406,700	356,400	384,400
Length of Moon's shadow cone (umbra)			
Miles	236,050	228,200	232,120
Kilometers	379,870	367,230	373,540

Significance

Most of the time the Moon's shadow is too short to reach the Earth. Therefore, total solar eclipses occur *less* often than annular solar eclipses.

Angular Size of the Sun and Moon (as seen from Earth)

	Maximum	Minimum	Mean
Angular diameter of the Sun	32′31.9″	31′27.7″	31′59.3″
Angular diameter of the Moon	33′31.8″	29′23.0″	31′05.3″

Significance

The Moon's angular diameter can *exceed* the Sun's angular diameter by as much as 6.6% (2.1 arc minutes), producing a *total* eclipse of the Sun.

The Sun's angular diameter can *exceed* the Moon's angular diameter by as much as 10.7% (3.1 arc minutes), producing an *annular* eclipse of the Sun.

Most of the time, the Moon's angular diameter is *smaller* than the Sun's angular diameter. Therefore, **total solar eclipses occur *less* often than annular solar eclipses.**

in the sky about once a month. The word *month* comes from this circuit of the Moon.

The actual time for the Moon to complete this cycle is 29.53 days, and it is called a synodic month, after the Greek *synodos*, "meeting"—the meeting of the Sun and the Moon.

The Moon gives off no light of its own. It shines only by reflected sunlight. So half the Moon is always lighted by the Sun. But as the Moon orbits the Earth, sometimes we see the Earth-facing side fully illuminated and sometimes we see only a thin crescent. As the days pass, the Moon changes phase—crescent, gibbous, full. . .

In 29.53 days, the Moon goes from new moon through full moon and back to new moon again. Solar eclipses can take place only at new moon (dark of the moon), and lunar eclipses may occur only at full moon.

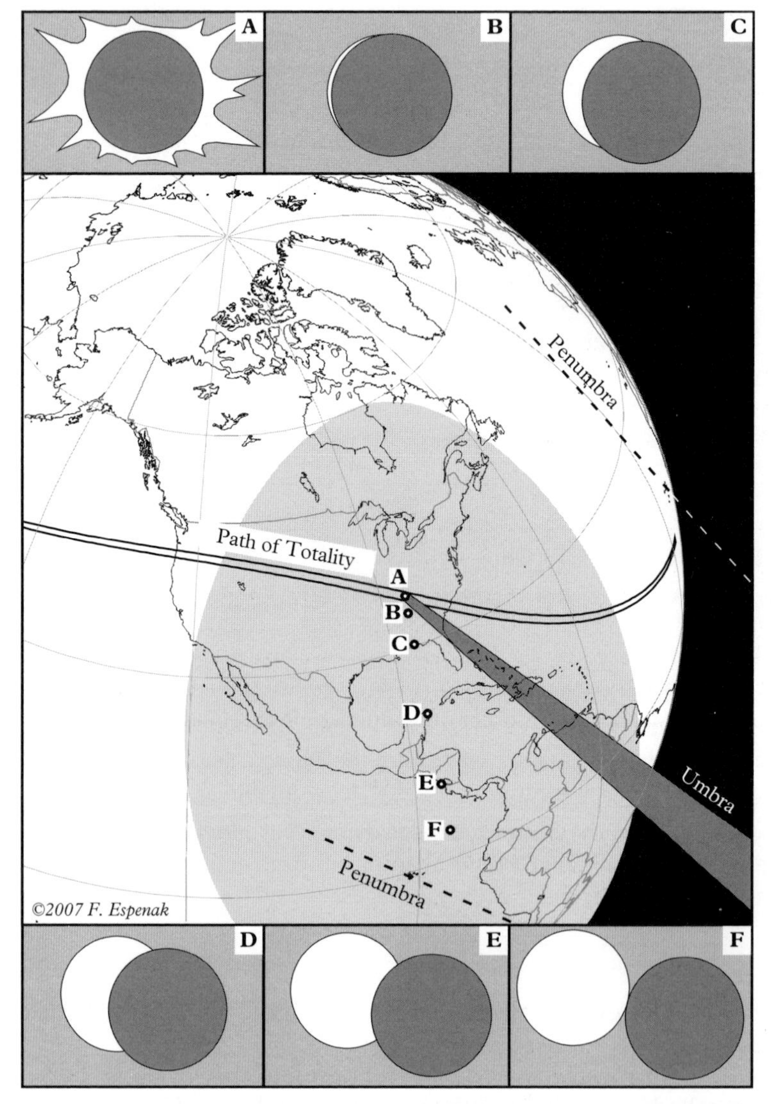

During a total eclipse of the Sun, the tip of the Moon's shadow touches the Earth and the Moon's orbital velocity carries the shadow rapidly eastward. Only along a narrow path is the eclipse total. Regions to the side of the path of totality experience varying degrees of partial eclipse. (This diagram illustrates the total eclipse of August 21, 2017.)

So why don't we have an eclipse of the Sun every 29.53 days—every time the Moon passes the Sun's position? The reason is that the Moon's orbit around the Earth is tilted to the Earth's orbit around the Sun by about 5°, so that the Moon usually passes above or below the Sun's position in the sky and cannot block the Sun from our view.

"Danger Zones"

The Moon's tilted orbit crosses the plane of the Earth's orbit at two places. Those intersections are called *nodes*. Node is from the Latin word *nodus*, meaning knot, in the sense of weaving, where two threads are tied together. The point at which the Moon crosses the plane of the Earth's orbit going northward is the ascending node. Going south, the Moon crosses the plane of the Earth's orbit at the descending node.

A solar eclipse can occur only when the Sun is near one of the nodes as the Moon passes. If the Sun stood motionless in a part of the sky away from the nodes, there would be no eclipses, and you would not be agonizing over this. But the Earth is moving around the Sun, and, as it does so, the Sun appears to shift slowly eastward around the sky, through all the constellations of the zodiac, completing that journey in one year. In that yearly circuit, the Sun must cross the two nodes of the Moon. Think of it as a street intersection at which the Sun does not pause and runs the stop sign every time. It is an accident waiting to happen. When the Sun nears a node, there is the "danger" that the Moon will be coming and—crash!

No. The Moon is 400 times closer to the Earth than the Sun, so the worst—the best—that can happen is that the Moon will pass harmlessly but stunningly right in front of the Sun. The Sun's apparent pathway in the sky is called the *ecliptic* because it is only when the Moon is crossing the ecliptic that eclipses can happen. Thus, twice a year roughly, there is

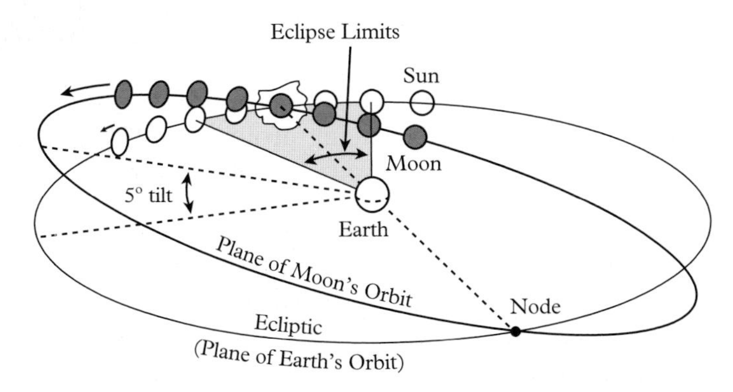

The paths of the Sun and Moon (as seen from Earth) illustrate why eclipses occur only when the Sun is near the intersection (node) where the Moon crosses the ecliptic. The plane of the Moon's orbit is tilted approximately 5° to the ecliptic plane.

Eclipse Limits ("Danger Zones")

	Maximum*	Minimum*	Mean
A solar eclipse of some kind will occur at new moon if the Sun's angular distance from a node of the Moon is less than:	18°31′	15°21′	16°56′
A central (total or annular) solar eclipse will occur at new moon if the Sun's angular distance from a node of the Moon is less than:	11°50′	9°55′	10°52′

Sun's apparent eastward movement in the star field each day (due to the Earth orbiting the Sun): about 1°

Moon's synodic period (from new moon to new moon: the time the Moon takes to complete its eastward circuit of the star field and catch up with the Sun again): 29.53 days

Significance

The Sun cannot pass a node of the Moon without at least one solar eclipse occurring, and two are possible. If one occurs, it can be either a partial or a central eclipse (total or annular). If two occur, both will be partials, about one month apart.

The Sun can pass a node without a central solar eclipse (total or annular) occurring.

* Limits vary due to changes in the apparent angular size and speed of the Sun and Moon caused by the elliptical orbits of the Earth and Moon.

a "danger period," called an eclipse season, when the Sun is crossing the region of the nodes and an eclipse is possible.

The Sun comes tootling up to the node traveling about 1° a day.[4] The hot-rod Moon, however, is racing around the sky at about 13° a day. Now if the Sun and Moon were just dots in the heavens, they would have to meet precisely at a node for an eclipse to occur. But the disks of the Sun and Moon each take up about half a degree in the sky. And the Earth, almost 8,000 miles (12,800 km) across, provides an extended viewing platform. Therefore, the Sun needs only to be *near* a node for the Moon to sideswipe it, briefly "denting" the top or bottom of the Sun's face. That will happen whenever the Sun is within 15⅓° of a node.

An "eclipse alert" begins when the Sun enters the danger zone 15⅓° west of one of the Moon's nodes and does not end until the Sun escapes 15⅓° east of that node. The Sun must traverse 30⅔°. Traveling at 1° a day, the Sun will be in the danger zone for about 31 days. But the Moon completes its circuit, going through all its phases, and catches up to the

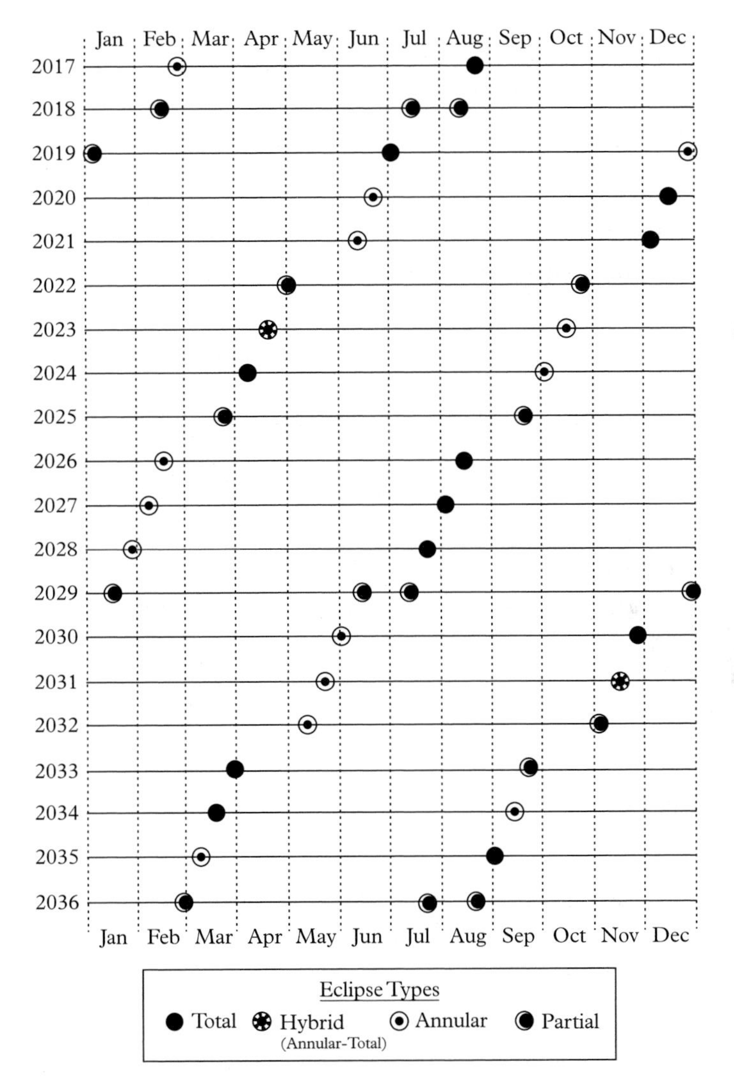

Solar eclipses 2017–2036 plotted to show eclipse seasons. Each calendar year, eclipses occur about 20 days earlier. Consequently, eclipse seasons shift to earlier months in the year. [Graph by Fred Espenak]

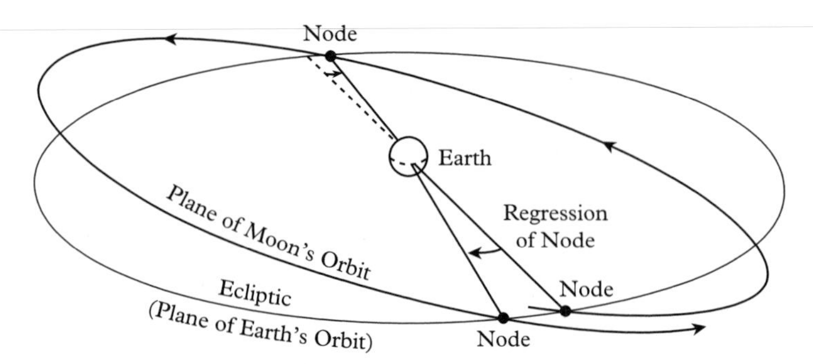

Each time the Moon completes an orbit around the Earth, it crosses the Earth's orbit at a point west of the previous node. Each year the nodes regress 19.4°, making a complete revolution in 18.61 years.

Solar Eclipses Outnumber Lunar Eclipses

There are more solar than lunar eclipses. That surprises most people because they have seen an eclipse of the Moon but not an eclipse of the Sun. The reason is simple. When the Moon passes into the shadow of the Earth to create a lunar eclipse, the event is seen wherever the Moon is in view, which includes half the planet. Actually, a lunar eclipse is seen from more than half the planet because during the course of a lunar eclipse (up to 4 hours), the Earth rotates so that the Moon comes into view for additional areas.

In contrast, whenever the Moon passes in front of the Sun, the shadow it creates—a solar eclipse—touches only a small portion of the Earth. On the average, your house will be visited by a *total* eclipse of the Sun only once in about 375 years.[a] To be touched by the dark shadow (umbra) of the Moon is quite rare.

But from either side of the path of a total eclipse, stretching northward and southward 2,000 miles (3,200 km) and sometimes more, an observer sees the Sun partially eclipsed. Even so, the zone of partial eclipse covers a much smaller fraction of the Earth's surface than a lunar eclipse. So more people have seen lunar eclipses than partial solar eclipses. Only a tiny fraction of people, about one in 50,000, have witnessed a total solar eclipse.

In his *Canon of Eclipses* (1887), Theodor von Oppolzer and his assistants computed with pen on paper all eclipses of the Sun and Moon from 1208 BCE to 2161 CE. He cataloged 8,000 solar eclipses and 5,200 lunar eclipses—about three solar eclipses for every two lunar eclipses.

This ratio can be misleading, however. Oppolzer counted all solar eclipses, whether they were total (the Moon's *umbra* touches the Earth) or partial (only the Moon's *penumbra* touches the Earth). But for lunar eclipses, Oppolzer counted only those in which the Moon was totally or partially immersed in the Earth's *umbra*. He did not count *penumbral* lunar eclipses because they are virtually unnoticeable. If he had included penumbral lunar eclipses in his census, thereby counting *all* forms of solar and lunar

Sun every 29.53 days. It is not possible for the Sun to crawl through the danger zone before the Moon arrives. A solar eclipse *must* occur each time the Sun approaches a node and enters one of these danger zones, about every half year.

In fact, if the Moon nips the Sun at the beginning of a danger zone (properly called an *eclipse limit*), the Sun may still have 30 days of travel left within the zone. But the Moon takes only 29.53 days to orbit the Earth and catch up with the Sun again. So it is possible for the Sun to be nipped by the Moon twice during a single node crossing, thereby creating two partial eclipses within a month of one another.

The closer the Sun is to the node when the Moon crosses, the more nearly the Moon will pass over the center of the Sun's face. In fact, if the

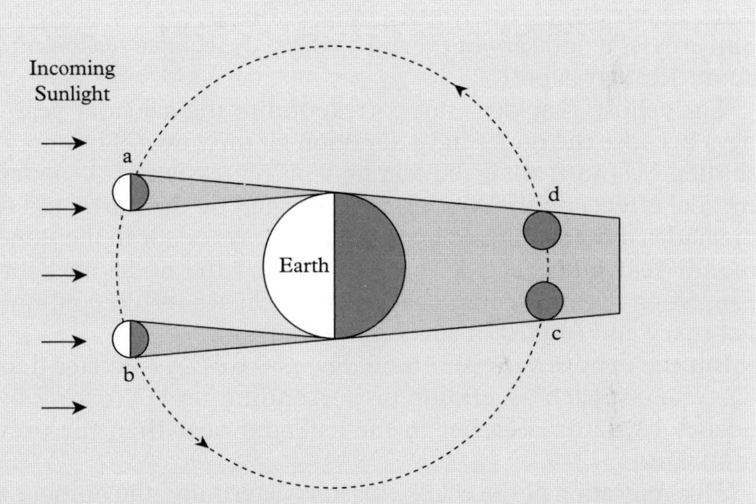

eclipses, the ratio would have been close to even, but solar eclipses would still have prevailed.

The reason why solar eclipses slightly outnumber lunar eclipses is most easily visualized if you imagine looking down on the Sun-Earth-Moon system and if you start by considering only total solar and total lunar eclipses.

The Moon will be totally eclipsed whenever it passes into the shadow of the Earth—between ***c*** and ***d*** on the diagram. At the Moon's average distance from Earth, that shadow is about 2.7 times the Moon's diameter. But there will be an eclipse of the Sun whenever the Moon passes between the Earth and Sun—between points ***a*** and ***b***. The distance between ***a*** and ***b*** is greater than between ***c*** and ***d***, so total solar eclipses must occur slightly more often.

[a] Jean Meeus: "The Frequency of Total and Annular Solar Eclipses at a Given Place," *Journal of the British Astronomical Association*, volume 92, April 1982, pages 124–126.

Sun is within about 10° of the node at the time of the Moon's crossing, a central eclipse will occur somewhere on Earth. Depending on the Moon's distance from the Earth and the Earth's distance from the Sun, this central eclipse will be total or annular.

Nodes on the March

There should be a solar eclipse or two every six months, whenever the Sun crosses one of the Moon's nodes. Actually, the Sun crosses the ascending node of the Moon, then the descending node, and returns to the ascending node in only 346.62 days—the *eclipse year*. Within this eclipse year there are two *eclipse seasons*, intervals of 30 to 37 days as the Sun approaches, crosses, and departs from a node. All eclipses will fall within eclipse seasons. There can be no eclipses outside this period of time. Because the Sun crosses a node about every 173 days, the eclipse seasons are centered about 173 days apart.

The eclipse year does not correspond to the calendar year of 365.24 days because the nodes have a motion all their own. They are constantly shifting westward along the ecliptic. This regression of the nodes is caused by tidal effects on the Moon's orbit created by the Earth and the Sun. If the nodes did not shift, eclipses would always occur in the same calendar month year after year. If the Sun crossed the nodes in February and August one year to cause eclipses, the eclipses would continue to fall in February and August in succeeding years.

But the eclipse year is 346.62 days, 18.62 days shorter than a calendar year, so each ascending or descending node crossing by the Sun occurs 18.62 days earlier in the calendar year than the previous one of its kind.

This migration of the eclipse seasons determines the number of eclipses that may occur each year. One solar eclipse must occur each eclipse season, so there have to be at least two solar eclipses each year (although both may be partial). But because of the width of the eclipse limits—up to 19 days on either side of the Sun's node crossing—and the slowness of the Sun's apparent motion, there can be two solar eclipses at each node passage (both partials). Thus, occasionally there will be four solar eclipses in one calendar year.

There can actually be five. Because the eclipse year lasts 346.62 days, almost 19 days less than a calendar year, if a solar eclipse occurs before or on January 18 (or January 19 in a leap year), that eclipse year could conceivably bring two solar eclipses in January and two more around July. That eclipse year would then end in mid-December, and a new eclipse year would begin in time to provide one final solar eclipse before the

Paths of the total eclipses of 2017 and 2024 through the United States. In a period of 6⅔ years, the people near the confluence of the Ohio and Mississippi Rivers will see two total solar eclipses. [Map ©2016 Michael Zeiler]

end of December. At most, therefore, there can be five solar eclipses in a calendar year.

Heavenly Rhythm

Eclipses, then, are like fresh fruit—available only in season. Ancient peoples who kept written records of eclipses, such as the Babylonians did from 750 BCE on, noticed after decades of observation that eclipses happen only at certain intervals. These eclipse seasons are separated from one another by 6 or occasionally 5 new moons.[5]

From the expanse of their empire, the Babylonians[6] could see only about half the lunar eclipses and only a small fraction of the solar eclipses, so, for them at first, eclipse seasons were times of "danger" when an eclipse was possible.

But gradually, as they studied their eclipse records, the Babylonians realized that if they took the date of one eclipse and counted forward a certain interval in months, they would find the record of another eclipse. Eclipses have a long-term rhythm of their own. So they took the records and counted forward into the future a certain interval in months—and *predicted* that an eclipse would occur. And it did.

By about 600 BCE, the Babylonians could accurately predict the next eclipse—and the one after that and the one after that.[7] They realized that eclipse seasons were periods during which one to three eclipses (1 or 2 solar; 0 or 1 lunar) would necessarily occur. One of the two great celestial lights might be partially or totally darkened. They could tell which one would be eclipsed and, by 300 BCE, even to what extent. The Babylonians also realized that if they couldn't see an eclipse, it didn't mean that their prediction was wrong. It just meant that the eclipse occurred somewhere else on Earth.

The Babylonians discovered and inscribed on clay tablets in their cuneiform writing several different eclipse cycles. The most famous and most useful of these eclipse rhythms is the *saros*.[8] The Babylonians noticed that 6,585 days (18 years 11 days) after virtually every lunar eclipse, there was another very similar one. If the first was total, the next was almost always total. And these eclipses, separated by 18 years, occurred in the same part of the sky, as if they were related to one another.

In a sense, they were. Imagine that a total solar eclipse occurs one day at the Moon's descending node. After 6,585 days, the Moon has completed 223 lunations (synodic periods) of 29.53 days each and returned to new moon at that same node. In that same period of 6,585 days, the Sun has endured 19 eclipse years of 346.62 days each and has returned to the descending node, forcing another solar eclipse to occur. And because 6,585 days is very close to an even 18 calendar years, this solar eclipse occurs at the same season of the year as its predecessor, and with the Sun very close to the same position in the zodiac that it occupied at the eclipse 18 years earlier. Even though 18 years have intervened, these two eclipses certainly seem to be relatives.

Frequency of Solar Eclipses

Solar eclipses by types – for 5,000 years: 2000 BCE to 3000 CE

Total:	26.7%
Annular:	33.2%
Hybrid (Annular/Total):	4.8%
Partial:	35.3%

Solar eclipses outnumber lunar eclipses almost 3 to 2
(excluding penumbral eclipses, which are seldom detectable visually)

Annular eclipses outnumber total eclipses about 5 to 4

Solar eclipses per century (average over 4,530 years): 238.9

Maximum number of solar eclipses per year: 5 (4 will be partial)

Minimum number of solar eclipses per year: 2 (both can be partial)

Maximum number of *total* solar eclipses per year: 2

Minimum number of *total* solar eclipses per year: 0

Maximum number of solar and lunar eclipses per year:
7 (4 solar and 3 lunar *or* 5 solar and 2 lunar)*

Minimum number of solar and lunar eclipses per year:
2 (2 solar and 0 lunar)*

Examples of years in which only 2 solar eclipses occur:
2017, 2020, 2021, 2022, 2023, 2024, 2025, 2026, 2027, 2028, 2030

Examples of years in which 4 solar eclipses occur: 2000, 2011, 2029, 2047

Examples of years in which 5 solar eclipses occur: 1805, 1935, 2206, 2709

Maximum diameter of the Moon's shadow cone (umbra)
as it intercepts the Earth to cause a total eclipse: 170 miles (273 km)

Maximum diameter of the Moon's "anti-umbra" as it intercepts
the Earth to cause an annular eclipse: 232 miles (374 km)

Source: Fred Espenak and Jean Meeus: *Five Millennium Canon of Solar Eclipses: –1999 to +3000* (2000 BCE to 3000 CE) (Greenbelt, Maryland: NASA Goddard Space Flight Center, 2006; NASA Technical Publication 2006–214141).

* Counts total and partial lunar eclipses but not penumbral lunar eclipses.

Total Eclipses—Duration of Totality

Longest duration (theoretical):	7 minutes 32 seconds
Longest duration in 10,000 years[a]:	7m 29s—July 16, 2186 CE
Longest duration by millennium:	
3000 BC to 2001 BC:	7m 21s—May 16, 2231 BCE
2000 BC to 1001 BC:	7m 05s—July 3, 1443 BCE
1000 BC to 1 BC:	7m 28s—June 15, 744 BCE
1 AD to 1000 AD:	7m 24s—June 27, 363 CE
1001 AD to 2000 AD:	7m 20s—June 9, 1062 CE
2001 AD to 3000 AD:	7m 29s—July 16, 2186 CE
3001 AD to 4000 AD:	7m 18s—July 24, 3991 CE
4001 AD to 5000 AD:	7m 12s—August 4, 4009 CE
Longest duration of the 20th century[b]: (6 minutes or longer)	6m 29s—May 18, 1901 6m 20s—September 9, 1904 6m 51s—May 29, 1919 7m 04s—June 8, 1937 7m 08s—June 20, 1955 7m 04s—June 30, 1973 6m 53s—July 11, 1991
Longest duration of the 21st century[c]: (6 minutes or longer)	6m 39s—July 22, 2009 6m 23s—August 2, 2027 6m 06s—August 12, 2045 6m 06s—May 22, 2096
Number of eclipses with 7 minutes or more of totality in 21st century:	0
Number of eclipses with 7 minutes or more of totality from July 1, 1098 CE to June 8, 1937 CE (839 years):	0

Note: All solar eclipses with long durations of totality have dates centered around July 4, the mean date for the Earth at aphelion (farthest from the Sun), when the Sun appears slightly smaller in size and is more easily covered by the Moon.

[a] Based on the 10,000-year period from 3000 BC to 7000 AD

[b] All the long total eclipses of the 20th and 21st centuries are members of saros series 136 except for September 9, 1904 and May 22, 2096.

[c] All the long total eclipses of the 20th and 21st centuries are members of saros series 136 except for September 9, 1904 and May 22, 2096.

All the more so because another lunar cycle crucial to eclipses has a multiple that also adds up to 6,585 days. That cycle is the *anomalistic month*—the time it takes the Moon in its elliptical orbit around the Earth

Annular Eclipses—Duration of Annularity

Longest duration (theoretical):	12 minutes 30 seconds
Longest duration in 10,000 years[a]:	12m 24s—December 6, 150 CE
Longest duration by millennium:	
3000 BC to 2001 BC:	11m 02s—November 24, 2037 BCE
2000 BC to 1001 BC:	12m 07s—December 12, 1656 BCE
1000 BC to 1 BC:	12m 08s—December 22, 178 BCE
1 CE to 1000 CE:	12m 24s—December 6, 150 CE
1001 CE to 2000 CE:	12m 09s—December 14, 1955 CE
2001 CE to 3000 CE:	11m 08s—January 15, 2010 CE
3001 CE to 4000 CE:	12m 09s—January 14, 3080 CE
4001 CE to 5000 CE:	11m 08s—January 20, 4885 CE
Longest duration of the 20th century[b]: (9 minutes or longer)	11m 01s—November 11, 1901
	11m 37s—November 22, 1919
	12m 00s—December 2, 1937
	12m 09s—December 14, 1955
	12m 03s—December 24, 1973
	11m 41s—January 4, 1992
Longest duration of the 21st century[c]: (9 minutes or longer)	11m 08s—January 15, 2010
	10m 27s—January 26, 2028
	9m 42s—February 5, 2046

Note: All solar eclipses with long durations of annularity have dates centered around January 3, the mean date for the Earth at perigee (closest to the Sun), when the Sun appears slightly larger in size and is harder for the Moon to cover.

Note: A long annularity means that the Moon is smaller than usual in apparent size, creating an annular eclipse that is farthest from being a total eclipse. Thus, a long annularity will provide less landscape and sky darkening, and less likelihood of a glimpse of Baily's beads.

[a] Based on the 10,000-year period from 3000 BCE to 7000 CE.

[b] All the long annular eclipses of the 20th and 21st centuries are members of saros series 141.

[c] All the long annular eclipses of the 20th and 21st centuries are members of saros series 141.

to go from *perigee* (closest to Earth) to *apogee* (farthest) and back to *perigee*.[9] When the Moon is near perigee, its angular size in the sky is just slightly larger, creating solar eclipses that are total rather than annular. The anomalistic month is 27.55 days long and 239 of these cycles add up to 6,585 days. If the previous eclipse occurred with the Moon at perigee,

Erratic Intervals Between Total Solar Eclipses

Location	Dates	Interval
An average town		Average of 375 years
Unusually long intervals		
London	October 29, 878 May 3, 1715	837 years
Jerusalem	August 2, 1133 August 6, 2241	1,108 years
United States mainland	February 26, 1979 August 21, 2017	38 years
Unusually short intervals—past		
New Guinea (southern)	June 11, 1983 November 22, 1984	1½ years
Ukraine	March 14, 190 BCE July 17, 188 BCE October 19, 183 BCE	3 in 7½ years
Sumatra	May 18, 1901 January 14, 1926 May 9, 1929	3 in 28 years
Spain	July 8, 1842 December 22, 1870 May 28, 1900 August 30, 1905	4 in 63 years
Jerusalem (region of)	March 1, 357 BCE July 4, 336 BCE April 2, 303 BCE	3 in 54 years
Stonehenge, England	May 3, 1715 May 22, 1724	9 years
Yellowstone National Park, United States of America	July 29, 1878 January 1, 1889	10½ years

the new eclipse will occur with the Moon near perigee, providing another total eclipse.

So, after a time interval of 6,585 days, the geometrical configuration of the Sun, Moon, and Earth is almost exactly repeated. Another solar

Erratic Intervals Between Total Solar Eclipses

Location	Dates	Interval
Mexico City	March 7, 1970 July 11, 1991	21 years
Wichita, Kansas, United States	September 17, 1811 November 30, 1834	23 years
Turkey (northern)	August 11, 1999 March 29, 2006	6⅔ years
Lobito, Angola (just north of)	June 21, 2001 December 4, 2002	1½ years
Unusually short intervals—future		
Confluence of Ohio and Mississippi Rivers, United States (southeast Missouri, southern Illinois, and western Kentucky)	August 21, 2017 April 8, 2024	6⅔ years
Australia	April 20, 2023 July 22, 2028 November 25, 2030 July 13, 2037 December 26, 2038	5 in 15 years
Florida panhandle, United States (Pensacola to Tallahassee)	August 12, 2045 March 30, 2052	6⅔ years
Paris	September 3, 2081 September 23, 2090	9 years
Bagé, Brazil (Brazil-Uruguay border)	January 8, 2103 February 8, 2111 September 15, 2118	3 in 15 years
New York City	May 1, 2079 October 26, 2144	65 years
Antwerp, Belgium	May 25, 2142 June 14, 2151	9 years

eclipse occurs at the same node, at the same season, in the same part of the sky. If it was a total eclipse before, it will almost certainly be a total eclipse again.

Do Your Own Eclipse Prediction

Take the date of any solar or lunar eclipse and add 6,585.32 days to it and you will accurately predict a subsequent eclipse of the same kind that will closely resemble the one 18 years earlier. Take the date of *every* solar and lunar eclipse that occurs and keep on adding 6,585.32 days to it and you will have, with rare exceptions, a quite reliable list of future eclipses.

Flaws in the Rhythm

But these eclipses 6,585 days apart are not identical twins. The match between unrelated periods—the Moon's cycle of phases, the Sun's eclipse year cycle, and the Moon's cycle from perigee to perigee—is not perfect.

223 synodic periods of the Moon (lunations) at 29.5306 days each	= 6,585.32 days
19 eclipse years of the Sun at 346.6201 days each	= 6,585.78 days
239 anomalistic months of the Moon (revolution from perigee to perigee) at 27.55455 days each	= 6,585.54 days

These synchronizing cycles are out of step with one another by fractions of a day. And those fractions of a day have their consequences.

Consider first that 223 synodic periods of the Moon amount to 6,585.32 days or, in calendar years, 18 years 11⅓ days (18 years 10⅓ days if five leap years intervene). Because the saros period is 18 years 11⅓ days, each subsequent eclipse occurs about one-third of the way around the world westward from the one before it. For a lunar eclipse, visible to half the planet, this westward shift usually does not push the eclipse out of view. For a solar eclipse, however, visible over a narrow swath of the Earth's surface, the subsequent eclipse in that saros series would almost never be visible from the initial site. The eclipse would be happening in an entirely different part of the world.

After three saros cycles—54 years 34 days—the eclipse would be back to its original longitude, but it would have shifted, on the average, about 600 miles (1,000 km) northward or southward, taking totality and even major partiality out of view for the original observer.[10]

A saros period truly does predict with accuracy that a solar eclipse will happen, but it would have been extremely hard for an ancient astronomer to confirm that the predicted eclipse had taken place and thus difficult for the astronomer or those he served to retain confidence in his solar eclipse predictions.

The Evolving Saros

Now consider that 223 lunations (synodic periods) amount to 6,585.32 days, while 19 eclipse years of the Sun consume 6,585.78 days. The difference after 18 years 11⅓ days is 0.46 day. The result is that the Sun is not exactly where it was before in relation to the Moon's node. It is 0.477° farther west. If the Moon just barely grazed the western part of the Sun before, it will clip a little more the next time. At each return of the saros, the eclipses become larger and larger partials until the Moon is passing across the center of the Sun's disk, yielding total or annular eclipses. Then, with the passing generations, as the Sun is farther west within the eclipse limit, the eclipses return to partials. Finally, after about 1,300 years, the Sun is no longer within the eclipse limits when the Moon, after 223 lunations, arrives. After about 1,300 years of adding 6,585.32 days to the date of an eclipse to predict the next, the prediction fails. No eclipse occurs. That saros has died.[11]

Finally, consider that 223 lunations amount to 6,585.32 days while 239 anomalistic months of the Moon (perigee to perigee) total 6,585.54 days. If the Moon was at perigee and hence largest in angular size during one central eclipse, the eclipse must be total. But with each succeeding eclipse in that saros series, the Moon is a little farther from perigee, until finally the Moon's angular size is too small to completely cover the Sun, and eclipses become annular.

The Saros Family

For people long ago, the discovery of the saros and other eclipse cycles probably brought some comfort that eclipses, however dire their interpretation, were part of nature's rhythms. Using these rhythms, it was possible to predict future eclipses without knowing anything about the mechanism that produced them. Even after people understood the causes of eclipses, it was far easier to predict their occurrence by the saros (or other cycles)

Creating the Ultimate Eclipse

What conditions would provide the longest total eclipse of the Sun?

First, the Moon should be near maximum angular size, which means it should be near perigee—the point in its orbit when it is closest to Earth. That happens once every 27.55 days (an anomalistic month).

Second, the Sun should be near minimum angular size, which means the Earth should be near aphelion—the point in its orbit when it is farthest from the Sun. That happens once every 365.26 days (an anomalistic year). At present, aphelion occurs in early July.

Third, to prolong the eclipse as much as possible, the Moon's eclipse shadow must be forced to travel as slowly as possible. Here the observer enters the formula. At the time of a solar eclipse, the Moon's shadow is moving about 2,100 miles an hour (3,380 km an hour) with respect to the center of the Earth. But the Earth is rotating from west to east, the same direction that the eclipse shadow travels. At a latitude of 40° north or south of the equator, the surface of the Earth turns at about 790 miles an hour (1,270 km an hour), slowing the shadow's eastward rush by that amount. At the equator, the Earth's surface rotates at 1,040 miles an hour (1,670 km an hour), slowing the shadow's speed to only 1,060 miles an hour (1,710 km an hour), thereby prolonging the duration of totality, Baily's beads, and all phases of the eclipse.

Finally, to prolong the eclipse just a few seconds more, the Moon should be directly overhead, so that we are using the full radius of the Earth to place ourselves about 4,000 miles (6,400 km) closer to the Moon than the limb of the Earth, thus maximizing the Moon's angular size and therefore its eclipsing power. The latitude where the Moon stands overhead varies with the seasons as the Sun appears to oscillate 23½° north and south of the equator. Thus the maximum duration of totality *always* occurs in the tropics but rarely occurs exactly on the equator.

Eclipse expert Jean Meeus calculates that the maximum possible duration of totality in a solar eclipse is currently 7 minutes 32 seconds. The value slowly changes due to variations in the eccentricity of the Moon's orbit.[a]

[a] Jean Meeus: "The Maximum Possible Duration of a Total Solar Eclipse," Journal of the British Astronomical Association, volume 113, number 6, 2003; pages 343–348.

than to calculate all the factors surrounding the varying motions and apparent sizes of the Sun, Moon, and Earth.

Today, with modern electronic computers to crunch orbits and tilts and wobbles into extremely accurate eclipse predictions, the saros would seem to be an anachronism. Yet it is interesting to view any single total eclipse as a member of an evolving family that had its origin in the distant past, has gradually risen from insignificance to great prominence, and will

Basic Sun and Moon Data Important to Eclipses	
Diameter	
Sun	864,989 miles (1,392,000 km)
Moon	2,160 miles (3,476 km)
Earth	7,927 miles (12,756 km)
Mean distance from Earth	
Sun	92,960,200 miles (149,598,000 km)
Moon	238,870 miles (384,400 km)
Ratio of Sun's diameter to Moon's diameter:	400.5
Ratio of Sun's mean distance to Moon's mean distance:	389.1
Orbital speed of Moon (mean):	2,290 miles per hour (3,680 km per hour)
Speed through space of Moon's shadow during a solar eclipse (not the same as the Moon's orbital speed because of the orbital motion of the Earth):	2,100 miles per hour (3,380 km per hour)
Orbital eccentricity of Moon:	0.05490
Inclination of Moon's orbit to plane of Earth's orbit (ecliptic) (varies due to tidal effects of Earth and Sun):	5°08′ (mean) 5°18′ (maximum) 4°59′ (minimum)
Regression (westward drift) of Moon's nodes:	19.4° per year
Period for Moon's nodes to regress all the way around its orbit:	18.61 years

inevitably decline into oblivion. Chapters 13 and 16 explore the genealogy of the 2017 and 2024 eclipses.

NOTES AND REFERENCES

1. Epigraph: Alfonso X, King of Castile as cited by Arthur Koestler: *The Sleepwalkers* (New York: Grosset & Dunlap, 1963), page 69.
2. Alan D. Fiala, U.S. Naval Observatory, personal communication, April 1990.

3. The principal cause of the seasons is the tilt of the Earth's axis, not the Earth's rather modest change in distance from the Sun.
4. The Sun takes 365¼ days to appear to go through the constellations of the zodiac once around the sky, so some ancient peoples measured it or rounded it off to 360 days. Each day, the Sun goes one step around its great sky circle, which is why the Babylonians considered that a circle has 360 degrees. The Chinese circle had 365¼°.
5. Clemency Montelle: *Chasing Shadows: Mathematics, Astronomy, and the Early History of Eclipse Reckoning* (Baltimore: Johns Hopkins University Press, 2011), pages 71–74.
6. *Babylonian* here refers to Mesopotamian culture. At various times during this period, Babylon was ruled by Assyrians and Chaldeans.
7. Clemency Montelle: *Chasing Shadows: Mathematics, Astronomy, and the Early History of Eclipse Reckoning* (Baltimore: Johns Hopkins University Press, 2011), page 78. The earliest prediction of an eclipse in Babylonian records is 651 BCE.
8. The Babylonians called this 223-lunar-month eclipse cycle "18 years." The use of the word *saros* to mean this 223-lunar-month eclipse cycle was erroneously introduced in 1691 by Edmond Halley when he applied it to the Babylonian eclipse cycle on the basis of a corrupt manuscript by the Roman naturalist Pliny. The Babylonian sign SAR has meaning as both a word and a number. As a word, it means (among other things) *universe*. As a number, it means 3,600, signifying a large number. But there is no evidence that the Babylonians ever applied the word *saros* to this 18-year eclipse cycle.
9. This period is called anomalistic because it derives from the anomaly—or irregularity—of lunar motion due to the Moon's elliptical orbit.
10. The reason for the change in latitude after 54 years 34 days is the 34-day difference from a calendar year, which changes the eclipse's position within the season, so the altitude of the Sun is significantly different and the shadow is cast farther north or south. The triple saros cycle of 54 years 34 days was known to the Babylonians, who passed it along to the Greeks, who called it the *exeligmos*.
11. In a sense, it has not vanished altogether. With each period of 6,585.32 days, the Sun's position with respect to the node continues to slip westward without experiencing or causing any eclipses until, after about 5,500 years, it encounters the eclipse limit of the opposite node, and that saros may be said to be reborn. George van den Bergh: *Periodicity and Variation of Solar (and Lunar) Eclipses* (Haarlem: H. D. Tjeenk Willink, 1955).

A MOMENT OF TOTALITY

A Lasting Impression

Glenn Schneider saw his first total eclipse of the Sun in 1970 at age 14. He traveled with his amateur astronomy club from New York City to Greenville, North Carolina. They observed from the football field of East Carolina University. Glenn was loaded with still cameras, a movie camera, binoculars, and a telescope. He had carefully rehearsed over and over again every action he would take to capture each precious moment of the 2 minutes 54 seconds of totality.

As totality neared, he stationed himself high in the stands to see the horizon and watch the approach of the Moon's shadow. Then he raced down the steps to his telescope and cameras, checked the sky as the Sun blinked out—and *froze.* He stood and stared at the hole in the sky that hid the Sun, at the luminescent prominences briefly visible at the Sun's rim, at the delicate white tracery of the corona. He was breathless. He couldn't budge to activate his cameras. For 2 minutes 54 seconds, he stared at the sight he had heard about and studied—and for which all the reading, all the pictures, and all the stories were utterly inadequate.

He was profoundly moved—and in that moment he promised himself he would see every total eclipse of the Sun for the rest of his life. Glenn is now Dr. Glenn Schneider, University of Arizona astronomer and principal scientist for many Hubble Space Telescope projects. He continues to pursue solar eclipses. 2017 will be his 34th total.[1]

[1] Glenn Schneider, interview, May 5, 2015.

3

Ancient Efforts to Understand

I look up. Incredible! It is the eye of God. A perfectly black disk, ringed with bright spiky streamers that stretch out in all directions.

Jack B. Zirker, solar astrophysicist (1984)[1]

Ancient peoples around the world have left monuments and artifacts demonstrating their reverence for the sky and their efforts to record celestial motions. More than 2,500 years ago, the Babylonians could predict solar and lunar eclipses. And the ability to anticipate eclipses may go back further still.

Stonehenge

The earliest and most famous of monuments that testify to a people's high level of astronomical knowledge is Stonehenge, in southern England. The builders of Stonehenge began work about 5,000 years ago.

The initial Stonehenge consisted of a circular embankment 360 feet (110 meters) in diameter, some postholes, and four markers set in a rectangle, later replaced by large standing stones.[2] To dig holes in the ground, the Stonehenge people used the antlers of deer. They had no metal tools, no wheeled vehicles, no draft animals—and no writing. Instead, they left their celestial knowledge set in stone.

About 2600 BCE, when the first pyramid was built in Egypt, the Stonehenge builders brought the 35-ton Heel Stone to the site, the first of the great boulders to be erected.[3] It stands outside the circular embankment to the northeast, 250 feet (77 meters) from the center of Stonehenge. But the Heel Stone was not alone. It had a twin that stood beside it, just to the north.[4]

As seen from the center of Stonehenge, the Sun at the beginning of summer—the longest day of the year—rose between the Heel Stone and its now-vanished companion. With this dramatic sunrise, the people of Stonehenge could celebrate the summer solstice. Using other alignments, they could also mark the beginning of summer and winter, creating an accurate solar calendar of great benefit to farmers and herders.

Aerial view of Stonehenge. [Courtesy of English Heritage]

For someone standing at the center of Stonehenge, the embankment around the monument served to level the horizon of rolling hills. Within the embankment, four stones—the Station Stones—outlined a rectangle within the circle. The sides of this rectangle offered interesting lines of sight. The short side of the rectangle pointed toward the same spot on the horizon that the two Heel Stones framed, the position where the Sun rose farthest north of east, marking the commencement of summer. Facing in the opposite direction along the short side of the rectangle, an observer would see the place where the Sun set farthest south of west, signaling the beginning of winter.

In contrast, the long sides of the rectangle provided alignments for crucial rising and setting positions of the Moon. Looking southeast along the length of the rectangle, an observer was facing the point on the horizon where the summer full moon would rise farthest south. In the opposite direction, looking northwest, this early astronomer's gaze was led to the spot on the horizon where the winter full moon would set farthest north. These positions marked the north and south limits of the Moon's motion.

The structure of Stonehenge offers additional testimony to its builders' efforts to understand the motion of the Moon. Evidence of small holes near the remaining Heel Stone strongly suggests that the users of Stonehenge observed and marked the excursion of the Moon as much as 5 degrees north and south of the Sun's limit.[5] This motion above and below the Sun's position is caused by the tilt of the Moon's orbit to the

Viewed from the center of Stonehenge, the Sun rises just to the left of the remaining Heel Stone at the beginning of summer, the longest day of the year. [Courtesy of English Heritage]

Earth's path around the Sun. Because of this tilt, the Moon does not pass directly in front of the Sun (a solar eclipse) or directly into the Earth's shadow (a lunar eclipse) each month.

In the years that followed the installation of the Heel Stones and Station Stones, new generations of ancient Britons added two sets of giant archways. Those arches, completed by 2450 BCE, provide the familiar silhouette of Stonehenge we know today.

The outer set—the Sarsen Circle[6]—formed a ring of 30 linked archways, approximating the days in a lunar month. One of the 30 uprights was only half the diameter of the others, as if to suggest 29½ days, a more accurate record of the time it takes the Moon to complete a cycle of phases.

Inside this Sarsen Circle was a horseshoe of five even larger freestanding archways, the trilithons, with uprights that weigh up to 50 tons—one pillar weighing as much as 25 cars. The trilithon archways may have framed extreme setting positions of the Sun and Moon.

The massive, shaped boulders of the Sarsen Circle and trilithons were dragged from a quarry 20 miles (32 kilometers) away, carefully positioned, and set upright to codify in stone the discoveries made by the people before them.

In the last phase of building at Stonehenge, about 1800 BCE, two concentric circles of holes were dug just outside the Sarsen Circle—one with 30 holes and the other with 29. These circles reinforce the evidence that astronomers at Stonehenge were counting off the 29½-day cycle of lunar phases, from new moon to full moon and back to new moon again. Eclipses of the Sun can only take place at new moon; lunar eclipses can only occur at full moon.

Because the builders of Stonehenge had discovered and accurately recorded the range in rising and setting positions of the Sun and Moon and had built a monument whose alignments marked these positions quite well, they may have been able to recognize when the Moon was on course to intercept the position of the Sun, to cause a solar eclipse. Perhaps they

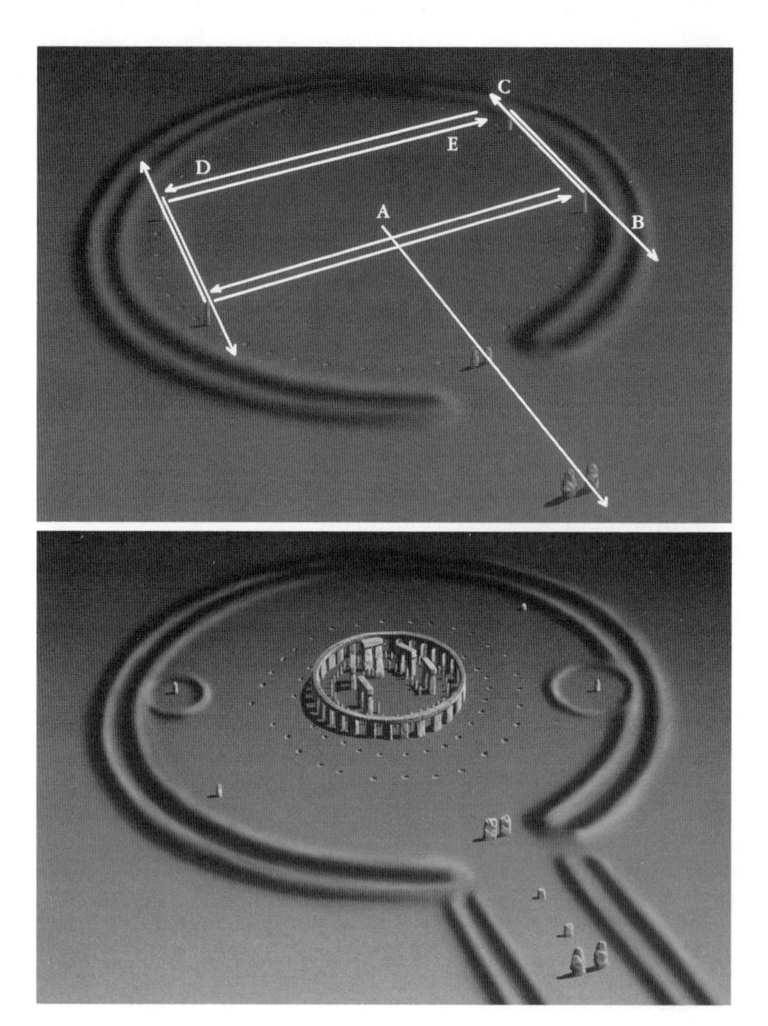

Stonehenge. *Top:* First phase of construction showing alignments A and B, northernmost sunrise: first day of summer; C, southernmost sunset: first day of winter; D, southernmost moonrise; E, northernmost moonset; *Bottom:* Final phase of construction. [©1983 Hansen Planetarium]

could tell when the Moon was headed for a position directly opposite the Sun, which would carry it into the shadow of the Earth for a lunar eclipse.[7] They almost certainly could not predict where or what kind of solar eclipse would be seen, but they might have been able to warn that on a particular day or night, an eclipse of the Sun or Moon was *possible*.

The builders of Stonehenge left no written records of their objectives or results, so we must judge from the monument and its alignments what they knew. Whatever that was, they thought it so worth celebrating that for 1,500 years the rulers and common people were willing to devote vast amounts of time, physical effort, and ingenuity to raising a lasting monument of awesome size, amazing precision, and haunting beauty.

Certainly Stonehenge was more than an astronomical observatory 4,500 years ago. For this Stone Age people, watching the Sun and Moon rise and set over the alignments of Stonehenge must have been a thrilling theatrical spectacle and a profoundly inspiring experience—to be so in touch with the rhythms of the universe.

China

A frequently recounted Chinese story says that Hsi and Ho, the court astronomers, got drunk and neglected their duties so that they failed to predict (or react to) an eclipse of the Sun. For this, the emperor had them executed. So much for negligent astronomers.

If this story were an account of an actual event, the dynasty mentioned would place the eclipse somewhere between 2159 and 1948 BCE, making it by far the oldest solar eclipse recorded in history. But all serious attempts to identify one particular eclipse as the source of this story have been abandoned as scholars have recognized that the episode is mythological.

In ancient Chinese literature, Hsi-Ho is not two persons but a single mythological being who is sometimes the mother of the Sun and at other times the chariot driver for the Sun. Later, in the *Shu Ching* (Historical Classic), parts of which may date from as early as the 7th or 6th century BCE, this single character is split, not into two, but into six. In the *Shu Ching* story, the legendary Chinese emperor Yao commissions the eldest of the Hsi and Ho brothers "to calculate and delineate the sun, moon, the stars, and the zodiacal markers; and so to deliver respectfully the seasons to the people."[8] In further orders, he sends a younger Hsi brother to the east and another to the south; he orders a younger Ho brother to the west and another to the north. Each is responsible for a portion of the rhythms of the days and seasons, to turn the Sun back at the solstices and to keep it moving at the equinoxes.

These mythological magicians are also charged with the prevention of eclipses, hence the story that appears later in the *Shu Ching* about the

The Hsi and Ho brothers receive their orders from Emperor Yao to organize the calendar.

emperor's anger with his servants for failing to *prevent* an eclipse, not just predict or respond ceremonially to it.[9]

The earliest Chinese word for eclipse, *shih*, means "to eat" and refers to the gradual disappearance of the Sun or Moon as if it were eaten by a celestial dragon.[10] It was a bad omen. The Chinese recorded more than a thousand sightings of solar and lunar eclipses—including more solar eclipses than any other civilization. Their accounts, mostly inscribed on animal bones, seldom contain the observational detail provided by the Babylonians. Perhaps there was more information in the original reports, but nearly all were lost long ago, leaving us with summaries in dynastic histories.

The earliest reliable accounts of Chinese eclipses come from *Spring and Autumn Annals* (*Ch'un-ch'iu*), recording eclipses from 772 to 481 BCE, including a total solar eclipse in 709 BCE.[11] As they recorded more and more eclipses, the Chinese began to notice that over time eclipses occur in patterns. So, like the Babylonians, without yet understanding what caused an eclipse, the Chinese discovered by the late 1st century BCE that they could predict eclipses. They could take the date of an eclipse, count forward a certain number of months, and reliably predict when another eclipse would occur, even if it was not visible from their city.[12]

By the early 3rd century CE, the Chinese understood how the motions of the Sun and Moon cause eclipses and were able to use this knowledge to predict eclipses more accurately.[13]

The Maya

In the New World, there were ancient people who, like the Babylonians and the Chinese, used writing to record eclipses and from these records detected a rhythm by which they could predict them or at least warn of their likelihood. Those people were the Maya and we know of their achievement through one of their books—one of only four that survived the Spanish conquest and its zealous destruction of the religious beliefs of the native peoples.

All that we know of Maya accomplishments in recognizing the patterns of eclipses comes from a manuscript called the Dresden Codex. Written in hieroglyphs and illustrated in color, the book was painted on processed tree bark with pages that open and shut in accordion folds. The Dresden Codex dates from the 11th century CE and is probably a copy of an older work.

We can only wonder what was lost when the conquering Spaniards destroyed by the thousands the books of the Maya and other Mesoamerican peoples. What remains is impressive enough. The Maya realized that discernible eclipses occur at intervals of 5 or 6 lunar months. Five or 6 full moons after a lunar eclipse, there was the *possibility* of another lunar eclipse. Five or 6 new moons after a solar eclipse, another solar eclipse was *possible*.

The Maya had discovered in practical, observable terms the approximate length of the eclipse year, 346.62 days, and the eclipse half year of 173.31 days. The interval for one complete set of lunar phases is 29.53 days. Six lunations amount to approximately 177.18 days, close enough to the eclipse half year (173.31 days) so that there is the "danger" of an eclipse at every sixth new or full moon, but not a certainty. After another 6 lunar months, the passing days have amounted to 354.36, nearly 8 days *too long* to coincide with the Sun's passage by the Moon's node. An eclipse is less likely. As the error mounts, the need increases to substitute a 5-lunar-month cycle into the prediction system rather than the standard 6-lunar-month count.

Some great genius must have noticed after recording a sizeable number of eclipses, that major eclipses were occurring only at intervals of 177 days (6 lunar months) or 148 days (5 lunar months). Using the date of an observed solar or lunar eclipse, it would then have been possible to predict the likelihood of another eclipse, even though in some cases an eclipse would not occur, and in others it would not be visible from Mesoamerica.

In the Dresden Codex there are eight pages with a variety of pictures representing an eclipse. Each depiction is different, but most show the glyph for the Sun against a background half white and half black. In two of the pictures, the Sun and background are being swallowed by a serpent. Leading up to each picture is a sequence of numbers: a series of 177s ending with a 148. Each sequence adds up to the number of days in well-known 3- to 5-year eclipse cycles. At the end of each burst of numbers stands the giant, haunting symbol of an eclipse.

Astronomer-anthropologist Anthony Aveni notes that "The reduction of a complex cosmic cycle to a pair of numbers was a feat equivalent to those of Newton or Einstein and for its time must have represented a great triumph over the forces of nature."[14]

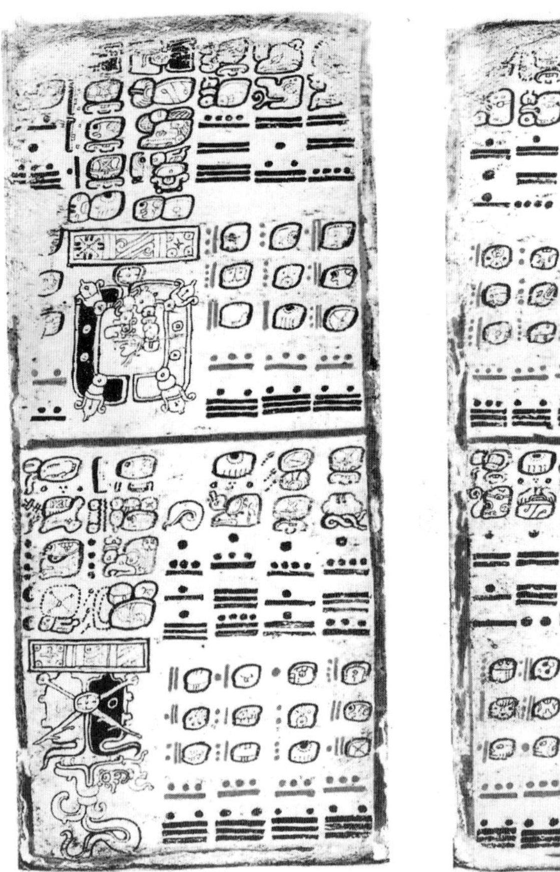

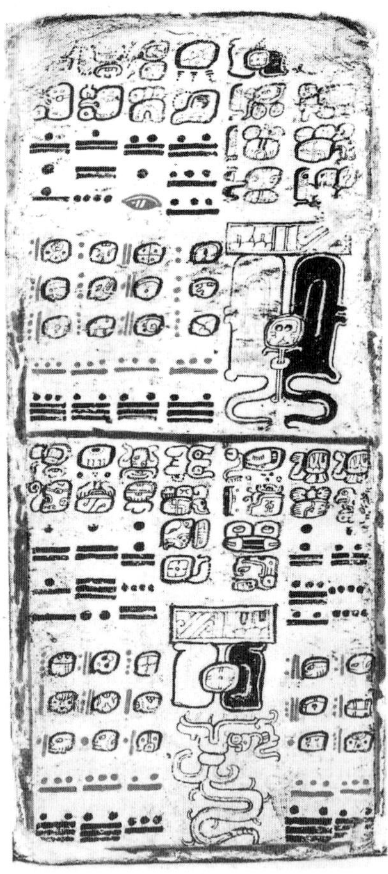

Portion of the Maya solar eclipse prediction tables from the Dresden Codex: at the bottom are the day counts that lead up to a solar eclipse, indicated, *bottom right*, by a serpent swallowing a symbol for the Sun. [Courtesy of Ancient Americas at LACMA]

From the Maya, we have the numbers that demonstrate one of the greatest of their many discoveries about the rhythms of the sky, but we have no account of the emotion the common folk felt when they observed an eclipse. Perhaps the closest we can come is a passage in the Florentine Codex of the Aztecs, who inherited and used the Mesoamerican calendar but apparently knew little of the astronomy discovered by the Maya a thousand years earlier.

> When the people see this, they then raise a tumult. And a great fear taketh them, and then the women weep aloud. And the men cry out, [at the same time] striking their mouths with [the palms of] their hands. And everywhere great shouts and cries and howls were raised.... And they said: "If the sun becometh completely eclipsed, nevermore will he give light; eternal darkness will fall, and the demons will come down. They will come to eat us!"[15]

NOTES AND REFERENCES

1. Epigraph: Jack B. Zirker: *Total Eclipses of the Sun* (New York: Van Nostrand Reinhold, 1984), page vi.
2. There were also 56 chalk-filled Aubrey Holes just inside the embankment, but their function is still not generally agreed upon.
3. Stonehenge construction dates come from Clive Ruggles: *Astronomy in Prehistoric Britain and Ireland* (New Haven, Connecticut: Yale University Press, 1999), pages 35–41; and Anthony Aveni: *Stairways to the Stars: Skywatching in Three Great Ancient Cultures* (New York: John Wiley & Sons, 1997), pages 57–91.
4. The hole for this stone was discovered in 1979 under the shoulder of the road that passes close to the Heel Stone. The original stone is gone. Michael W. Pitts: "Stones, Pits and Stonehenge," *Nature*, volume 290, March 5, 1981, pages 46–47.
5. The limits of the Moon's motion north and south of the ecliptic can differ by as much as 10 minutes of arc from the mean inclination of 5 degrees 8 minutes because of gravitational perturbations on the Moon caused by the Sun and the equatorial bulge of the Earth.
6. In medieval times, people living near Stonehenge called these stones "sarsen," short for Saracen," meaning Muslim—hence foreign, strange, pagan.
7. A lunar eclipse prediction is much easier to verify because lunar eclipses are seen over half the Earth.
8. James Legge, editor and translator: *The Chinese Classics*, volume 3, *The Shoo King* [Shu Ching] (Hong Kong: Hong Kong University Press, 1960), part 1, chapter 2, paragraphs 3–8 (pages 18–22). Legge romanizes Hsi's name as He. In some retellings, Hsi's name appears as Hi. The *Shu Ching* is one of five books in the *Wu Ching* (Five Classics), the sourcebooks of the Confucian tradition.

9. The Hsi and Ho story appears in a chapter that is an exhortation by the Prince of Yin, commander of the armies, to government officials to fulfill their duties to the administration, thereby making the emperor "entirely intelligent." If anyone neglects this requirement, "the country has regular punishments for you."

   > Now here are Hsi and Ho. They have entirely subverted their virtue and are sunk and lost in wine. They have violated the duties of their office and left their posts. They have been the first to allow the regulations of heaven to get into disorder, putting far from them their proper business. On the first day of the last month of autumn, the sun and moon did not meet harmoniously in Fang. The blind musicians beat their drums; the inferior officers and common people bustled and ran about. Hsi and Ho, however, as if they were mere personators of the dead in their offices, heard nothing and knew nothing – so stupidly went they astray from their duty in the matter of the heavenly appearances and rendering themselves liable to the death appointed by the former kings. The statutes of government say, "When they anticipate the time, let them be put to death without mercy; when they are behind the time, let them be put to death without mercy."

   We never hear whether Hsi and Ho were ever tracked down and executed.
10. Joseph Needham and Wang Ling: *Science and Civilisation in China*, volume 3, *Mathematics and the Sciences of the Heavens and the Earth* (Cambridge: At the University Press, 1959), page 409. Most of this information is also available in Colin A. Ronan's abridgment of Needham's work, *The Shorter Science and Civilisation in China*, volume 2 (Cambridge: Cambridge University Press, 1981).
11. F. Richard Stephenson: *Historical Eclipses and Earth's Rotation* (Cambridge: Cambridge University Press, 1997), pages 58–59, 213, 220–221. Chinese eclipse recordkeeping continued through 1500 CE.
12. The Chinese used a 135-month (11-year) eclipse period rather than the 223-month (18-year) saros period favored by the Babylonians.
13. Joseph Needham and Wang Ling: *Science and Civilisation in China*, volume 3, *Mathematics and the Sciences of the Heavens and the Earth* (Cambridge: At the University Press, 1959), page 421.
14. Anthony F. Aveni: *Skywatchers of Ancient Mexico* (Austin: University of Texas Press, 1980), page 181.
15. Bernardino de Sahagún: *Florentine Codex; General History of the Things of New Spain*, book 7, *The Sun, Moon, and Stars, and the Binding of the Years*, translated from the Aztec by Arthur J. O. Anderson and Charles E. Dibble (Santa Fe, New Mexico: School of American Research; Salt Lake City: University of Utah, 1953), pages 36 and 38.

A MOMENT OF TOTALITY

The Value of Totality

In 2003, Xavier Jubier, a computer scientist who lives in Paris and created the interactive eclipse maps on Google, traveled with a group of 80 to a total eclipse in Antarctica. They made arrangements to fly on a Russian cargo plane from Cape Town, South Africa to a Russian research station in Antarctica—the equivalent of a coast-to-coast flight across the United States. In Antarctica, on eclipse day, they traveled 9 miles (15 kilometers) further inland by snowcat. For the eclipse, the sky was perfectly clear and the temperature was −13°F (−25°C).

One of the people who accompanied Jubier into Antarctica was a 70-year-old American woman, recently widowed, who was living in Botswana. To raise money to travel to this eclipse she sold her house. After the eclipse, she told Jubier she loved it and had absolutely no regrets.[1]

[1] Xavier Jubier, interview, July 31, 2014.

4

Eclipses in Mythology

There was at the same time something in its singular and wonderful appearance that was appalling: and I can readily imagine that uncivilised nations may occasionally have become alarmed and terrified at such an object . . .

Francis Baily, on seeing his first total eclipse (1842)[1]

Two great lights brighten the heavens. Life depends on them. Eclipses take that light away. Through the ages, most cultures responded to eclipses of the Sun and Moon with stories to explain the eerie events.

The mythology of solar eclipses might be divided into several themes, and these themes are found scattered throughout the world:

- A celestial being (usually a monster) attempts to destroy the Sun.
- The Sun fights with its lover the Moon.
- The Sun and Moon make love and discreetly hide themselves in darkness.
- The Sun-god grows angry, sad, sick, or neglectful.[2]

Within these myths is a great truth. The harmony and well-being of Earth rely on the Sun and Moon. Abstract science cannot convey this profound realization as powerfully as a myth in which the celestial bodies come to life.

The Sun for Lunch

Most often in mythology a solar eclipse is considered to be a battle between the Sun and the spirits of darkness. The fate of Earth and its inhabitants hangs in the balance. With so much at stake, the people enduring an eclipse are anxious to help the Sun in this struggle if they can.

In the mythology of the Norse tribes, Loki, an evil enchanter, is put in chains by the gods. In revenge, he creates giants in the shape of wolves. The mightiest is Hati, who causes an eclipse by swallowing the Sun. Sköll, the other wolf-like giant, follows the Moon, always seeking a chance to devour it.[3] Old French and German chants echo this belief: "God protect the Moon from wolves."

In the mythology of India, the demons and their younger brothers, the gods, fight over the possession of amrita, the nectar of immortality. Rahu, a demon, disguises himself and attends a gathering of the gods to steal the amrita. But the Sun and Moon recognize him and alert chief god Vishnu. Just as Rahu grabs the amrita and begins to drink it, Vishnu hurls a discus that slices off Rahu's head. Not a drop of amrita gets past Rahu's throat, so his body dies. But Rahu has a mouthful of amrita, so his head is immortal. Rahu's head flies off into the heavens to take revenge on the Sun and Moon. Whenever he catches them, he swallows them—and that's why we have eclipses. But Rahu has only a head and no body, so soon after he swallows them, the Sun and Moon reappear.

In some versions of the myth, the headless body of Rahu has a name: Ketu. One retelling says that Ketu falls to Earth with a colossal thud, causing quakes. Another says that Ketu remains in the heavens as the constellations.[4]

The Indian myth of Rahu spread into China, Mongolia, and southern Siberia. Rahu took on different names but his severed head went on chasing and chewing on the Sun and Moon.

When the Sun and Moon scream for help, the people on the ground respond by hollering and by throwing stones and shooting arrows into the sky to scare away the monster.

In southern Siberia, a Buryat myth says the eclipse-maker is Arakho (Rahu), a beast who formerly lived on Earth. In those days long ago, human beings were quite hairy. Arakho roamed the Earth munching the hair off their bodies until people became the nearly hairless creatures we are today. This annoyed the gods, who chopped Arakho in two. So Arakho no longer grazes on human hair. Instead, his head, which is still alive, chomps on the Sun and Moon, causing eclipses.

Rahu appears again in Indonesia and Polynesia as Kala Rau, all head and no body, who eats the Sun, burns his tongue, and spits the Sun out.

Other Sun-Eaters and Sun-Beaters

From around the world come stories of many different monsters intent on devouring the Sun and Moon. In China, it was a heavenly dog who causes eclipses by eating the Sun.[5] In South America, the Mataguaya Indians of the pampas saw eclipses as a great bird with wings outspread, assailing the Sun or Moon.[6] In Armenian mythology, eclipses were the work of dragons who sought to swallow the Sun and Moon. In contrast, another Armenian myth says that a sorcerer can stop the Sun or Moon in their courses, deprive them of light, and even force them down from the

Egyptian emblem of the winged Sun at the top of the Gateway of Ptolemy at Karnak. [William Tyler Olcott: *Sun Lore of All Ages*]

skies. Despite the Moon's size, once it has been brought down to Earth, the sorcerer can milk it like a cow.[7]

In an Egyptian myth, Ra, the Sun god, travels across the sky each day in a boat. His eternal enemy Apep is always watching for his opportunity to attack. Apep, god of chaos, hater of sunlight, is a whale-size serpent who lives in the depths of the Nile. At just the right moment, he leaps from the water, opens his jaws, and swallows the Sun—boat and all. But traveling with Ra are his defenders. They are always able—so far—to rip the Sun from Apep and cast the vicious serpent back into the depths. Ra, the Sun, continues his daily boat ride westward across the heavens. Apep hungers for another try.[8]

In North Africa, a Berber myth explains a solar eclipse like this: A huge, winged ifrit (evil jinni) zooms skyward from its underground lair and swallows the Sun. But the meal gives him a cosmic case of heartburn. He vomits up the Sun and it resumes shining.[9]

The Pomo Indians of northern California gave a solar eclipse a name that was also an explanation: Sun-got-bit-by-bear. This bear was walking in the sky when it bumped into the Sun. Stand aside, said the Sun. Get out

Drawing of the corona by Samuel P. Langley from the top of Pike's Peak, July 29, 1878. The elongated corona resembles the Egyptian emblem of the winged Sun. [Mabel Loomis Todd: *Total Eclipses of the Sun*]

of my way, said the bear. So, initiating a road rage tradition that lives on in America, they fought about it, with the bear chewing on the Sun, creating an eclipse. Then the bear and the Sun continued their journeys—until the bear bumped into the Moon, which led to another argument, which led to a Moon-got-bit-by-bear. Then the bear and the Moon continued on their way . . . [10]

On rare occasions in mythology, the Sun and Moon are not totally innocent victims. In a variant Hindu myth, the Sun and Moon once borrowed money from a member of the savage Dom tribe and failed to pay it back. In retribution, a Dom occasionally devours the two heavenly bodies.[11]

The Original Black Holes

In two instances, the hungry monster that swallows the Sun or Moon becomes quite scientifically sophisticated in character. A western Armenian myth, said to be borrowed from the Persians, tells of two dark bodies, the children of a primeval ox. These dark bodies orbit the Earth closer than the Sun and Moon. Occasionally they pass in front of the Sun or Moon and thereby cause an eclipse.[12]

The sudden appearance of the corona during totality would have been a terrifying sight to ancient civilizations. The corona was photographed from northern China during the total eclipse of August 1, 2008. [Nikon D200 DSLR, Sigma 170–500 mm zoom at 500 mm, f/6.3, 1/45 second, ISO 1000. ©2008 Patricia Totten Espenak]

Still more remarkable is a Hindu myth that speaks of the Navagrahas, the Nine Seizers. These nine "planets" that wander through the star field include the usual seven bodies familiar to the Greeks—Sun, Moon, Mercury, Venus, Mars, Jupiter, and Saturn—plus Rahu and Ketu. Rahu (eater of the Sun and Moon) and Ketu (Rahu's lower half) are the ascending and descending nodes of the Moon, the shifting points in the sky where the Moon crosses the apparent path of the Sun.[13] Thus, quite correctly, the Sun would be at risk of an eclipse whenever it passes by Rahu or Ketu.

Buddhism carried Rahu and Ketu from India to China in the 1st century CE, where they became Lo-Hou and Chi-Tu. They were imagined as two invisible planets positioned at the nodes in the Moon's path: Lo-Hou at the ascending node and Chi-Tu at the descending node. These "dark stars" were numbered among the planets and were considered to be the cause of eclipses.[14]

Love, Marriage, and Domestic Violence

A Germanic myth explains eclipses differently. The male Moon married the female Sun. But the cold Moon could not satisfy the passion of his fiery bride. He wanted to go to sleep instead. The Sun and Moon made a bet—whoever awoke first would rule the day. The Moon promptly fell asleep, but the Sun, still irritated, awoke at 2 a.m. and lit up the world. The day was hers; the Moon received the night. The Sun swore she would never spend the night with the Moon again, but she was soon sorry. And the Moon was irresistibly drawn to his bride. When the two come together, there is a solar eclipse, but only briefly. The Sun and Moon begin to reproach one another and fall to quarreling. Soon they go their separate ways, the Sun blood-red with anger.

The native Awawak and Gê people of South America also have myths in which the Sun and Moon fight each other, creating an eclipse.[15] It's hard to know the antiquity of such stories, but if they are ancient, they may indicate a realization that the Sun and Moon *together* cause eclipses, which was beyond the understanding of most ancient people.

It is not always a fight between the Sun and Moon that causes an eclipse. Sometimes it is love and modesty. The Tlingit Indians of the Pacific coast in northern Canada explained a solar eclipse as the Moon-wife's visit to her husband. Across the continent, in southeastern Canada, the Algonquin Indians also envisioned the Sun and Moon as loving husband and wife. If the Sun is eclipsed, it is because he has taken his child into his arms.[16]

The Fon people of Benin in western Africa say the male Sun rules the day and female Moon rules the night. They love one another, but they are very busy, always in motion. They meet whenever they can, which is not often. But when they do, they modestly turn off the light.[17] For the Tahitians, the Sun and Moon were lovers whose union creates an eclipse. In that darkness, they lose their way and create the stars in order to light their return.

An Angry Sun

Sometimes, as in a folktale from eastern Transylvania, it is the perversion of mankind that brings on an eclipse. The Sun shudders, turns away in disgust, and covers herself with darkness. Stinking fogs gather. Ghosts appear. Dogs bark strangely and owls scream. Poisonous dews fall from the skies, a danger to man and beast. Neither humankind nor animals should consume water or eat fresh fruit or vegetables. Such beliefs persisted into the 19th century. This poisonous dew that supposedly accompanied eclipses could be the source of an outbreak of the plague or other epidemics. If people had to leave their homes, they wrapped a towel around their mouths and noses to strain out the noxious vapors. Clothes caught drying outdoors during a solar eclipse were considered to be infected.

Europeans were not alone in their belief that a solar eclipse brought a dangerous form of precipitation. Eskimos in southwestern Alaska believed that an unclean essence descended to Earth during an eclipse. If it settled on

A partial solar eclipse on April 29, 2014, was imaged at sunset in this multiple exposure composite. Clouds during sunrise/sunset eclipses would have made it easier for ancient people to see the partial phases. [Canon 50D, 70–200 mm zoom, Baader solar filter, ISO 200, exposure 1/100 second every 3 minutes. ©2014 Stephen Mudge]

utensils, it would produce sickness. Therefore, when an eclipse began, every Eskimo woman turned all her pots, buckets, and dishes upside down.[18]

Yet it was not always an angry Sun that brought darkness to the Earth. When US Coast Survey scientist George Davidson observed the eclipse of August 7, 1869, from Kohklux, Alaska, he found that the Native Americans there attributed the eclipse to an illness of the Sun. He had alerted them to the impending eclipse, but they doubted it. Halfway through the partial phase, the Indians and their chief quit work and hid in their houses: "They looked upon me as the cause of the Sun's being 'very sick and going to bed.' They were thoroughly alarmed, and overwhelmed with an indefinable dread."[19]

Sometimes in mythology, an eclipse is not a monster devouring the Sun, not a sickness of the Sun, not a fight between the Sun and Moon, not even the result of the always abundant sins of mankind. Sometimes an eclipse is what in sports would be called an unforced error. The Nuxalk (Bella Coola) Indians of the Pacific coast in Canada have a myth that begins with a remarkable observational description of the Sun's apparent annual pathway though the sky. The trail of the Sun, they say, is a bridge whose width is the distance between the summer and winter solstices, the northernmost and southernmost positions of the Sun. In the summer, the Sun walks on the right side of this bridge; in the winter, he walks on the left. The solstices are where the Sun sits down. Accompanying the Sun on his journey are three guardians who dance about him. Sometimes the Sun simply drops his torch, and thus an eclipse occurs.[20]

Warding Off the Evil

Corruption and death are a frequent theme of eclipse myths. Evil spirits descend to Earth or emerge from underground during eclipses.

On June 16, 1406, says an enlightened and bemused French chronicler, "between 6 and 7 a.m., there was a truly wonderful eclipse of the Sun which lasted nearly half an hour. It was a great shame to see the people withdrawing to the churches and believing that the world was bound to end. However the event took place, and afterward the astronomers gathered and announced that the occurrence was very strange and portended great evil."[21] (Half an hour is too long for totality, but the right length for conspicuously reduced light surrounding totality.)

For Hindus, the place to be during an eclipse was in the water, especially in the purifying current of the Ganges.[22] This Hindu practice of immersing oneself in water was known to the French philosopher and popularizer of science Bernard Le Bovier de Fontenelle, as recorded in his *Entriens sur la pluralité des mondes* (Conversations on the Plurality of Worlds) in 1686.

> All over India, they believe that when the sun and moon are eclipsed, the cause is a certain dragon with very black claws which tries to seize those two bodies, wherefore at such times the rivers are seen covered with human heads, the people immersing themselves up to the neck, which they regard as a most devout position, and implore the sun and moon to defend themselves well against the dragon.[23]

A Participatory Event

From around the world come reports of people trying to assist the Sun and Moon against the peril of eclipse. Screaming, crying, and shouting are supposed to encourage the Sun and Moon to escape the clutches of the evil spirit. Historians recorded that Germans watching a lunar eclipse in the Middle Ages chanted in unison, "Win, Moon."

The Sun and Moon were the supreme gods of the Indians of Colombia. When these gods were threatened by an eclipse, the people seized their weapons and made warlike sounds on their musical instruments. They also shouted to the gods, promising to mend their ways and work hard. To prove it, they watered their corn and worked furiously with their tools during the eclipse.[24]

People frequently augmented their voices with the clanging of metal pots, pans, and knives. The Chippewa Indians in the northeastern United States and southeastern Canada went even further. Seeing the Sun's light being extinguished, they shot flaming arrows into the sky, hoping to rekindle the Sun.[25] The Sencis of eastern Peru also shot fire arrows toward the Sun, but not to rekindle it. They were trying to scare off a savage beast that was attacking the Sun.

Ethnographers descended on the Kalina tribe in Suriname to collect their folklore and watch their behavior as the total eclipse of June 30, 1973, neared. In Kalina mythology, the Sun and Moon are brothers. They usually get along well together, but occasionally they have sudden and ferocious quarrels that endanger mankind. At such times, it is important to separate the combatants by making a maximum of noise: banging on tools, hollow objects, and instruments. The fading of the Sun or Moon means that one has been knocked unconscious. When this happens, the tribesmen yell, "Wake up, Papa!" Papa, here, is a term of respect, not an indication that the Kalina consider themselves descended from the Sun or Moon. After the 1973 eclipse, invisible because of clouds but noticeable because of darkening, the old women (the pot makers) rounded up the children and used branches to spread white clay all over them, somewhat too vigorously for the children's enjoyment. Then they smeared the women and finally the men from head to foot with white clay. The white clay was the blood of the injured Moon that had dripped onto the ground. It was necessary to wash oneself with the Moon's blood to restore purity in man and whiteness to the Moon. After an hour, the tribesmen washed themselves off in the river.[26]

In ancient Mexico and Central America, the most important god was represented as a plumed serpent. For the Maya, he was Kukulcán; for the Aztecs, Quetzalcóatl. At an eclipse of the Moon, and most especially during an eclipse of the Sun, a special snake was killed and eaten.[27]

In prehistoric times, screams, cries, banging noises, and prayer may not have been deemed adequate to ward off eclipses or their effects. In many places around the world, human sacrifice was performed at the appearance of unexpected and confusing sights, such as an eclipse or a comet. Yet few eclipse myths refer to human sacrifice, suggesting that this practice had largely been abandoned before most eclipse myths were preserved. An exception was in Mexico and Central America, where the Spanish invaders saw the Aztecs and their neighbors carry out human sacrifice in the early 16th century. For the Aztecs, almost any natural or political event was commemorated with a sacrifice. On the occasion of an eclipse, the Sun was in need of help from people, just as he had constant help from the dog Xolotl (Sho-LOT-uhl). Xolotl was the god of human monstrosities (including twins), so it was humpbacks and dwarfs who were sacrificed to the Sun to help him prevail.[28]

Solar eclipses boded ill for everyone, it seems, except prospectors. In Bohemia, people believed that a solar eclipse would help them find gold.

NOTES AND REFERENCES

1. Epigraph: Francis Baily: "Some Remarks on the Total Eclipse of the Sun, on July 8th, 842," *Memoirs of the Royal Astronomical Society*, volume 15, 1846, page 6.
2. Viktor Stegemann: "Finsternisse" in Hanns Bächtold-Stäubli, editor: *Handwörterbuch des Deutschen Aberglaubens*, Bände 2 (Berlin: W. de Gruyter, 1930), columns 1509–1526. Germanic eclipse lore described in this chapter comes from this article unless otherwise noted.
3. In some versions of the myth, Sköll eats the Sun and Hati eats the Moon. See E. C. Krupp: *Beyond the Blue Horizon: Myths and Legends of the Sun, Moon, Stars, and Planets* (New York: HarperCollins, 1991), page 162.
4. Jan Knappert: *Indian Mythology: An Encyclopedia of Myth and Legend* (London: HarperCollins, 1991), page 203. Arthur Berriedale Keith: *Indian Mythology*, volume 6 of *The Mythology of All Races* (Boston: Marshall Jones, 1917), page 151.
5. John C. Ferguson: *Chinese Mythology*, volume 8 of *The Mythology of All Races* (Boston: Marshall Jones, 1928), page 84.
6. Hartley Burr Alexander: *Latin-American Mythology*, volume 11 of *The Mythology of All Races* (Boston: Marshall Jones, 1920), page 319.
7. Mardiros H. Ananikian: *Armenian Mythology*, volume 7 of *The Mythology of All Races* (Boston: Marshall Jones, 1925), page 48.
8. Harold Scheub: *A Dictionary of African Mythology: The Mythmaker as Storyteller* (Oxford: Oxford University Press, 2000), page 216.
9. Harold Scheub: *A Dictionary of African Mythology: The Mythmaker as Storyteller* (Oxford: Oxford University Press, 2000), page 74.
10. E. C. Krupp: *Beyond the Blue Horizon: Myths and Legends of the Sun, Moon, Stars, and Planets* (New York: HarperCollins, 1991), page 162.
11. Arthur Berriedale Keith: *Indian Mythology*, volume 6 of *The Mythology of All Races* (Boston: Marshall Jones, 1917), pages 232–233.

12. Mardiros H. Ananikian: *Armenian Mythology*, volume 7 of *The Mythology of All Races* (Boston: Marshall Jones, 1925), page 48.
13. Arthur Berriedale Keith: *Indian Mythology*, volume 6 of *The Mythology of All Races* (Boston: Marshall Jones, 1917), page 233.
14. Joseph Needham and Wang Ling: *Science and Civilisation in China*, volume 3, *Mathematics and the Sciences of the Heavens and the Earth* (Cambridge: At the University Press, 1959), page 228.
15. E. C. Krupp: *Beyond the Blue Horizon: Myths and Legends of the Sun, Moon, Stars, and Planets* (New York: HarperCollins, 1991), page 162.
16. Hartley Burr Alexander: *North American Mythology*, volume 10 of *The Mythology of All Races* (Boston: Marshall Jones, 1916), pages 25, 277.
17. Harold Scheub: *A Dictionary of African Mythology: The Mythmaker as Storyteller* (Oxford: Oxford University Press, 2000), page 168.
18. James George Frazer: *Balder the Beautiful*, volume 1; *The Golden Bough*, volume 10 (London: Macmillan, 1930), page 162.
19. Mabel Loomis Todd: *Total Eclipses of the Sun*, revised edition (Boston: Little, Brown, 1900), page 131.
20. Hartley Burr Alexander: *North American Mythology*, volume 10 of *The Mythology of All Races* (Boston: Marshall Jones, 1916), page 255.
21. Paul Yves Sébillot: *Le folk-lore de France*, tome 1, *Le ciel et la terre* (Paris: Librairie orientale & américaine, 1904), page 52. The chronicler was Jean Juvénal des Ursins.
22. Arthur Berriedale Keith: *Indian Mythology*, volume 6 of *The Mythology of All Races* (Boston: Marshall Jones, 1917), page 234.
23. A free translation from the "Second Soir." Compare Bernard Le Bovier de Fontenelle: *A Plurality of Worlds*, translated by John Glanvill (England: Nonesuch Press, 1929), with the excerpt in François Arago: *Popular Astronomy*, volume 2, translated by W. H. Smyth and Robert Grant (London: Longman, Brown, Green, Longmans, and Roberts, 1858), page 349.
24. Hartley Burr Alexander: *Latin-American Mythology*, volume 11 of *The Mythology of All Races* (Boston: Marshall Jones, 1920), pages 277–278.
25. James George Frazer: *The Magic Art*, volume 1; *The Golden Bough*, volume 1 (London: Macmillan, 1926), page 311.
26. Patrick Menget: "30 juin 1973: station de Surinam," *Soleil est mort; l'éclipse totale de soleil du 30 jin 1973* (Nanterre, France: Laboratoire d'ethnologie et de sociologie comparative, 1979), pages 119–142.
27. Hartley Burr Alexander: *Latin-American Mythology*, volume 11 of *The Mythology of All Races* (Boston: Marshall Jones, 1920), page 135.
28. Hartley Burr Alexander: *Latin-American Mythology*, volume 11 of *The Mythology of All Races* (Boston: Marshall Jones, 1920), page 82.

A MOMENT OF TOTALITY

My Favorite Eclipse

by Ken Willcox

Every total eclipse is different—and wonderful—but if you see more than one, you probably have a favorite. Mine was the eclipse of November 3, 1994, on the Altiplano in Bolivia.

The Altiplano is a huge desert high in the Andes Mountains. It was there we had come, 110 of us from all over the United States and Europe, to a small plot of land 12,516 feet (3,815 meters) above sea level. I selected the Altiplano because it provided the best chance of clear skies along the eclipse path. But the price of clear skies was a major logistical problem: transporting 110 people safely and comfortably to a remote high-altitude viewing site in a foreign land.

On Wednesday, November 2, we boarded our own private train in La Paz, with a dining car and also a baggage car to accommodate all our equipment to record three precious minutes of time. At 4:50 p.m., our chartered narrow-gauge train began its spiraling climb out of the trench in which La Paz lies, then turned south toward our destination 200 miles (320 kilometers) away, a spot on the central line, 7½ miles (12 kilometers) south of Sevaruyo. In the middle of the night, using the global positioning system (GPS), Jim Zimbleman of the Smithsonian Institution and I navigated the train to a precise spot along the tracks selected two years earlier on a site inspection trip.

As we approached our site about 4 a.m., we were met by 70 gun-toting soldiers provided for our protection by the Bolivian army. Under floodlights, before dawn, in the middle of nowhere high in the Andes, amateur and professional astronomers, and Bolivian soldiers dragged equipment from the train and prepared to record an extraordinary celestial event.

One hour after sunrise, the clouds began breaking up. Only high, scattered cirrus remained at first contact at 7:19 a.m., and those too were vanishing.

A family of Aymara Indians, descendants of the Incas, had been invited to join us. We gave them solar filters to view the partial phases. This Bolivian family huddled around their father as totality approached. We kept glancing at them to see what their response to the disappearing Sun and the onset of totality would be. A few minutes before totality, the father made his three boys look down at the ground until the eclipse was over. The father feared the souls of the boys would be unalterably affected.

"Here it comes!" someone shouted. "Where?" "Over there," pointing to the northwest. "Oh yes! I see it!" "It's getting dark now . . . it's getting real dark now . . . it's really getting . . . OH MY GOD!" Cheers and gasps accompanied the beginning of totality. It took everything I had to keep my mind focused on the task at hand, and even that didn't work when the lady behind me broke down crying. Her husband wrapped his arms around her.

As totality ended, I shot a last few exposures blinded by my own tears, gave up, and turned around and photographed the couple behind me.

The Aymara Indians were right. All our souls were unalterably affected by that eclipse. But it was—and is—a sublime experience, one that should be sought, not feared.[1]

[1] Adapted from Mark Littmann, Fred Espenak, and Ken Willcox: *Totality: Eclipses of the Sun* (New York: Oxford University Press, 2008; updated 2009), pages 142–143.

5

The Strange Behavior of Man and Beast—Long Ago

The Sun . . .
In dim eclipse disastrous twilight sheds
On half the nations, and with fear of change
Perplexes monarchs.

John Milton (1667)[1]

The Human Reaction

One of the most dramatic responses in history to a total solar eclipse is presented by Herodotus, the first Greek historian, writing around 430 BCE.

> War broke out between the Lydians and the Medes [major powers in Asia Minor], and continued for five years . . . The Medes gained many victories over the Lydians, and the Lydians also gained many victories over the Medes. . . . As, however, the balance had not inclined in favour of either nation, another combat took place in the sixth year, in the course of which, just as the battle was growing warm, day was on a sudden changed into night. This event had been foretold by Thales, the Milesian, who forewarned the Ionians of it, fixing for it the very year in which it actually took place. The Medes and Lydians, when they observed the change, ceased fighting, and were alike anxious to have terms of peace agreed upon.

A treaty was quickly made and sealed by the marriage of the daughter of the Lydian king to the son of the Median king.[2] This story indicates the awe that ancient people felt when confronted with a total eclipse of the Sun.[3]

Modern astronomers, armed with the dates of the kings described in the account and a knowledge of the dates and paths of ancient eclipses, have generally settled upon May 28, 585 BCE as the eclipse to which the story refers, if Herodotus, given to fanciful embellishment, can be trusted about an event that occurred a century before he was born.[4]

In his account, Herodotus credits Greek philosopher Thales with predicting this eclipse. If so, Thales would have been the first person *known* to have calculated a future solar eclipse. But the Babylonians, not the Greeks, were the leaders in eclipse prediction. Their cuneiform writing on clay tablets from the period of the 585 BCE eclipse shows recognition of an 18-year-11-day rhythm in eclipses—the saros.[5] Might Thales have borrowed this eclipse rhythm from the Babylonians as it was being developed?

However, such a rhythm predicts not just the year but the month and precise day of the eclipse. Yet Herodotus seems amazed that Thales could be accurate to "the very year in which it actually took place." Was Herodotus so surprised that Thales could predict an eclipse accurate to the day that he simply could not believe that degree of precision and used the more conservative "year" instead? That would be out of character for the flamboyant Herodotus. Yet predicting a solar eclipse accurate to a year is not much of a trick since there are a minimum of two solar eclipses a year. The problem is to predict a total eclipse for a particular location on Earth. Could Thales have accomplished this? It is doubtful.

The saros period is actually 18 years 11⅓ days, so the Earth has spun through an extra 8 hours. Thus, each subsequent eclipse falls about one-third of the way around the world westward from the one before it. The result is that successive eclipses in a saros series are almost never visible from the same site.

The eclipse in the same saros series that preceded 585 BCE occurred on May 18, 603 BCE, with an early morning path from the northern portion of the Red Sea to the northern tip of the Persian Gulf, about 600 miles (1,000 kilometers) distance from the end of the path of the May 28, 585 BCE eclipse. Thales could have heard reports of the 603 BCE eclipse and used it to calculate the date for the 585 BCE eclipse. But the saros projection would not have told him where the eclipse would be visible. Thales, then, first of the great Greek philosophers, could have warned of the *possibility* of a solar eclipse, but he could not predict from the saros period that it would be visible in Asia Minor. And there is no evidence that he had the celestial knowledge or the mathematics to calculate it from orbital considerations.

Of course the key to appreciating the story of the solar eclipse that stopped a war is the realization that people long ago were stunned by a total eclipse of the Sun and incredulous that someone could predict such an event. Quite often in ancient history, eclipses are reported to have played a decisive role in the turn of events.

Herodotus tells of another turning point in world history that he says hinged on a solar eclipse. Xerxes and his Persian army were about to march from Sardis to Abydos on their advance toward Greece.

During a battle between the Lydians and the Medes on May 28, 585 BCE, a total eclipse of the Sun occurred. It scared the soldiers so badly that they stopped fighting and signed a treaty. [Mabel Loomis Todd: *Total Eclipses of the Sun*]

At the moment of departure, the sun suddenly quitted his seat in the heavens, and disappeared, though there were no clouds in sight, but the sky was clear and serene. Day was thus turned into night; whereupon Xerxes, who saw and remarked the prodigy, was seized with alarm, and sending at once for the Magians, inquired of them the meaning of the portent. They

> replied—"God is foreshadowing to the Greeks the destruction of their cities; for the sun foretells for them, and the moon for us." So Xerxes, thus instructed, proceeded on his way with great gladness of heart.[6]

To disaster! He reached and burned Athens, but his navy was destroyed by the Greeks and his forces had to withdraw. Twice more Xerxes invaded Greece, but each time his armies were crushed. After his last defeat, his nobles assassinated him.

Xerxes' first march against Greece actually occurred in 480 BCE, but the only major eclipse visible in the region near that date was the total eclipse of February 17, 478 BCE. Thus, the story tells us less about observational astronomy in that era than about the power exercised by eclipses over the minds of men and the effectiveness of their use to heighten the drama of a story.

A final story from Greece illustrates an advance from superstitious dread of eclipses to an understanding of what causes them. On August 3, 430 BCE, Pericles and his fleet of 150 warships were about to sail for a raid upon their enemies.

> But at the very moment when the ships were fully manned and Pericles had gone onboard his own trireme, an eclipse of the sun took place, darkness descended and everyone was seized with panic, since they regarded this as a tremendous portent. When Pericles saw that his helmsman was frightened and quite at a loss what to do, he held up his cloak in front of the man's eyes and asked him whether he found this alarming or thought it a terrible omen. When he replied that he did not, Pericles asked, "What is the difference, then, between this and the eclipse, except that the eclipse has been caused by something bigger than my cloak?" This is the story, at any rate, which is told in the schools of philosophy.[7]

The eclipse was a large partial at Athens and annular about 600 miles (1,000 kilometers) to the northeast. This eclipse had also been recorded by Thucydides, without the didactic story, but exhibiting an increased awareness of the cause of eclipses: "The same summer, at the beginning of a new lunar month, the only time by the way at which it appears possible, the sun was eclipsed after noon. After it had assumed the form of a crescent and some stars had come out, it returned to its natural shape."[8]

Almost 2,000 years later, however, superstition still clouded the public's view of eclipses. On June 3, 1239, Thomas, Archdeacon of Split, Croatia, chronicled "the wonderful and terrible eclipse of the Sun." "Such great fear overtook everyone," he wrote, "that just like madmen they ran about to and fro shrieking, thinking that the end of the world had come."

On October 6, 1241, only 2½ years later, Split experienced a second total solar eclipse—a rare event—and Thomas once again recorded "great terror among everyone."[9]

Four centuries later, not much had changed. Paris was a center of education, yet on August 12, 1654, "at the mere announcement of a total eclipse, a multitude of the inhabitants of Paris hid themselves in deep cellars."[10]

No wonder then that the sight of a total eclipse on July 29, 1878, had a powerful effect on Native Americans near Fort Sill, Indian Territory (now Oklahoma). A non-Indian described it this way:

> It was the grandest sight I ever beheld, but it frightened the Indians badly. Some of them threw themselves upon their knees and invoked the Divine blessing; others flung themselves flat on the ground, face downward; others cried and yelled in frantic excitement and terror. Finally one old fellow stepped from the door of his lodge, pistol in hand, and, fixing his eyes on the darkened Sun, mumbled a few unintelligible words and raising his arm took direct aim at the luminary, fired off his pistol, and after throwing his arms about his head in a series of extraordinary gesticulations retreated to his own quarters. As it happened, that very instant was the conclusion of totality. The Indians beheld the glorious orb of day once more peep forth, and it was unanimously voted that the timely discharge of that pistol was the only thing that drove away the shadow and saved them from ... the entire extinction of the Sun.[11]

The Animal Response

Noting their own primal response to the daytime darkening of the Sun, people through the ages have been fascinated by the reaction of animals to a total eclipse. Reports go back more than 750 years. In describing the eclipse of June 3, 1239, Ristoro d'Arezzo wrote: "We saw the whole body of the Sun covered step by step ... and it became night ... and all the animals and birds were terrified; and the wild beasts could easily be caught ... because they were bewildered."[12]

As a university student in Portugal, astronomer Christoph Clavius saw the total eclipse of August 21, 1560: "Stars appeared in the sky, and (miraculous to behold) the birds fell down from the sky to the ground in terror of such horrid darkness."[13]

In 1706 at Montpellier in southern France, observers reported that "bats flitted about as at the beginning of night. Fowls and pigeons ran precipitately to their roosts." In 1715, the French astronomer Jacques Eugène d'Allonville, Chevalier de Louville, traveled to London for the eclipse and observed that at totality "horses that were laboring or employed on the high roads lay down. They refused to advance."[14]

By 1842, some people were even conducting behavioral experiments on their pets. "An inhabitant of Perpignan [France] purposely kept his dog without food from the evening of the 7th of July. The next morning, at the instant when the total eclipse was going to take place, he threw a piece of bread to the poor animal, which had begun to devour it, when the sun's last rays disappeared. Instantly the dog let the bread fall; nor did he take it up again for two minutes, that is, until the total obscuration had ceased; and then he ate it with great avidity."[15]

William J. S. Lockyer, son of the pioneering solar spectroscopist, traveled to Tonga for an eclipse in 1911. The weather conditions were miserable and the insects numerous and very hungry. He and his colleagues caught only a brief view of the corona through thin clouds and the scientific results were meager. The only members of his team with good results were those studying animal behavior. The horses did not seem to notice the darkening, but fowl ran home to roost and pigs lay down. Flowers closed. But most memorable of all were the insects, which had been completely silent until the moment of totality and then sang as if it were night. "The noise," recalled Lockyer, "was most impressive, and will remain in my memory as a marked feature of that occasion."[16]

A Cure for Eclipses

In 1834, the Indians of the Kiowa Tribe gathered together at the End of the Mountains, west of the Wichita Mountains in Oklahoma, for their annual Sun Dance ceremony. On November 30, right at noon, when people were about to eat dinner, the sky began to darken, but there were no clouds blocking the Sun. Darker and darker it got. "The Sun is dying," the people yelled. "A snake has come up from the Underworld," screamed others. "It is swallowing the Sun." "What can we do?" they cried.

A great medicine man came forward, his face painted as a bobcat—the strongest power there is. The medicine man began to shake his rattle and dance and sing, screeching like a bobcat—as if the Sun was sick and he was treating a sick person.

Soon, where the Sun had vanished, there was a sliver of light, like the edge of a fingernail. The medicine man danced and sang even harder. The light grew and grew. Birds flew from their nests and began to sing. The world was alive again. Young men and women laughed. Mothers went back to serving dinner. The Sun Dance ceremony continued.

People showered the medicine man with gifts. They didn't know what made the Sun go away, but they did know who brought the Sun back.[17]

To many early people, the eerie twilight sky of totality was interpreted as the end of the world. A wide-angle view of totality was shot from Dundlod, India, on October 24, 1995. [Nikon FE SLR, Nikkor 50 mm, Ektachrome 100, f/5.6, 1/4 second. ©1995 Fred Espenak]

The Shawnee Prophet Uses an Eclipse

Tenkskwatawa, the Shawnee Prophet (1775?–1837?), was an important Indian religious leader in Ohio and Indiana in the early 19th century. He saw great danger for his people as they increasingly adopted the customs of the European settlers, especially alcohol. He urged them to return to traditional Indian ways and to unite into a single Indian nation under the leadership of his brother Tecumseh to resist the encroachment of white men with their fraudulent treaties.

General William Henry Harrison, later president of the United States, was at that time the governor of Indiana Territory, where the Shawnee Prophet was successfully recruiting converts to his Indian religious revival. Seeking to undermine the credibility of the Shawnee Prophet as a shaman, Harrison urged Indians to demand proof from the Prophet that he could perform miracles. Thinking in biblical terms, Harrison asked if Tenkskwatawa could "cause the Sun to stand still, the Moon to alter its course, the rivers to cease to flow, or the dead to rise from their graves."

The followers of the Shawnee Prophet did not need such displays, but Tenkskwatawa was a canny politician. He proclaimed that on June 16, 1806, he would blot the Sun from the sky as a sign of his divine powers. Whether he knew of this total eclipse from a British agent or from an almanac is uncertain, but a great many Indians gathered at the Shawnee Prophet's camp as the appointed day dawned clear.

At the proper moment, the Prophet, in full ceremonial regalia, pointed his finger at the Sun, and the eclipse began. When the Prophet called out to the Good Father of the Universe to remove his hand from the face of the Sun, the light gradually returned to the Earth. Response to the Prophet's performance was overwhelming and his fame spread rapidly and widely. Harrison's condescension had backfired, to his embarrassment.

But the westward migration of European settlers was unstoppable. In the Battle of Tippecanoe in 1811, Harrison destroyed the Shawnee Prophet's religious center, killing many Indians, and breaking the power of Tenkskwatawa.[a]

[a] See especially Laurence A. Marschall: "A Tale of Two Eclipses," *Sky & Telescope*, volume 57, February 1979, pages 116–118.

NOTES AND REFERENCES

1. Epigraph: John Milton: *Paradise Lost*, book 1, lines 594 and 597–599. See John Milton: *The Complete Poems* (New York: Crown Publishers, 1936), page 24.
2. Herodotus: *The History*, volume 1, translated by George Rawlinson; Everyman's Library, volume 405 (London: J. M. Dent, 1910), book 1, chapter 74, pages 36–37.
3. A similar battle-stopping eclipse occurred over southern Japan on November 11, 1183 CE. Two clans, the Minamoto and the Taira, were in the midst of a 5-year war. They were preparing for battle when an annular eclipse intervened. The Minamoto soldiers were scared out of their wits and fled. The Taira won the day. Unlike the war between the Medes and the Lydians, no treaty was signed. Hostilities continued. Two years later the Minamoto won the war. Shigeru Nakayama: *A History of Japanese Astronomy—Chinese Background and Western Impact*. (Cambridge, Massachusetts: Harvard University Press, 1969), page 51. Thomas Crump also mentions this story in *Solar Eclipse* (London: Constable, 1999), page 192.
4. Robert R. Newton lists three annular eclipses seen in the region during a 50-year period that he feels are equally likely to have given rise to this story, although an annular eclipse is not nearly as spectacular as one that is total. *Ancient Astronomical Observations and the Accelerations of the Earth and Moon* (Baltimore: Johns Hopkins Press, 1970), pages 94–97.
5. Clemency Montelle: *Chasing Shadows: Mathematics, Astronomy, and the Early History of Eclipse Reckoning* (Baltimore: Johns Hopkins University Press, 2011), page 78.
6. Herodotus: *The History*, volume 2, translated by George Rawlinson; Everyman's Library, volume 406 (London: J. M. Dent, 1910), book 7, chapter 37, page 136.

7. Plutarch: *The Rise and Fall of Athens: Nine Greek Lives*, translated by Ian Scott-Kilvert (Baltimore: Penguin Books, 1960), pages 201–202. Ironically, Pericles' raid was disastrous for the Athenian forces; they fell victim to the plague. Pericles was fined and temporarily stripped of power.
8. Thucydides: *History of the Peloponnesian War*, translated by Richard Crawley (New York: E. P. Dutton, 1910), book 2, paragraph 28. Stars would not have been visible at Athens. Perhaps Thucydides heard reports from where the eclipse was annular and incorporated them into his account.
9. F. Richard Stephenson: *Historical Eclipses and Earth's Rotation* (Cambridge: Cambridge University Press, 1997), page 401. Thomas Crump: *Solar Eclipse* (London: Constable, 1999), page 185.
10. From the chapter "Second soir" in Bernard Le Bovier de Fontenelle: *Entretiens sur la pluralité des mondes* (Paris: Chez la veuve C. Blageart, 1686), cited in François Arago: *Popular Astronomy*, volume 2, translated by W. H. Smyth and Robert Grant (London: Longman, Brown, Green, Longmans, and Roberts, 1858), pages 359–360.
11. Mabel Loomis Todd: *Total Eclipses of the Sun*, revised edition (Boston: Little, Brown, 1900), pages 141–142.
12. F. Richard Stephenson and David H. Clark: *Applications of Early Astronomical Records*, Monographs on Astronomical Subjects, number 4 (New York: Oxford University Press, 1978), page 9. Also F. Richard Stephenson: *Historical Eclipses and Earth's Rotation* (Cambridge: Cambridge University Press, 1997), page 397.
13. F. Richard Stephenson and David H. Clark: *Applications of Early Astronomical Records*, Monographs on Astronomical Subjects, number 4 (New York: Oxford University Press, 1978), page 14.
14. François Arago: *Popular Astronomy*, volume 2, translated by W. H. Smyth and Robert Grant (London: Longman, Brown, Green, Longmans, and Roberts, 1858), page 359.
15. François Arago: *Popular Astronomy*, volume 2, translated by W. H. Smyth and Robert Grant (London: Longman, Brown, Green, Longmans, and Roberts, 1858), page 362.
16. William J. S. Lockyer: "The Total Eclipse of the Sun, April 1911, as Observed at Vavau, Tonga Islands," in Bernard Lovell, editor: *Astronomy*, volume 2, The Royal Institution Library of Science (Barking, Essex: Elsevier Publishing, 1970), pages 190–191.
17. Alice Marriott and Carol K. Rachlin: *Plains Indian Mythology* (New York: Thomas Y. Crowell, 1975), pages 131–132. There is an inconsistency in this story: The Sun Dance is held in the summer and Marriott and Rachlin introduce the story as occurring in the summer. The total solar eclipse occurred on November 30, 1834. The Kiowa also recorded another terrifying event that had occurred a year earlier—the night when the stars fell. That was the great Leonid meteor storm of November 13, 1833. Marriott and Rachlin recount that story too, pages 129–131.

A MOMENT OF TOTALITY

Fashionably Late to an Eclipse

The great astronomy popularizer Camille Flammarion, quoting astronomer François Arago, tells of the eclipse of May 22, 1724, visible from Paris. Supposedly, Jacques Cassini, director of the Paris Observatory, invited a marquis and his aristocratic lady friends to observe the eclipse with him at the observatory. However, the ladies fussed so long with their gowns and hair styles that the party arrived late, a few minutes after totality had ended. "Never mind, ladies," said the marquis, "we can go in just the same. M. Cassini is a great friend of mine and he will be delighted to repeat the eclipse for you."[1]

[1] Camille Flammarion: *The Flammarion Book of Astronomy*, edited by Gabrielle Camille Flammarion and André Danjon, translated by Annabel and Bernard Pagel (New York: Simon and Schuster, 1964), page 147.

6

The Sun at Work

As long as the Sun shall shine,
As long as the rivers shall flow,
As long as the Moon shall rise
As long as the grass shall grow.

Anonymous, North American Indian expression
for how long a promise should last[1]

At the Core

At this moment, deep in the core of the Sun, the nucleus of a hydrogen atom—a proton—is colliding and fusing with another hydrogen nucleus, and the collisions and fusions proceed until four hydrogen nuclei have become the nucleus of one helium atom. In this nuclear reaction, a tiny amount of mass has been destroyed: not lost, but converted into energy. It is this reaction that powers the Sun and all stars, creates their light and heat, for more than 90% of their active lives.

In the Sun, tens of trillions of these reactions take place every *second.* Every second 600 million tons of hydrogen become 596 million tons of helium, and 4 million tons of mass become energy, in accordance with Einstein's famous equation $E = mc^2$. A little bit of mass yields a vast amount of energy.[2]

Even though the Sun is actually losing mass at the rate of 4 million tons every second, this weight-reduction plan is far from a crash diet. At 4 million tons a second, it would take the Sun 14.8 trillion years to consume itself entirely, if it could. But it can't. The heat to run this nuclear fusion comes from the gravitational force of all the mass of the Sun pressing inward on the core, and it is only within this central 25% of the Sun's diameter that the temperatures—up to 27 million °F (15 million °C)—are hot enough to generate and sustain this reaction.[3]

Over billions of years, the hydrogen at the core is converted into helium until the core is too clogged with helium for hydrogen fusion to continue. It is then that stars begin to die. Our Sun has been shining for 4.6 billion years and has enough hydrogen at its core to continue shining much as it does now for about another 5 billion.

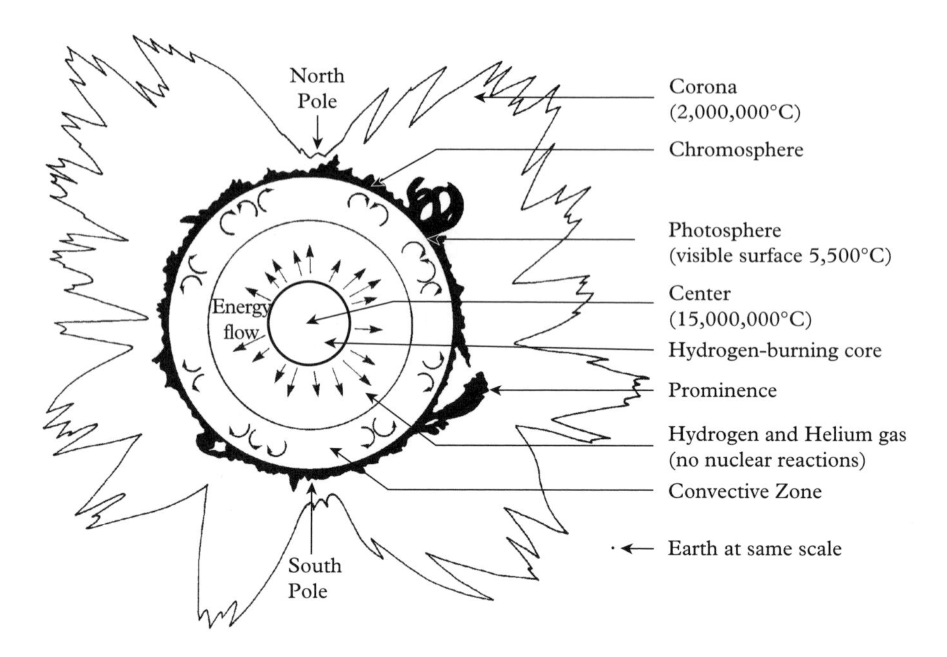

Cross section of the Sun and its atmosphere. [Drawing by Josie Herr after William K. Hartmann: *Cosmic Voyage Through Time and Space*, ©1990 Wadsworth Publishing, Inc., by permission of the publisher]

The fiercely hot reaction at the heart of our Sun is concealed from our view by 432,500 miles (696,000 kilometers) of opaque gases. And a good thing too. The principal radiation generated at the Sun's core is not visible light but gamma rays. There would be no life on Earth if the Sun radiated large quantities of this high-energy radiation in our direction.

Fortunately, it does not. High-energy photons radiate outward from the Sun's core but are absorbed in the crush of other atomic particles, which in turn reradiate this energy. But the energy emitted is not the same. With each absorption and emission, some of the photon's original energy is lost. A photon that started as a highly energetic gamma ray, if tracked from absorption to absorption, would gradually become an x-ray, then an ultraviolet ray, and then visible light as it bounced around randomly inside the Sun.

At about 130,000 miles (200,000 kilometers) below the Sun's surface, the temperature and density have fallen enough so that energy is conveyed upward less by radiation than by convection—the rise of gases heated by this energy from below.

If energy created at the Sun's core could leave the Sun directly traveling at the speed of light, it would emerge from the Sun's surface in 2⅓ seconds. But because of the countless absorptions and reemissions in random directions, a typical photon requires 10 million years to reach the Sun's surface. There, at last, it is free to move directly away from the Sun at the speed of light. At that pace, it travels the distance from the Sun to the Earth in 8⅓ minutes.

Layers Above

The Sun's internal layers of radiation and convection are hidden from our eyes. What we see of the Sun, when we or the atmosphere provide adequate filters, is an apparent disk, glaring white-hot at a temperature of about 10,000°F (5,500°C). This disk is an optical illusion, since the Sun is gaseous throughout. We are really seeing the layer of the Sun in which the density and ionization of atoms is so great that the gas becomes opaque. This region that provides the Sun with the appearance of a surface is called the *photosphere* ("light sphere"). It is only about 200 miles (300 kilometers) deep. It is there that we notice sunspots, areas of

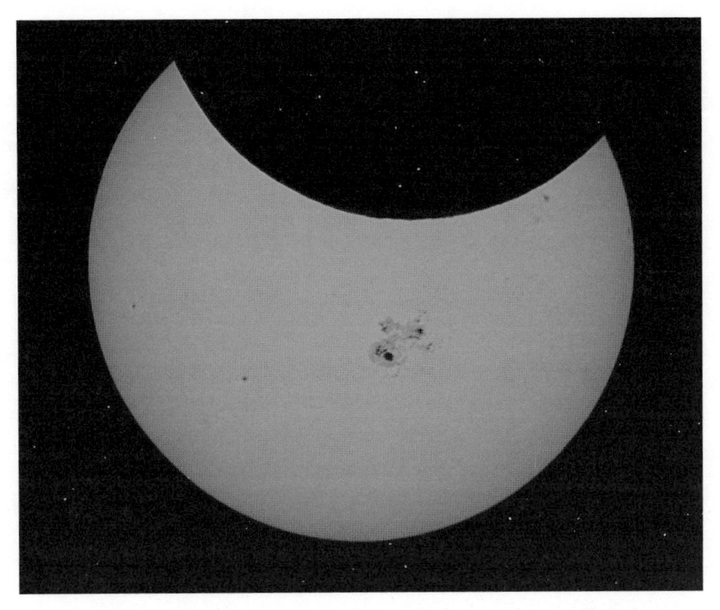

The photosphere of the Sun has sunspots that vary in number and size over an 11-year period. The biggest sunspot to grace the face of the Sun in more than two decades just happened to be visible during the partial solar eclipse of October 23, 2014. The sunspot was as big as the planet Jupiter. The enormous amount of energy stored in its twisted magnetic fields was responsible for producing several major solar flares. [Vixen 90 mm fluorite refractor, fl = 810 mm, f/9, 1/2500 second, ISO 100, Thousand Oaks Type III filter, ©2014 Fred Espenak].

magnetic disturbance on the Sun, the size of Earth, or Jupiter, or even larger, appearing and disappearing and riding along in the photosphere with the Sun's rotation. The sunspots increase and decrease in number over a period of about 11 years.

If we examine the photosphere more closely, we see that it has a mottled appearance, created by rising columns of hot, bright gas surrounded by darker haloes where the gas has cooled and is descending, to be heated again. The photosphere is boiling. These *granulations* of upwelling and downfalling gases are typically 500 miles (800 kilometers) in diameter and last for 5 or 10 minutes. The gases in the granulations, the topmost layer of the convection zone, carry with them magnetic fields from deep within the Sun. As these gases tumble up and down, the magnetic lines of force twist and snap, and this magnetic turbulence governs the behavior of gases in the photosphere and in the solar atmosphere above it.

Above the photosphere is the *chromosphere* ("color sphere"), aptly named for its vibrant reddish color. Seen with a telescope at the rim of the Sun, it looks like a fire-ocean.[4] Its lower levels are cooler than the white-hot photosphere, with temperatures of about 7,200°F (4,000°C). Its upper

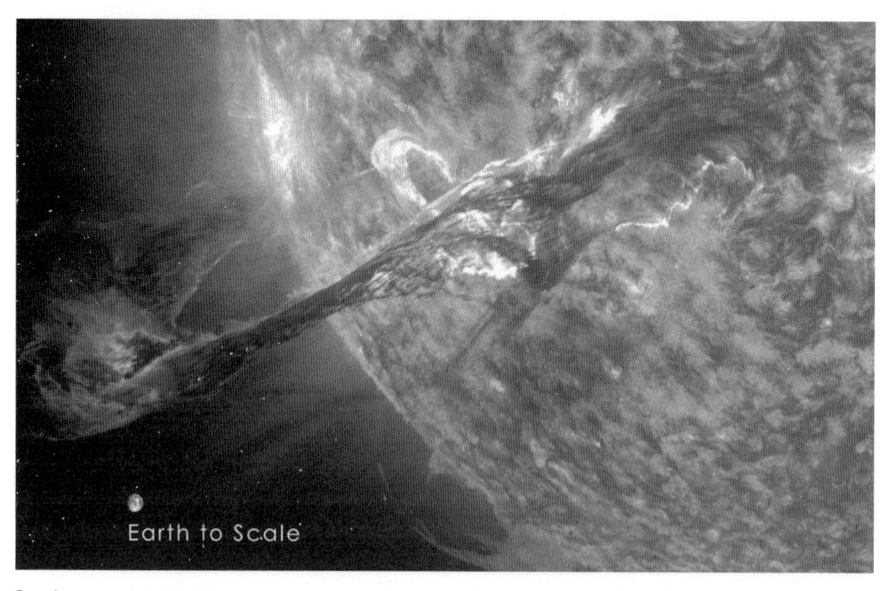

On August 31, 2012, an enormous solar prominence erupted out into space. The event triggered a coronal mass ejection, or CME, which traveled at over 900 miles per second (1,400 kilometers per second). Several days later the plasma reached Earth's magnetosphere, causing the aurora to appear in the night sky. [Photograph courtesy of NASA/SDO/AIA/GSFC]

layers, however, are much hotter than the photosphere—about 18,000°F (10,000°C). This temperature causes hydrogen atoms to emit a red wavelength that is primarily responsible for the chromosphere's fiery color.[5] The chromosphere is a thin atmospheric layer, only about 1,600 miles (2,500 kilometers) thick, although there are no sharp boundaries above or below.

Seen under high magnification along the rim of the Sun, the chromosphere is not a smooth layer of gas. It looks like an erratic forest of thin topless trees. These spiked features that compose the chromosphere are known as *spicules*. They are less than 400 miles (700 kilometers) in diameter but may tower thousands of miles high, reaching out of the chromosphere and into the corona. These spicules may be not only columns of rising gas but also the outlines of magnetic flux tubes that transfer magnetic fields from the photosphere to the corona.

It is in the photosphere and chromosphere that prominences and flares are rooted and stretch upward into the corona. They too are transient features of the Sun that vary with the rhythm of the sunspots below them. *Prominences* are the same temperature as the upper chromosphere and glow with the same red color of excited hydrogen. They are condensed clouds of solar gas, but much cooler and denser than the surrounding corona. The prominences are bent and twisted by local magnetic fields. Magnetic forces keep the gases in the prominences from all falling back to the surface, which would take only about 15 minutes if gravity were the only force acting. Most often these clouds, the prominences, do slowly rain material back toward the surface, but occasionally they erupt outward. *Flares* are much stronger eruptions, also triggered by the magnetic activity of the Sun, that launch great torrents of mass and energy from the Sun at millions of miles an hour.

The chromosphere is also the birthplace of *solar tornados*, almost the size of Earth, whirling at speeds up to 300,000 miles per hour (500,000 kilometers per hour), spiraling up and through the corona, probably contributing fast-moving atomic fragments to the solar wind of particles flowing spaceward from the Sun.[6] How these solar tornados form and function is not certain, but, as with virtually every feature seen on the Sun's surface and in its atmosphere, the explanation almost certainly involves twisted magnetic fields.

Upward and Outward

In the upper reaches of the chromosphere and extending outward into the corona is a realm called the *transition region*, so named because the temperature there suddenly climbs to about 2 million °F (1 million °C).

Why should the temperature of the *corona* ("crown"), the outer atmosphere of the Sun, be so high? It is much hotter than the photosphere and chromosphere, yet those layers are closer to the core where the Sun generates its energy. The answer again lies in the Sun's magnetic fields. The Sun not only behaves like a giant bar magnet (like the Earth with its magnetic poles), but the Sun is also pocked by many local magnetic regions with intensities greater than the Sun's polar magnetism. These surface sites of magnetic activity, the largest marked by sunspots, are induced by varying magnetic fields in the Sun's interior. These magnetic fields are borne to the surface by the columns of hot gases seen as granulations in the photosphere. The random motion of the rising and falling gases contort the lines of magnetic force. These magnetic fields stretch into the chromosphere and corona as loops and arches, as if they were invisible electric wires attached to terminals in the photosphere. The twisted and coiled magnetic fields induce electrical currents into the corona, which heat the gases there. But explaining exactly how this heating is accomplished in the near vacuum of the corona remains a challenge to solar astronomers.[7]

Solar flares are violent explosions in the Sun's atmosphere that release huge amounts of energy and subatomic particles. This close-up image from NASA's Solar Dynamics Observatory (SDO) shows the eruption of a flare on July 6, 2012. The flare caused a radio blackout as well as a solar energetic particle event in which fast particles traveling behind the flare impacted Earth's magnetosphere. [Photograph courtesy of NASA/SDO/AIA]

The gases rising from the surface of the Sun are all so hot that their atoms are missing some or all of their electrons. This plasma (ionized gas) expands as it rises from the surface of the Sun. As it expands, its density declines and these charged gases are more easily warped by the Sun's magnetic fields. The patterns of this magnetic control can be seen in the loops and arches of the prominences that extend high into the corona. Here and there a magnetic loop in the corona is stressed so greatly that it snaps and perhaps 10 billion tons of million-degree plasma is slung into space as a huge expanding bubble traveling as fast as 4.5 million miles per hour (2,000 kilometers per second). Such outpourings are called *coronal mass ejections*.[8]

Coronal mass ejections, eruptive prominences, and flares are probably different parts of the same phenomenon. They involve the expulsion of hot gases and twisted ropes of magnetic fields from the Sun, they often occur together, and all are more frequent at sunspot maximum.[9]

Coronal mass ejections enhance and create shock waves in the normal solar wind with charged particles (primarily protons and electrons) that are escaping the Sun's gravitational hold at speeds of 900,000 miles per hour (400 kilometers per second) or more. Coronal mass ejections also carry with them some of the coronal magnetic field, and distort the existing field in the solar wind. When the altered magnetic fields and these especially energetic subatomic particles strike the Earth, they create colorful displays of the northern and southern lights—the aurora.

But solar windstorms can also damage electronic equipment on satellites in Earth orbit; disrupt telephone, radio, and television transmission; and overload electric power lines, causing blackouts (as one did for 9 hours to 6 million Canadians in Quebec on March 13, 1989).

Scientists now think that coronal mass ejections, rather than flares, are the principal cause of the aurora and other geomagnetic events.

In visible light, the corona shows graceful, delicate streamers and brush-like features.[10] However, in x-ray pictures, which better capture the activity of high-temperature gases, the corona is a riot of everchanging loops, plumes, eruptions, and contrasting light and dark regions. These dark regions of lesser activity are called *coronal holes*. It is primarily through these coronal holes, where magnetic fields are weaker, that the solar wind escapes into space. Slower and more variable solar wind flows from coronal streamers.

The coronal holes and streamers seen during a total eclipse mark the visible departure from the Sun of the solar wind. The Sun loses about 10 million tons of material a year in the solar wind. But such a loss is insignificant compared to the Sun's total mass and to the rate at which the Sun is converting mass into energy at its core.

Coronal loops are fountains of multimillion-degree electrified gas in the atmosphere of the Sun that are 300 times hotter than the Sun's visible surface. The shape and structure of the coronal loops map out the magnetic field in the lower corona. [NASA Transition Region and Coronal Explorer (TRACE) spacecraft]

The solar wind blows outward in all directions. The Earth is a small target 93 million miles (150 million kilometers) away, so only two out of every billion particles in the solar wind will reach the Earth, to cause mischief here. But they are enough. Scientists are studying "space weather" with the hope of predicting when the Earth will be struck by a stormy blast of enhanced solar wind.

To the eye, the white flame brushes of the corona can extend outward from the Sun 2 million miles (3 million kilometers) or more until they are so tenuous that they are no longer visible. But the corona is still measurable out to the Earth and beyond in the form of the solar wind. The Earth orbits the Sun within the Sun's rarefied outer atmosphere.

The Gift of Totality

We do not normally see the chromosphere or the corona. They are concealed from our view by the overwhelming glare of the photosphere, half a million times brighter than the corona. To study the Sun, early scientists had only its surface to rely on—just the photosphere and sunspots. Progress was slow.

We do not see the chromosphere and the corona, the prominences, and the coronal mass ejections unless something blocks the glare of the photosphere so that the faint atmosphere of the Sun is revealed. In the 19th century, astronomers discovered that the Moon was their scientific collaborator, obscuring the Sun's surface from time to time so that they might see a part of the Sun that had never been studied before.

In less than a century, total solar eclipses, and the scientific instruments and theories they helped to stimulate, revealed the composition of the Sun, the structure of the Sun's interior, and the wonder of how the Sun shines.

A coronal mass ejection blasts immense bubbles of hot plasma into the solar system. This January 8, 2002, image shows a widely spreading coronal mass ejection *(right half of picture)* hurling more than a billion tons of electrons and protons into space at millions of kilometers per hour. [Photograph courtesy of ESA/NASA SOHO LASCO]

NOTES AND REFERENCES

1. Epigraph listed as Anonymous: North American Indian in *Bartlett's Familiar Quotations*.
2. Solar physicists Charles Lindsey, Joseph Hollweg, and Jay Pasachoff calculated or reviewed the statistics and information in this chapter. Personal communication, October 20–November 3, 1998. Lindsey calculates that 4.26 million metric tons per second (4.69 million English tons per second) of mass are converted to energy in the core of the Sun. The figure of 14.8 trillion years is based on the Sun converting *all* its mass to energy, which, of course, it cannot do. Only 0.7% of the hydrogen mass is converted into energy; the rest becomes helium. If only the total amount of hydrogen available is considered, the Sun's life span falls to about 77 billion years. Of course, only the hydrogen close to the center of the Sun is under sufficient pressure so that the temperatures are great enough for fusion to occur. Only about 10% of the Sun's hydrogen will undergo fusion, reducing the Sun's lifetime to 8–10 billion years.
3. To generate and sustain a hydrogen-to-helium fusion reaction, the core of a star must have a temperature of at least 18 million °F (10 million °C). To have enough gravity to generate sufficient pressure to obtain this high a core temperature, a star must have at least 8% of the mass of the Sun.
4. The expressions "fire-ocean" to describe the chromosphere and "flame-brushes" to describe features in the corona were used by Agnes M. Clerke: *A Popular History of Astronomy during the Nineteenth Century*, 4th edition (London: A. and C. Black, 1902), pages 68, 175.
5. Spectroscopically, the red of the chromosphere is produced by the hydrogen-alpha line.
6. Solar tornados were discovered in 1998 by David Pike and Helen Mason using the Solar and Heliospheric Observatory (SOHO), a collaboration of the European Space Agency (ESA) and NASA (United States).
7. The temperature of the corona can be misleading. The Sun's magnetic fields cause the electrically charged atoms of the corona to move at great speeds (high temperature), but the density of these ions is so low that the corona has relatively little heat (energy in a given volume). The corona is so rarefied that if you had a box there 100 miles (160 kilometers) on each side (1 million cubic miles; 4.1 million cubic kilometers), you would entrap less than a pound (0.4 kilogram) of matter. The corona is a good vacuum by laboratory standards on Earth.

   Solar physicist Joe Hollweg points out that it is also hard to explain the heating of the chromosphere, which requires as much energy as the corona. Personal communication, August 3, 1998.
8. Coronal mass ejections were discovered using NASA's Orbiting Solar Observatory 7 between 1971 and 1973 and confirmed using NASA's Solar

Maximum Mission satellite in 1980 and after 1984, following repair in orbit by Space Shuttle astronauts. In the 1990s, a new generation of solar spacecraft studied corona mass ejections: Yohkoh (Japan), SOHO (*So*lar and *H*eliospheric *O*bservatory—ESA and NASA), and TRACE (*T*ransition *R*egion *a*nd *C*oronal *E*xplorer—NASA). In the 2000s, the next generation of solar observatories continues the quest: Hinode (Japan) and STEREO (*S*olar-*Te*rrestrial *Re*lations *O*bservatory—NASA).

9. At sunspot minimum, about one coronal mass ejection a week is observed. Near sunspot maximum, two or three coronal mass ejections are observed each day on the average.
10. Coronal structures trace out the magnetic field lines, like iron filings trace out the field around a bar magnetic.

A MOMENT OF TOTALITY

The Audio of the Video

Gary Ropski and Barbara Schleck make a video recording of each eclipse, complete with sounds of people's reactions as they watch the approach of totality and the sudden revelation of the Sun's corona as totality begins. They were playing an eclipse video for friends one evening, when one said, "Play it again, but this time close your eyes and just listen." So they did. The oohs and aahs built to a crescendo at the start of totality. "Eclipse?" the friend said. "Sounds to me like an orgy."[1]

[1] Gary Ropski and Barbara Schleck live in Chicago. He is an lawyer specializing in intellectual property. She is a journalist. Interviewed May 22, 2015.

7

The First Eclipse Chasers

I did not expect, from any of the accounts of preceding eclipses that I had read, to witness so magnificent an exhibition as that which took place.

Francis Baily (1842)[1]

An Unlikely Beginning

Francis Baily, the man who might be said to have founded the field of solar physics, received only an elementary education, was not trained in science, and did not get around to astronomy until the age of 37. Like his father, a banker, he entered the commercial world as an apprentice when he was 14. But adventure called. When his seven years of apprenticeship expired, he sailed for the New World and spent the next two years, 1796–1797, exploring unsettled parts of North America, narrowly escaping from a shipwreck, flatboating down the Ohio and Mississippi Rivers from Pittsburgh to New Orleans, and then hiking nearly 2,000 miles back to New York through territory inhabited mostly by Indians. He liked the United States so much that he planned to marry and become a citizen, but he finally abandoned those plans and returned home in 1798.

Back in England, he began efforts to mount an expedition to explore the Niger River in Africa. He could not raise enough money, however, so he became a stockbroker. To dedication and enthusiasm, he quickly added a reputation for intelligence and integrity, and he made a fortune. He exposed stock exchange fraud and helped clean it up. He published a succession of explanations of life insurance methods and comparisons of insurance companies, which became wildly popular. He also published a chart of world history that was equally popular, confirming the nickname given to him in his apprentice days: the Philosopher of Newbury (his birthplace).

His first astronomical paper (1811) tried to identify the solar eclipse allegedly predicted by Thales. In 1818, he called attention to an annular eclipse of the Sun coming in 1820, and he observed it from

Francis Baily. [Royal Astronomical Society]

southeastern England. That same year, he became one of the founders of the Astronomical Society of London, later the Royal Astronomical Society.

In 1825, he retired from the stock market to devote all his time to his new profession. He was 51 years old. His revisions of a series of old star catalogs were considered so valuable that the Royal Astronomical Society twice awarded him its Gold Medal and four times elected him president. Although he was not renowned as an observer, he had an abiding fascination with eclipses, a good eye for detail, and the ability to express what he saw.

Thus, it was in 1836 that a few words from Francis Baily sparked the immediate, intense, and unending study of the physical properties of the Sun that had been generally ignored or discounted until then. He traveled to an annular eclipse of the Sun in southern Scotland and watched on May 15, 1836, as mountains at the Moon's limb occulted the face of the Sun but allowed sunlight to pour through the valleys between them, so that the ring of sunlight around the rim of the Moon was broken up into "a row of lucid points, like a string of bright beads."[2] With those words, Baily generated fervor for solar physics and founded the industry of eclipse chasing.

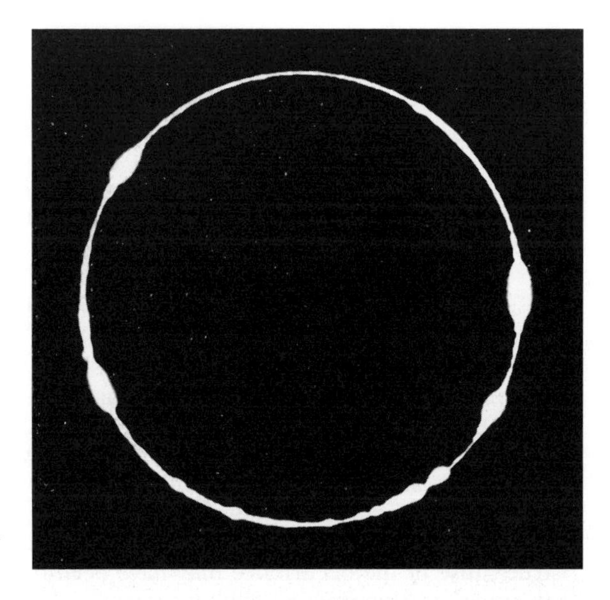

Baily's beads during the annular eclipse of April 28, 1930. [Courtesy of Lick Observatory]

The Surprise of Totality

At the next accessible eclipse, July 8, 1842, a high percentage of the astronomers of Europe migrated to southern France and northern Italy to see "Baily's beads." Baily, now 68 years old, went too.

This was not an annular eclipse, as Baily had seen twice before. It was total. No European astronomer then alive had ever seen a total eclipse.

Baily set up his telescope at an open window in a building at the university in Pavia, Italy. Again Baily's beads were visible—to Baily at least. George Airy, England's astronomer royal, observing from Turin, Italy, did not see them. Baily was just jotting down the time of appearance and duration of the beads.

> . . . when I was astounded by a tremendous burst of applause from the streets below, and at the *same moment* was electrified at the sight of one of the most brilliant and splendid phenomena that can well be imagined. For, at that instant the dark body of the moon was *suddenly* surrounded with a *corona*, or kind of bright *glory*, similar in shape and relative magnitude to that which painters draw round the heads of saints, and which by the French is designated an *auréole.*[3]

Baily was not the first to use the word *corona* to designate the glowing outer atmosphere of the Sun visible during a total eclipse, but his striking description of it caught everyone's attention as it never had before and forever united the word with the phenomenon.

> When the total obscuration took place, which was *instantaneous*, there was an universal shout from every observer . . . I had indeed anticipated the appearance of a luminous circle round the moon during the time of total obscurity: but I did not expect, from any of the accounts of preceding eclipses that I had read, to witness so magnificent an exhibition as that which took place. . . . It riveted my attention so effectually that I quite lost sight of the string of *beads*, which however were not completely closed when this phenomenon first appeared.

There was so much to see, said Baily, that in future eclipses, each observer should be assigned a single observing task.

> Splendid and astonishing, however, as this remarkable phenomenon really was, and although it could not fail to call forth the admiration and applause of every beholder, yet I must confess that there was at the same time something in its singular and wonderful appearance that was appalling: and I can readily imagine that uncivilised nations may occasionally have become alarmed and terrified at such an object, more especially in times when the true cause of the occurrence may have been but faintly understood, and the phenomenon itself wholly unexpected.

It was the last eclipse Francis Baily was to see. Two years later he died. Of him, historian Agnes M. Clerke wrote: "He was gentle as well as just; he loved and sought truth; he inspired in an equal degree respect and affection . . . . Few men have left behind them so enviable a reputation."[4]

Drawings of corona and prominences at different eclipses. *Left*: July 8, 1842; *right*: July 28, 1851. [François Arago: *Popular Astronomy*, edited and translated by W. H. Smyth and Robert Grant]

Before Baily

Never again would a total eclipse over an inhabited land mass go unattended by professional and amateur astronomers, even when the observers had to travel to remote sites at the farthest ends of the world. The corona, Baily's beads, prominences, shadow bands—all had been seen before, many times, throughout the world. But now they commanded attention and explanation.

The first written record of the corona may be Chinese characters inscribed on oracle bones from about 1307 BCE, which say "three flames ate up the Sun, and a great star was visible."[5] Or were these flames prominences instead? The first unequivocal description of the corona comes from a chronicler named Leo more than 2,000 years later, observing from Constantinople the eclipse of December 22, 968 CE:

> Everyone could see the disc of the Sun without brightness, deprived of light, and a certain dull and feeble glow, like a narrow headband, shining around the extreme parts of the edge of the disc.[6]

The great astronomer Johannes Kepler did not see a total eclipse himself, but from the reports he read, he concluded that the corona must be material around the Sun and not the Moon. Giacomo Filippo Maraldi, an Italian-born French astronomer, provided evidence that the corona is part of the Sun because the Moon traverses the corona during a solar eclipse; the corona does not move with the Moon but stays fixed around the Sun.

The Spanish astronomer José Joaquín de Ferrer, well ahead of his time, traveled to the New World to observe total eclipses in Cuba in 1803 and Kinderhook, New York, in 1806. He was probably the first to use the word "corona" to describe the glow of the outer atmosphere of the Sun seen during a total eclipse.

> The disk [of the Moon] had round it a ring of illuminated atmosphere, which was of a pearl colour . . . From the extremity of the ring, many luminous rays were projected to more than 3 degrees distance.—The lunar disk was ill defined, very dark, forming a contrast with the luminous corona . . .[7]

Ferrer quite correctly attributed the corona to the Sun. If this glow belonged to the Moon, he calculated, the lunar atmosphere would extend upward 348 miles (560 kilometers)—50 times more extensive than the atmosphere of Earth. Thus, he concluded, the corona "must without any doubt belong to the Sun." Baily, too, at the 1842 eclipse, attributed the corona to the Sun.

The first sure report of solar prominences came from Julius Firmicus Maternus in Sicily, who noticed them during the annular eclipse of July

17, 334 CE.[8] Edmond Halley, the great English astronomer, saw them clearly as bright red protrusions during the total eclipse of May 3, 1715.

Baily had pointed the way. But what was the significance of the corona and the prominences? New techniques—photography and spectroscopy—were emerging with the capability to record and explore these features of the Sun. In photography and spectroscopy, observational astronomy gained its two most powerful tools to augment the telescope.

The stage was set for the eclipse of July 28, 1851. The astronomical world gathered along its path of totality through Scandinavia and Russia. They solved one mystery and uncovered another.

The Debut of Photography

The first successful photograph of the Sun in total eclipse was a daguerreotype taken on July 28, 1851, by a professional photographer named Berkowski, assigned to the task by August Ludwig Busch, director of the observatory in Königsberg, Prussia.[9] The inner corona and prominences are clearly visible. Yet, for now, photography was just a curiosity, a promising experiment. Hardcore science was still carried out by visual observations made through telescopes.

At the 1851 eclipse, two teams of astronomers provided proof of what most astronomers suspected: that prominences were part of the Sun. Robert Grant and William Swan from the United Kingdom and Karl

First photograph of the Sun in total eclipse, July 28, 1851. Only the last name of the photographer is known: Berkowski. [Courtesy of Dorrit Hoffleit]

Ludwig von Littrow from Austria documented how the eastward motion of the Moon across the face of the Sun covered prominences along the east rim of the Sun as totality began and then uncovered prominences along the west rim as totality ended, demonstrating that the prominences belonged to the Sun and that the Moon was only passing in front of them.

Often in science, observations that lead to the solution of one problem simultaneously reveal a fascinating new problem. As George B. Airy, England's astronomer royal, was observing this eclipse, he noticed a jagged edge to the solar atmosphere just above the edge of the Moon. He called it the *sierra*, thinking that he might be looking at mountains on the Sun.

Airy thus became the first to call attention to the chromosphere, the lowest level of the transparent solar atmosphere immediately above the opaque photosphere that creates the appearance of a surface for the Sun. The jaggedness that Airy observed was actually innumerable small jets of rising gas called spicules.

By 1860, in the field of photography, daguerreotypes were obsolete, superseded by the faster wet plate (collodion) photographic process. Before photography, astronomers could only describe or sketch what

Drawing by Lilian Martin-Leake from a telescopic view of the chromosphere and the corona, May 28, 1900, showing spicules in the chromosphere that George Airy had thought were mountains. [Annie S. D. Maunder and E. Walter Maunder: *The Heavens and Their Story*]

they saw. The early photographic emulsions were not as sensitive to detail as desired, but they were more objective than the human eye.

On July 18, 1860, a total eclipse was visible from Europe and found astronomers waiting with improved cameras. Prominences remained a matter of high priority and they encountered the resourcefulness of British astronomer Warren De La Rue and Italian astronomer and Jesuit priest Angelo Secchi.

De La Rue's well-to-do family provided him with a solid education, after which he entered his father's printing business. There he demonstrated a great affinity for machinery. He could make any instrument or device run better. He was also a fine draftsman, and it was this talent that lured him into astronomy. He could produce drawings of the planets, Moon, and Sun that were better than those of astronomers. But no sooner had he produced his first excellent drawings than he found out about a new invention called photography. He never saw a mechanical device that did not fascinate him, so he was soon making improvements in cameras, inventing specialized cameras for solar photography, and photographing the Moon and Sun stereographically so that lunar features appeared in relief and sunspots revealed themselves as depressions, not mountains, in the photosphere.[10]

Angelo Secchi. [Mabel Loomis Todd: *Total Eclipses of the Sun*]

Observing the 1860 eclipse 250 miles (400 kilometers) away from De La Rue was Angelo Secchi. He too had a remarkable aptitude for inventing and coaxing instruments. Secchi was from a poor family and received his education through the Catholic Church in the demanding Jesuit tradition. From the beginning, he showed brilliance in mathematics and astronomy. When the Jesuits were expelled from Italy by a liberal, anticlerical government in 1848, Secchi spent a year in the United States as assistant to the director of the Georgetown University observatory. When the ban against Jesuits was lifted in 1849, he returned to Italy to become director of the Pontifical (now Vatican) Observatory of the Collegio Romano (or Gregorian University). He transformed it into a modern, well-equipped center for research in the new field of astrophysics. Secchi was one of the pioneers in applying spectroscopy to astronomy. He surveyed more than 4,000 stars and realized that stellar spectra could all be grouped into a handful of classifications.

De La Rue and Secchi ambushed the 1860 eclipse with improved cameras using the wet-plate process, which greatly reduced exposure time and thereby increased the clarity with which objects in motion could be seen. They captured the prominences and compared them. The prominences looked the same from the photographers' widely separated sites, so Secchi and De La Rue could conclude that they were indeed part of the Sun. If they had been features on the Moon, so much closer than the Sun, the difference in viewing angles (parallax) from Secchi's and De La Rue's separate sites would have given them a different appearance.

The Debut of Spectroscopy

On August 18, 1868, a great eclipse touched down near the Red Sea and swept across India and Malaysia. Once again, the international scientific community had assembled.

From the United Kingdom to the path of the eclipse had come, among others, James Francis Tennant and John Herschel (son of John F. W. Herschel, grandson of William Herschel, both renowned astronomers). Norman Pogson, born in England, represented India and the observatory he directed there. The French delegation included Georges Rayet and Jules Janssen. Each of them carried a new weapon just added to the scientific arsenal for prying secrets from the Sun during an eclipse. That tool was a spectroscope. It would prove as indispensable to eclipse studies of the structure of the Sun as it was rapidly demonstrating itself to be in other realms of astronomy and in all other physical and biological sciences. By passing the light of the corona or the prominences through a prism, it could be broken down into a spectrum of lines and colors. From this spectrum, a scientist could identify the chemical elements present and even their temperature and density.

As the eclipse sped along its course, spectroscopes pointed upward toward the prominences. They showed that the prominences emitted bright lines, and most of these were quickly identified with hydrogen. More and more it seemed that the Sun must be composed primarily of gas and that hydrogen must be a major constituent.[11] The spectroscopists, properly pleased with their results, packed their equipment and headed home.

All but one. His name was Jules Janssen, and the brightness of the prominences and the strength of their spectral lines had given him an idea. He wanted to look for them again when the eclipse was over, when the Moon was not blocking the intense glare of the Sun from view. Might it be possible for him to see the prominences and their spectrum in broad daylight?

The weather was cloudy the rest of that day. He would have to wait until tomorrow.

That Extra Step

Pierre Jules César Janssen was 12 years old when Baily called attention to the beads visible during the annular eclipse of 1836. A childhood accident had left Janssen lame and he never attended elementary or high school. His family was cultured, but his father was a struggling musician, so Jules had to go to work at a young age.

While employed at a bank, he earned his college degree in 1849. He then went on to gain a certificate as a science teacher and served as a substitute teacher at a high school. In 1857, he traveled to Peru as part of a government team to determine the position of the magnetic equator. There he became severely ill with dysentery and was sent home. At age 33, he seemed destined for a quiet life in teaching, if he could get a job. He became the tutor for a wealthy family in central France. At their steel mills, he noticed that the eye could watch molten metal without fatigue or injury, while the skin had to be protected from the heat. He wrote a careful study of how the eye protects itself against heat radiation, which earned him his doctorate in 1860.

The glow of molten metal had led him to spectroscopic analysis, which he then applied to the Sun and, in 1859, identified several Earth elements present in the Sun. He moved to Paris in 1862 to dedicate himself to solar physics and scientific instrument-making. His work had already made him a leader in solar spectroscopy. He used the changing spectrum of the Sun in its daily journey from horizon to horizon to separate spectral lines caused by the Earth's atmosphere from those originating in the Sun and

Jules Janssen *(1824–1907)*

Two years after his breakthrough at the 1868 eclipse, Jules Janssen again planned to apply spectral analysis to a solar eclipse, this one on December 22, 1870, in Algeria. When it came time for departure, however, the Franco-Prussian War was in progress and Paris was under siege. Colleagues in Britain had obtained from the Prussian prime minister safe passage for Janssen from Paris, but Janssen wouldn't accept favors from his country's enemies. He had a different plan: "France should not abdicate and renounce taking part in the observation of this important phenomenon . . . . An observer would be able, at an opportune moment, to head toward Algeria by the aerial route . . ."[a]

Although he had never been in a balloon before, on December 2, with a sailor as an assistant and himself as pilot, he ascended from Paris and headed west. Despite violent winds, he landed safely near the Atlantic coast. He reached Algeria in time—only to have the eclipse clouded out. From that experience, however, Janssen designed an aeronautical compass and ground speed indicator and prophesied methods of air travel that would "take continents, seas, and oceans in their stride." In 1898 and subsequently, he used balloons to study meteor showers from above the clouds, pioneering high altitude astronomy and foreseeing the advantages of space observations.

Janssen was also a leader in astrophotography. "The photographic plate is the retina of the scientist," he wrote. It was for spectroscopy, however, that Janssen was most renowned. His methods opened up the Sun's atmosphere to continuous study. The French government tried to find him an observatory position, but the director of the Paris Observatory did not want him. So Janssen was allowed to pick a site near Paris for a new observatory dedicated to astrophysics. He chose Meudon and directed the observatory from its founding in 1876 until his death in 1907.

"There are very few difficulties that cannot be surmounted by a firm will and a sufficiently thorough preparation," he wrote. But he was too modest in his self-assessment. To everything he investigated, he brought imagination and insight. Jules Janssen was one of the most creative scientists of any era.

[a] Quotations are from the sketch of Janssen by Jacques R. Lévy in the *Dictionary of Scientific Biography*.

to demonstrate the composition and density of the Earth's atmosphere through which the sunlight passed. Janssen then applied spectroscopy to the other planets and, in 1867, discovered water in the atmosphere of Mars.

He traveled to Guntur, India, for the total eclipse of August 18, 1868, to use his spectroscope on solar prominences. The great contemporary English spectroscopist Norman Lockyer spoke ringingly of that pivotal moment: "Janssen—a spectroscopist second to none— . . . was so struck with the brightness of the prominences rendered visible by the eclipse that, as the sun lit up the scene, and the prominences disappeared, he exclaimed, '*Je reverrai ces lignes la!*'"[12] [I will see those lines again!] The next morning he succeeded. He had found a way to study the atmosphere of the Sun without waiting for a total eclipse, traveling halfway around the world to see it, and hoping for good weather at the critical moment.

For two weeks Janssen continued to map solar prominences by this technique and continued to perfect it on his circuitous way home, with a stop in the Himalayas to observe at high altitude. He proved that prominences change considerably from one day to the next.

A standard spectroscope breaks down the light of a glowing object into the characteristic colors of its spectrum. Janssen modified the spectroscope by blocking unwanted colors so that the observer could view an object in the light of one spectral line at a time. He had invented the spectrohelioscope. The Sun could now be analyzed in detail on a daily basis.

A month after the eclipse, on his way home to France, Janssen wrote up his findings and sent them to the Academy of Sciences in Paris. His paper arrived a few minutes after one from England that reported precisely the same discovery.

Jules Janssen.

Coincidence

Joseph Norman Lockyer came from a well-to-do family with scientific interests. He received a classical education, traveled in Europe, and then entered civil service. So wide were his interests that he wrote on everything from the construction dates and astronomical purposes of Egyptian pyramids and temples, to Tennyson, to the rules of golf. "The more one has to do, the more one does," was his motto.[13]

When Gustav Kirchhoff and Robert Bunsen showed in 1859 how spectroscopy could be used to determine the chemical composition of objects in space, Lockyer saw the discovery as a key to what had seemed the locked door of the universe. Had not French philosopher Auguste Comte confidently asserted only 24 years earlier that never, by any means, would we be able to study the chemical composition of celestial bodies, and every notion of the true mean temperature of the stars would always be concealed from us.[14] It was clear that Comte was wrong. Lockyer bought a spectroscope, attached it to his 6¼-inch (16-centimeter) refracting telescope, and began his observations.

Although he had never seen a solar eclipse, it occurred to him that, since prominences were probably clouds of hot gas, he should be able to use a spectroscope to analyze prominences without waiting for an eclipse. This idea struck Lockyer two years before the 1868 eclipse that inspired Janssen to the same realization. Lockyer tried the experiment in 1867 but found his spectroscope inadequate to the task. So he ordered a new spectroscope to his specifications. Because of construction delays,

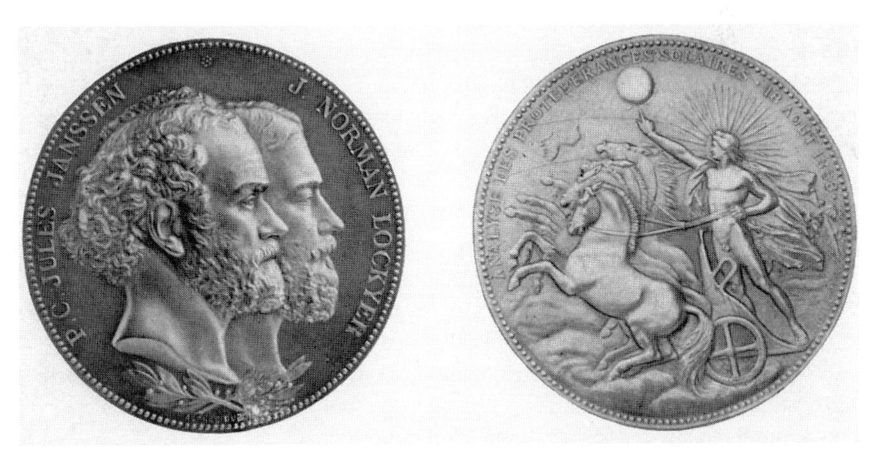

Medallion created by the Academy of Sciences in Paris to honor Janssen and Lockyer for their independent discovery of how to observe prominences without waiting for an eclipse. The front of the medal shows the heads of the two scientists. The reverse shows the Sun god Apollo pointing to prominences on the Sun.

however, it did not arrive until October 16, 1868, two months after the eclipse that Janssen saw in India. Lockyer rapidly and excitedly calibrated his new instrument and on October 20, 1868, he trained it on the rim of the Sun and recorded bright lines typical of hot gases under very little pressure. He wrote up his findings and sent them to the Academy of Sciences in Paris for presentation by his friend Warren De La Rue.

Just minutes before De La Rue was to speak, Janssen's letter arrived and both papers were read at the same session of the Academy to great acclaim for both scientists. A special medal was struck to honor them. It showed the heads of Janssen and Lockyer side by side.[15]

A New Element

Lockyer continued to examine the spectrum of the gases at the rim of the Sun. He recognized that the lower atmosphere of the Sun, what Airy

J. Norman Lockyer *(1836–1920)*

In 1869, the year after he independently showed how the atmosphere of the Sun could be analyzed without the benefit of an eclipse and discovered the element helium, Norman Lockyer founded the scientific journal *Nature*. He edited it for 50 years, until just before his death, keeping it alive through many crises.

While the French government was establishing a special observatory for Janssen, the British government likewise recognized the importance of Lockyer's contribution and set about creating a solar physics observatory for him. For its opening in 1875, Lockyer collected old and modern instruments and placed them on display. The display became permanent and grew. Lockyer had founded London's world famous Science Museum.

Lockyer was not shy in interpreting his findings to form startling theories. Often he was wrong, but always he provided useful data and often there was a nucleus of truth in his grand speculations. He thought that all atoms shared certain spectral lines and were therefore made of smaller common constituents. He was wrong about the spectra, but right about the composition of atoms.

He offered dates for the construction of ancient Egyptian temples based on their alignments with the rising and setting positions of the Sun and certain stars. His dates were wrong, but he was right that many of the temples had astronomical orientations and his work helped to establish the field of archeoastronomy.

When he died, a colleague wrote of him: "Lockyer's mind had the restless character of those to whom every difficulty is a fresh inspiration. His enthusiasm never failed him, despite repeated disappointments and opposition."[b]

[b] Alfred Fowler: "Sir Norman Lockyer, K.C.B., 1836–1920," *Proceedings of the Royal Society of London*, series A, volume 104, 1923, pages i–xiv.

J. Norman Lockyer in 1895, the year helium, the element he discovered on the Sun, was finally found to exist on Earth as well.

had called the sierra, was decidedly reddish in color, so he named it the *chromosphere*, and it has been known by that name ever since.

Lockyer was not done yet. In examining the spectrum of the prominences, he noticed a yellow line that he could not identify. It did not seem to belong to any element known on Earth. So he announced the existence of a new element and proposed the name *helium* for it, because it had been found in the Sun—*helios* in Greek. Most scientists rejected the idea of a new element, suggesting that this line was produced by a known element under unusual physical conditions. But Lockyer clung tenaciously to his interpretation.

Finally, in 1895, William Ramsay found trapped in radioactive rocks on Earth, an unknown gas that exhibited the mysterious spectral line that Lockyer had discovered on the Sun. Helium was an element. Lockyer had been right.

In 1869, just after he discovered helium, Lockyer had urged: "Let us . . . go on quietly deciphering one by one the letters of this strange hieroglyphic language which the spectroscope has revealed to us—a language written in fire on that grand orb which to us earth-dwellers is the fountain of light and heat, and even of life itself."[16]

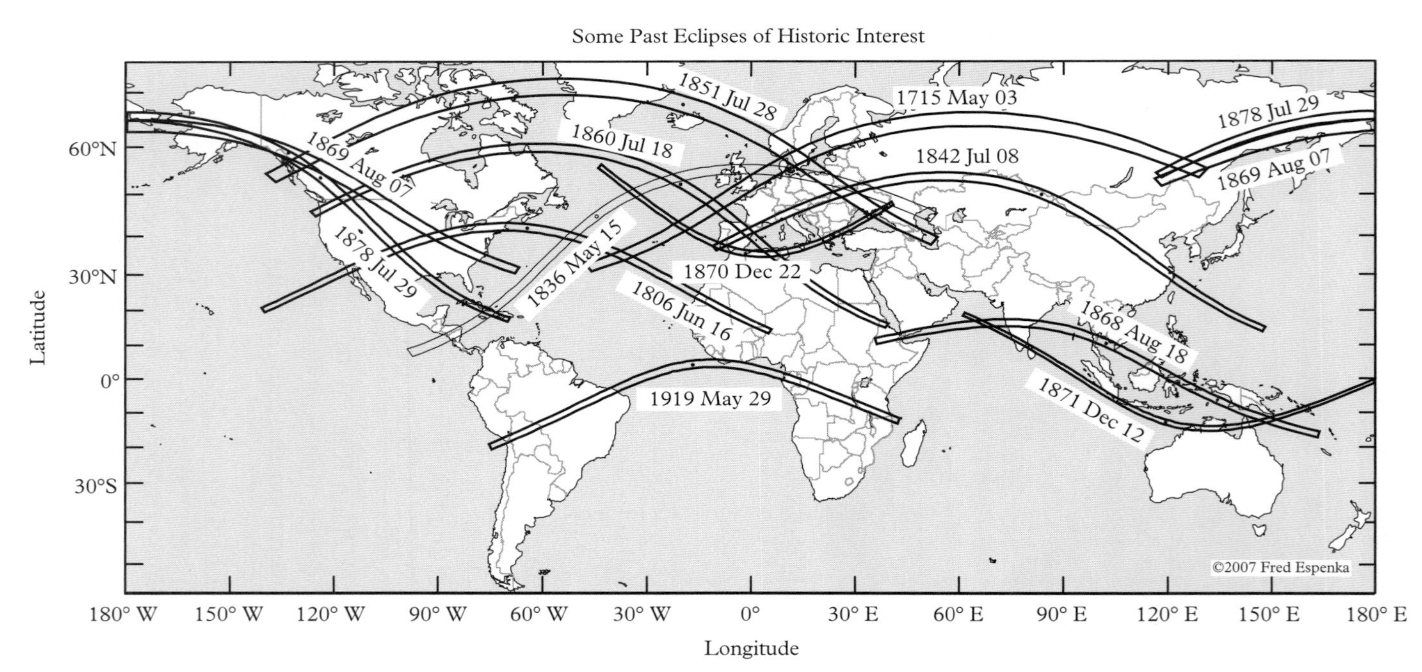

Paths of totality for the solar eclipses of 1715, 1806, 1836, 1842, 1851,1860, 1868, 1869, 1870, 1871, 1878, and 1919. [Map and eclipsecalculations by Fred Espenak]

The total eclipse of 1868 had raised the strong possibility of the existence of a new element—and, for the first time, one discovered not on Earth but in the heavens. One year later, on August 7, 1869, the United States lay on the path of a total eclipse. Two American astronomers, Charles A. Young and William Harkness, working separately, observed the event with spectroscopes. Each noticed a green line in the spectrum of the corona that defied identification with known elements. This suspected new element was called *coronium*.

In 1895, Lockyer's helium was identified in rocks on Earth, but coronium remained a spectral presence seen only on the Sun during total eclipses. As time passed, more elements were discovered on Earth until the periodic table of stable chemical elements was nearly complete. There was no room left for coronium to be an element. What could it be?

Might it be an already known element under such unusual conditions that it emitted a spectrum never before seen in a laboratory? Walter Grotrian of Germany in 1939 pointed the way and Bengt Edlén of Sweden in 1941 identified the green line of coronium as the element iron with 13 electrons missing—a "gravely mutilated state."[17] To ionize iron so greatly, the temperature of the corona had to be about 2 million °F (1 million °C) and its density had to be less than a laboratory vacuum. Because the conditions necessary for the production of such lines cannot be achieved in a laboratory, they are known as forbidden lines.

The Reversing Layer

Astronomy was a family tradition for Charles Augustus Young. His maternal grandfather and his father had been professors of astronomy at Dartmouth College. Charles entered Dartmouth at age 14, and four years later graduated first in his class. He immediately began teaching—classics!—at an elite prep school, and commenced studies at a seminary to become a missionary. But in 1856 he changed his plans and became professor of astronomy at Western Reserve College, with a break in his duties to serve in the Civil War. He returned to Dartmouth in 1866 at the age of 32 as professor of astronomy, holding the same chair as his father and grandfather before him. There he pioneered in spectroscopy, especially applied to the Sun.

At the eclipse of December 22, 1870, which he observed at Jerez, Spain, Young noticed that the dark lines in the Sun's spectrum become bright lines for a few seconds at the beginning and end of totality. He had discovered the *reversing layer*, the lowest 600 miles (1,000 kilometers) of the chromosphere, which is cooler than the photosphere and thus absorbs radiation of specific wavelengths, producing the ordinary

Charles A. Young. [Courtesy of the Mary Lea Shane Archives of the Lick Observatory]

dark-line spectrum of the Sun. However, when the Moon blocks the photosphere from view and the reversing layer of the chromosphere can be seen momentarily before it too is eclipsed, the bright-line spectrum of its glowing gases briefly flashes into view. Here at last was the layer responsible for the dark-line spectrum of the Sun seen on ordinary days. A long-missing piece in the puzzle of the structure, composition, and density of the solar atmosphere was fitted into place.

In 1877, Young was lured away from Dartmouth by the offer of more equipment and research time at the College of New Jersey, now Princeton University. Not only was he a great researcher, but a revered teacher and acclaimed writer. His textbooks were the standard of his day.

The Legacy of Eclipses

Throughout the final three decades of the 19th century, Janssen, Lockyer, and Young led expeditions to the major total eclipses and, weather permitting, always contributed useful data and often new discoveries.

The shape of the corona changes with sunspot activity. Near sunspot minimum, the corona is elongated at the Sun's equator. Near maximum, the corona is more symmetrical. These photographs were taken with a radial density filter developed by Gordon Newkirk that captures faint detail in the outer corona without overexposing the inner corona—similar to what the eye sees. The corona never appears the same twice. [Photographs courtesy of the High Altitude Observatory/National Center for Atmospheric Research]

Many scientists had noticed that the corona changes its appearance from one eclipse to the next. But it was Jules Janssen who first spotted a pattern to those variations. He compared the coronas of the 1871 and 1878 eclipses and concluded that the shape of the corona varies according to the sunspot cycle. In 1871, the Sun was near sunspot maximum and the corona was round. In 1878, near sunspot minimum, the corona was more concentrated at the Sun's equator.

As the 20th century began, solar eclipses were still the principal means of gathering information about the workings of the Sun. Every total eclipse over land was attended by scientists willing to travel great distances, endure hostile climates, and risk complete failure because of clouds for a few minutes' view of the corona—vital for the systematic study of the Sun launched more than half a century earlier by Francis Baily in his report on the annular eclipse of 1836.

One could see the Sun best when it was obscured.

What Eclipses at Jupiter Taught Us

by Carl Littmann

As the moons in the solar system revolve around their planets, they too create and undergo periodic solar and lunar eclipses. These events allowed the Danish astronomer Ole Römer in 1676 to prove, contrary to prevailing opinion, that light travels at a finite speed. He even succeeded in making the first good estimate of the speed of light. Such luminaries as Aristotle, Kepler, and Descartes had been certain that the speed of light was infinite. Gian Domenico Cassini, director of the Paris Observatory where Römer made his discovery, refused to believe the results, and Römer's triumph was not fully appreciated for half a century.

In observations of Io, innermost of Jupiter's four large moons, Römer noticed discrepancies between the observed times of its disappearance into the shadow of the planet and the calculated times for these events. He correctly explained that these discrepancies were due to the travel time of light between Jupiter and the Earth. When the Earth is approaching Jupiter, the interval between satellite eclipses is shorter because the distance light must travel is decreasing. When the Earth is moving farther from Jupiter, the interval between eclipses lengthens because the distance light must travel is increasing. Römer determined that light from Io took about 22 minutes longer to reach the Earth when the Earth was farthest from Jupiter than when it was closest. Thus, light required about 22 minutes to cross the orbit of Earth. The diameter of the Earth's orbit was not known at the time. Modern measurements show that light actually requires about 16⅔ minutes to make this journey.

Solar eclipse on Saturn caused by its ring system and a moon. At the bottom are moons Tethys and Dione. Tethys' shadow on Saturn can be seen at the upper right, just below the rings. [NASA/Jet Propulsion Laboratory Voyager 1*]*

Solar eclipse on Jupiter. The black dot left of center is the shadow cast by Io. Io, slightly larger than Earth's Moon, is visible over Jupiter's clouds in the upper right center. [NASA Hubble Space Telescope—WFPC 2]

*Saturn's shadow eclipses its rings [NASA/Jet Propulsion Laboratory—*Cassini *mission]*

When Römer returned to Denmark, the king gave him a succession of appointments, including master of the mint, chief judge of Copenhagen, chief tax assessor (everyone said he was fair!), mayor and police chief of Copenhagen, senator, and head of the state council of the realm—all this and more while he served as director of the Copenhagen observatory and astronomer royal of Denmark. He discharged all his duties with distinction.

Carl Littmann is a physicist and historian.

NOTES AND REFERENCES

1. Epigraph: Francis Baily: "Some Remarks on the Total Eclipse of the Sun, on July 8th, 1842," *Memoirs of the Royal Astronomical Society*, volume 15, 1846, page 4.
2. Francis Baily: "On a Remarkable Phenomenon That Occurs in Total and Annular Eclipses of the Sun," *Memoirs of the Royal Astronomical Society*, volume 10, 1838, pages 1–42. Baily, searching back through the records, realized that Edmond Halley in 1715 and many other observers had seen and reported this bead-like apparition before him. Among the previous observers of the beads, he named José Joaquin de Ferrer, who also described the corona and first called it by that name. Perhaps it was Ferrer's account and his use of the word corona that came to Baily's mind when he saw the eclipse of 1842.
3. This and the following Baily quotations are from Francis Baily: "Some Remarks on the Total Eclipse of the Sun, on July 8th, 1842," *Memoirs of the Royal Astronomical Society*, volume 15, 1846, pages 1–8.

4. Agnes M. Clerke: "Baily, Francis," *The Dictionary of National Biography*, volume 1 (London: Oxford University Press, 1921), page 903.
5. Joseph Needham and Wang Ling: *Science and Civilisation in China*, volume 3, *Mathematics and the Sciences of the Heavens and the Earth* (Cambridge: At the University Press, 1959), page 423.
6. F. Richard Stephenson: *Historical Eclipses and the Earth's Rotation* (Cambridge: Cambridge University Press, 1997), page 390.
7. José Joaquín de Ferrer: "Observations of the Eclipse of the Sun, June 16th, 1806, Made at Kinderhook, in the State of New-York," *Transactions of the American Philosophical Society*, volume 6, 1809, pages 264–275.
8. J. M. Vaquero and M. Vázquez: *The Sun Recorded Through History* (Heidelberg, Germany: Springer, 2009), page 203.
9. Dorrit Hoffleit: *Some Firsts in Astronomical Photography* (Cambridge, Massachusetts: Harvard College Observatory, 1950). The first successful photograph of the uneclipsed Sun, also a daguerreotype, was achieved by the French scientists Hippolyte Fizeau and Léon Foucault in 1845.
10. De La Rue made that discovery in 1861. Earlier evidence for sunspots as depressions had come from observations by Scottish astronomer Alexander Wilson in 1774. He noted that the geometry of sunspots seemed to change as they were seen from different angles as the Sun's rotation carried them across the solar disk and toward the limb.
11. The Sun as a sphere of hot gas had been proposed independently by Angelo Secchi and John Frederick William Herschel (William's son) in 1864.
12. J. Norman Lockyer: "On Recent Discoveries in Solar Physics Made by Means of the Spectroscope," in Bernard Lovell, editor: *Astronomy*, volume 1, The Royal Institution Library of Science (Barking, Essex: Elsevier Publishing, 1970) page 90.
13. Alfred Fowler: "Sir Norman Lockyer, K. C. B., 1836–1920," *Proceedings of the Royal Society of London*, series A, volume 104, December 1, 1923, pages i–xiv.
14. Auguste Comte: *The Essential Comte, Selected from Cours de philosophie positive*, translated by Margaret Clarke (London: Croom Helm, 1974), pages 74, 76.
15. A. J. Meadows: *Science and Controversy: A Biography of Sir Norman Lockyer* (London: Macmillan, 1972), page 53.
16. J. Norman Lockyer: "On Recent Discoveries in Solar Physics Made by Means of the Spectroscope," in Bernard Lovell, editor: *Astronomy*, volume 1, The Royal Institution Library of Science (Barking, Essex: Elsevier Publishing, 1970), pages 101–102.
17. "A gravely mutilated state" is the picturesque description found in Gabrielle Camille Flammarion and André Danjon, editors: *The Flammarion Book of Astronomy*, translated by Annabel and Bernard Pagel (New York: Simon and Schuster, 1964), page 227.

A MOMENT OF TOTALITY

Stumbling onto a Total Eclipse

Sheridan Williams and his fellow observers had set up their equipment on Antigua for the total eclipse of 1998. A young couple walking by approached him. Why the telescope? they asked. They were on their honeymoon and were utterly unaware that a total eclipse of the Sun would be visiting the island that day. Antigua had done little to inform its citizens and tourists about this rare event.

"Stay with us for the next hour and a half and you will see the most amazing sight you've ever seen," said Williams. He gave them solar filters and shared views through his telescope. He talked them through what was happening and what was going to happen as the Moon blocked the Sun. They were awed and grateful. It gave new meaning to the word *honeymoon*.[1]

[1] Sheridan Williams is a computer scientist who lives in London. Interviewed May 23, 2015.

8

The Eclipse that Made Einstein Famous

Oh leave the Wise our measures to collate.
One thing at least is certain, LIGHT has WEIGHT
One thing is certain, and the rest debate—
Light-rays, when near the Sun, DO NOT GO STRAIGHT.

Arthur S. Eddington (1920)[1]

Of all the lessons that scientists learned from eclipses, the most profound and momentous was the confirmation of Einstein's general theory of relativity. That lesson was provided by the total eclipse of the Sun on May 29, 1919.

The Birth of Relativity

In 1905, an obscure Swiss patent examiner third class named Albert Einstein published three articles in the same issue of the leading German scientific journal *Annalen der Physik* that utterly changed the course of physics. One proved the existence of atoms. The second laid the cornerstone for quantum mechanics. The third was a revolutionary view of space and time known as the special theory of relativity. It required only high-school algebra, yet its implications were so profound that this theory baffled many of the leading scientists of the day.

In the decade that followed the publication of the special theory of relativity, Einstein labored mightily to expand his concept to accelerated systems. In 1907, he formulated his principle of equivalence: there is no way for a participant to distinguish between a gravitational system and an accelerated system. For an observer in a closed compartment, does a ball fall to the floor by gravity because the compartment is resting on a planet or does the ball fall because the compartment is accelerating toward the ball? In both cases, the ball falls to the floor. In both cases, the

Albert Einstein in 1922. [Albert Einstein collection, The Huntington Library, San Marino, CA]

Einstein's principle of equivalence. A ball falls in a compartment. Does it fall by gravity because the compartment is resting on a planet or does it fall because the compartment is accelerating upward? Einstein realized that an observer in the compartment could not distinguish whether gravity or acceleration caused the ball to fall.

observer feels weight. It is impossible to tell whether that weight is from gravity or acceleration.

Einstein then realized from his principle of equivalence that relativity required gravity to bend light rays, much as it bends the paths of particles. In a closed accelerating compartment, a light on one wall is aimed directly at the opposite wall, across the line of motion. The beam travels at a finite speed, so in the time it takes to traverse the compartment, the opposite wall has moved upward. For an observer in the compartment, the light beam has struck the opposite wall below where it was aimed. The observer concludes that light has been bent.

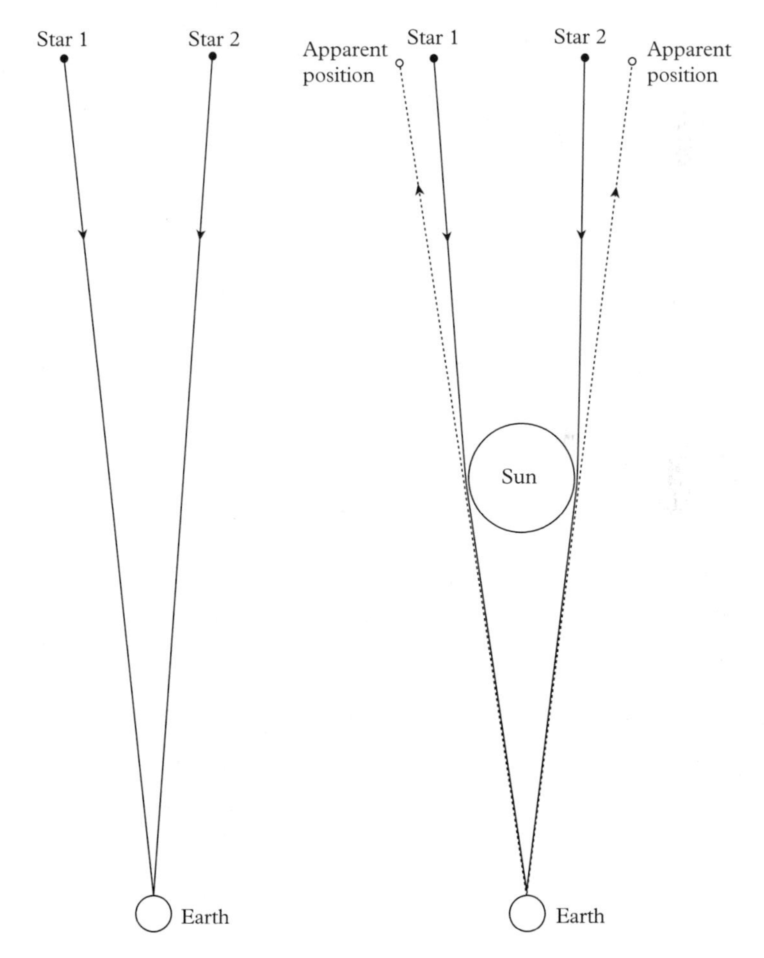

In a perfect vacuum with no gravity, light from distant stars travels in a straight line *(left)*. The gravitational field of the Sun bends light *(right)*, making the stars appear slightly farther apart than they actually are.

What about that same experiment performed in a compartment at rest on a planet? According to Einstein's principle of equivalence, there can be no difference between phenomena measured in the two compartments. Therefore, for the observer in the gravitational environment, light must be bent by gravity. But the effect is very small. It takes a lot of mass to bend light enough to be measured. In 1911, Einstein realized that this peculiar idea might be tested during a total solar eclipse.

The Sun is sufficiently massive that light from distant stars passing close to its surface would be deflected just enough to be measurable. This bending of the light from such stars causes them to appear displaced slightly outward from the Sun. The displacement could be recognized by comparing a photograph of the star field around the Sun with a photograph of the same star field when the Sun was not present.

When the Sun is visible, however, the stars are much too faint to be seen. If only the Sun's brightness could be shut off for just a few minutes—as happens during a total solar eclipse! Using his special theory of relativity, Einstein initially predicted that starlight just grazing the Sun would be bent 0.87 second of arc.[2]

The Search for Proof

A talented young scientist named Erwin Freundlich was the first to attempt to test this aspect of relativity.[3] Fascinated by the theory, he obtained solar eclipse photographs from observatories around the world that might show the displacement of stars. But photographs of previous eclipses were not adequate for the purpose.[4] The theory would have to be tested at a future eclipse, such as the one that would occur in southern Russia on August 21, 1914.[5] Freundlich was an assistant at the Royal Observatory in Berlin and tried to interest his colleagues in mounting a scientific expedition to test the theory. His superiors, however, were uninterested. Freundlich was allowed to go if he took unpaid leave and raised his own money. With youthful enthusiasm, he made his plans and informed Einstein of his intentions.

Einstein had just moved from Switzerland to Berlin to take a distinguished position created for him at the Kaiser Wilhelm Institute, the most prestigious research center in the world's science capital. Even among the giants there, Einstein stood out. Physicist Rudolf Ladenburg recalled, "There were two kinds of physicists in Berlin: on the one hand was Einstein, and on the other all the rest."

It was at this time, the spring of 1914, that Einstein was becoming increasingly withdrawn and oblivious to conventions of social behavior. He was confident in his new general theory of relativity but was deeply engrossed in its final formulation and its implications. Freundlich's wife Käte told of inviting Einstein to dinner one evening. At the conclusion of the meal, as the two scientists talked, Einstein suddenly pushed back his plate, took out his pen, and began to cover their prized tablecloth with equations. Years later, Mrs. Freundlich lamented, "Had I had kept it unwashed as my husband told me, it would be worth a fortune."[6]

That summer Freundlich took his scientific equipment and headed for the Crimea and the eclipse. On August 1, Germany declared war on Russia, commencing World War I. Freundlich was a German behind Russian lines. He and his team members were arrested and their equipment impounded. Within a month, Freundlich and crew were exchanged for high-ranking Russian officers, but they had missed the eclipse.[7]

Einstein deplored the war and German militarism, an attitude that drew the ire of many of his colleagues in Berlin. He ignored the hostility and concentrated all his energies on his research. In 1915 Einstein announced the completion of his general theory of relativity, a radically new theory of gravitation. It is perhaps the most prodigious work ever accomplished by a human being. Not only were its implications profound, but the mathematics needed to understand it were formidable. "Compared with [the general theory of relativity]," said Einstein, "the original theory of relativity is child's play."[8]

Einstein offered three tests to confirm or reject his theory. The first was the peculiar motion of Mercury. The entire orbit of Mercury was turning (precessing) more than Newton's law of universal gravitation

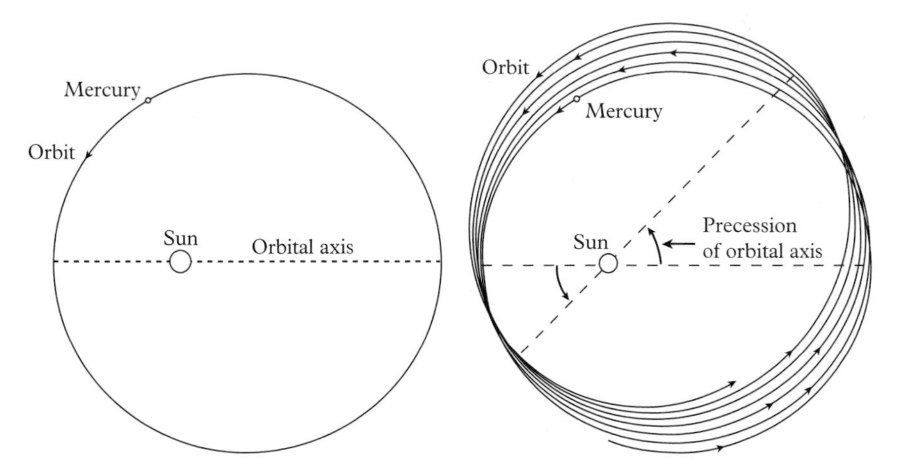

Precession of Mercury's orbital axis. *Left*: According to Newtonian theory, a planet orbiting the Sun follows a fixed elliptical path. *Right*: Einstein's general theory of relativity predicts that the axis of the ellipse will gradually rotate.

could explain. It was a tiny but measurable amount: 43 seconds of arc a *century*. This unexplained advance in Mercury's *perihelion* (its closest point to the Sun) had given rise to a suspicion that one or more planets lay between Mercury and the Sun. The suspicion was so strong that this suspected planet, scorched by the nearby Sun, had already been given a name: Vulcan, after the Roman god of fire. Many observers had tried to spot it as a tiny dot passing across the face of the Sun, and some even claimed success. But when an orbit for the planet was calculated and its next passage across the face of the Sun was due, the planet never kept the appointment. It did not exist. But it was not until Einstein formulated the general theory of relativity that the 43 seconds per century anomaly could be explained.

Einstein was more than just pleased when he realized that his theory could account for this discrepancy in the motion of Mercury. "I was beside myself with ecstasy," he wrote.[9] This explanation of a perplexing problem gave general relativity high credibility. But the power of the general theory would be even more evident if it could predict something never before contemplated or detected.

Einstein offered two such predictions: that starlight passing close to the Sun would be bent, and that light leaving a massive object would have its wavelengths stretched so that the light would be redder. This *gravitational redshift* was so small an effect that it could not be detected in the Sun with the equipment available at the time, so this proof of general relativity had to wait many years. It was finally detected in 1959 by Robert V. Pound and Glen A. Rebka, Jr., using the recently discovered Mössbauer effect in which the gamma rays emitted by atomic nuclei serve as the most precise of clocks. One of these atomic clocks was placed in the basement of a building and moved up and down so that its depth in the Earth's gravitational field varied minutely. The deeper in the basement the clock was, the longer the wavelength of its radiation. It had taken 45 years, but the gravitational redshift predicted by Einstein had at last been confirmed.[10]

In contrast, the gravitational deflection of starlight predicted by Einstein could be tested at most total eclipses of the Sun. Between 1911 and 1915, Einstein revised his calculation, using his new general theory of relativity. He found the deflection to be twice the initially assigned value. Starlight passing near the Sun would be bent 1.75 arc seconds. (Einstein and the world were fortunate that his initial prediction was not tested before it was revised; otherwise his later figure, although rigorously honest, might have seemed to be a manipulation to make the numbers come out right. There would have been far less drama in the confirmation of relativity.)

The diamond ring effect just before second contact was photographed from Libya during the total solar eclipse of March 29, 2006. [Nikon D200 DSLR, 170–500 mm zoom at 500 mm, f/5.0, 0.8s, ISO 400. ©2006 Patricia Totten Espenak]

The solar corona's enormous dynamic range and wealth of fine structure requires multiple exposures and special computer processing as illustrated in this image of the total eclipse of August 1, 2008. [Canon 450D and 300 mm, Canon 35OD an b200 mm, ISO 100, large range of exposures, image processing by Hana Druckmullerova. ©2008 Arne Danielsen]

Observers gaze in wonder at the August 1, 2008 total eclipse as seen in this time lapse composite shot from the shore of the Novosibirsk Reservoir in Russia. [Nikon D200, 17 mm lens, stacked multiple exposures at five-minute intervals. ©2008 Ben Cooper]

The totally eclipsed Sun stands above threatening clouds on July 21, 2009. Taken from Northern Cook Islands, at sea on M/S Paul Gauguin. [Canon 20Da, 28–105 mm zoom at 60 mm, ISO 100, f/5, 1.3 seconds. ©2009 Alan Dyer]

The silhouette of an intrepid eclipse chaser stands high above the lights of EL Calafate and Lake Argentino during the total eclipse of July 11, 2010. This challenging eclipse barely cleared the distant Andes as totality began. [Olympus SP570UZ, lens at 8.9 mm, f/3.4 at 1/2 second. ©2010 Janne Pyykkö]

From the frigid landscape of Antarctica, an eclipse watcher is dramatically captured against the total eclipse of November 23, 2003. [Canon EOS D60 DSLR, 100–400 mm Canon F4.5–5.6 L IS, f/8, 1/60s ISO 100. ©2003 Fred Bruenjes]

The wondrous sight of the diamond ring effect hanging 10° above the horizon is greeted by frozen eclipse chasers in the −20°C (−4°F) landscape of Svalbard on March 20, 2015. [Nikon D800, Nikon 14–24 zoom at 24 mm, ISO400, f/2.8, 1/8 second. ©2015 Stan Honda]

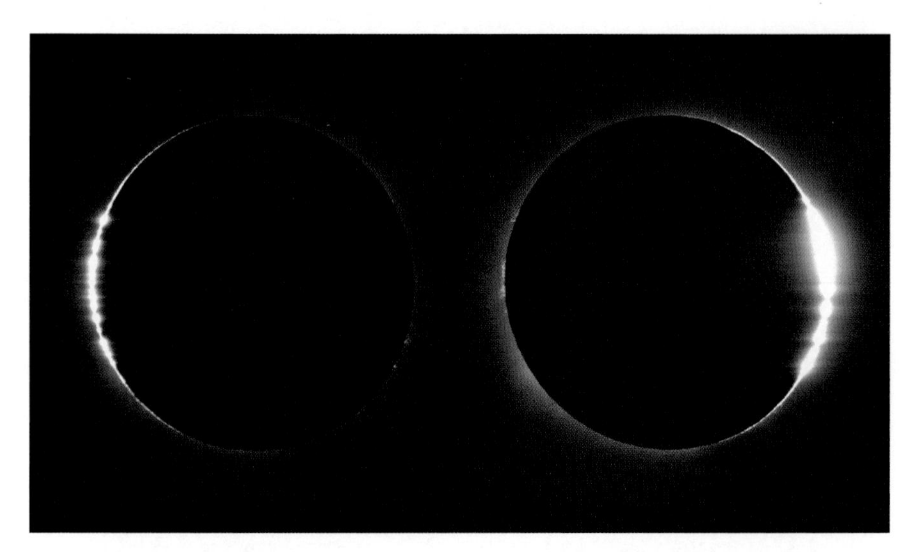

Short eclipses have fewer seconds of totality, but they offer fantastic views of Baily's beads. These two images were shot 34 seconds apart during the total solar eclipse of December 4, 2002 from Koolymilka, Australia. [Canon EOS D60 DSLR, Takahashi FS-60C w/ Extender-Q, fl = 568 mm f/9.5, exposures: 1/4000s and 1/1000s, ISO 800. ©2002 Arne Danielsen]

Just minutes after annularity, an oddly shaped crescent Sun sets behind giant wind turbines of a wind farm near Elida, New Mexico on May 20, 2012. [Nikon D90, Sigma 170–500 zoom at 500 mm, no solar filter, ISO 200, f/8, 1/500 second. ©2012 Fred Espenak]

A wide-angle view of totality captures eclipse chasers in a sugarcane field in North Queensland, Australia on November 14, 2012 [Canon 550D, Tokina 11–16mm zoom, ISO 3200, f/2.8, 1/25 second. ©2012 Stephen Mudge and Adam Poplawski]

A stunning view of the March 20, 2015 total eclipse from Svalbard was created from 29 individual exposures combined with custom software. In addition to the wealth of fine structure seen in the corona, lunar surface features are also revealed. [Nikon D810, TS Photo Line refractor, fl = 800 mm, f/6.9, exposures: 1/1600 to four seconds. © 2015 Miloslav Druckmüller, Shadia Habbal, Peter Aniol, Pavel Starha]

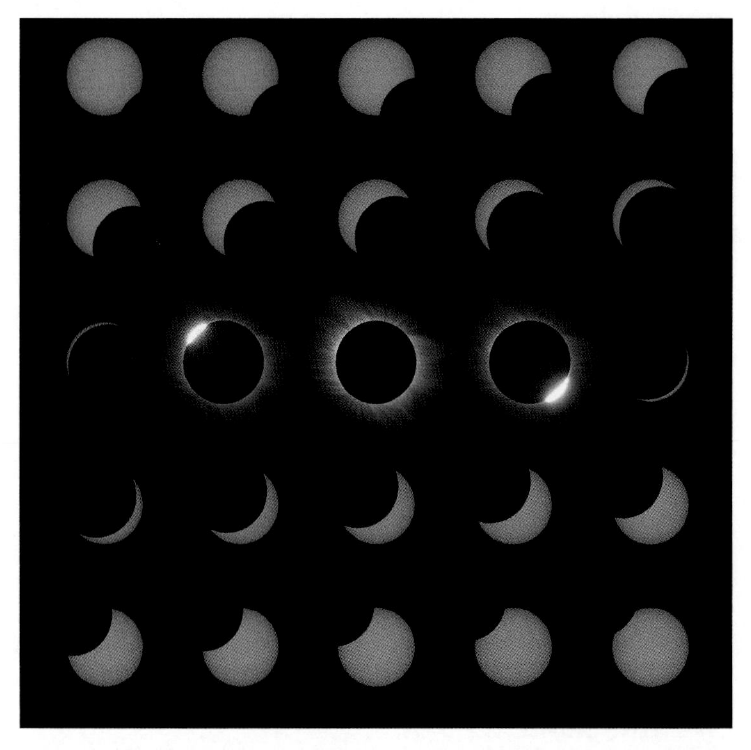

All stages of the March 29, 2006 total eclipse are captured in a mosaic of 25 images taken every seven minutes throughout the event. The diamond ring at each contact is included while the corona has been computer enhanced to show subtle details and prominences. [Nikon D200, Vixen 90 mm fluorite refractor, fl = 820 mm, f/9, ISO 400. ©2006 Fred Espenak]

A multiple-exposure time sequence records the partial phases and totality during the total solar eclipse of March 29, 2006 from Side, Turkey. [Zenith, 50 mm lens, Kodak Ektachrome, multiple exposures at f4. ©2006 Alex Tudorica]

The Moon's disk was surrounded by a diamond necklace during the beaded annular eclipse of May 30, 1984 from Pendleton, SC. At beaded annular eclipses, the Moon is very nearly the same size as the Sun so the Moon's higher mountains break the annular ring into a series of beads and crescent segments. [35 mm SLR, 3 inch F/15 Jaegers refractor, fl = 1140 mm, no solar filter, f/15, 1/500, ISO 64 slide film. ©1984 Johnny Horne]

Eclipse chasers gaze up at the totally eclipsed Sun from the Libyan desert on March 29, 2006. [Nikon N90s, Nikkor 16 mm Fisheye lens, f/5.6, auto (ISO 200). ©2006 Fred Espenak]

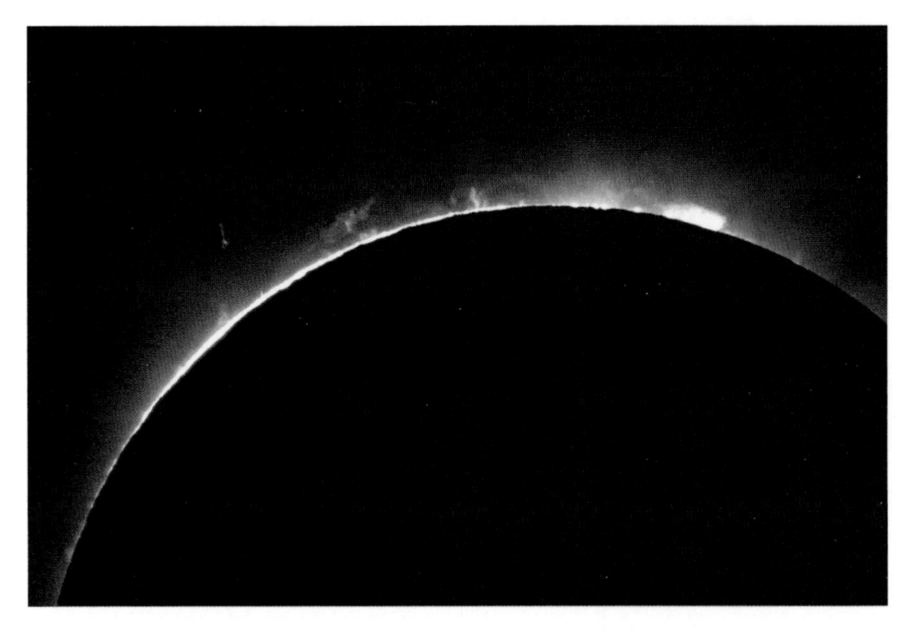

A forest of red prominences appears before third contact during the total solar eclipse of August 11, 1999 from Lake Hazar, Turkey. [Pentax ZX-M SLR, 94 mm Brandon refractor, f/30, 1/60s, Ektachrome V100 film pushed to ISO 200. ©1999 Greg Babcock]

In 1916 Einstein published his complete general theory of relativity. But the world hardly noticed. For two years, nations had been locked in World War I. Feelings against Germans ran strong in France and Great Britain, just as the Germans hated the French and British. Einstein sent his paper to a friend, scientist Willem de Sitter in The Netherlands. De Sitter, in turn, forwarded the paper to Arthur Eddington in England. At the age of 34, Eddington was already famous for his pioneering work in how stars emit energy. Eddington instantly recognized the significance of Einstein's discoveries and was deeply impressed by its intellectual beauty. He shared the paper with other scientists in Britain.

The 1919 Test

Frank Dyson, the astronomer royal of England, began planning for a British solar eclipse expedition in 1919 to test relativity. It was the perfect eclipse for the purpose because the Sun would be standing in front of the Hyades, a nearby star cluster, so there would be a number of stars around the eclipsed Sun bright enough for a telescope to see. But the timing could hardly have been worse. In 1917 Britain was in the midst of a terrible war whose issue was still very much in doubt. Yet somehow Dyson managed to persuade the government to fund the expedition, despite the fact that its purpose was to test and probably confirm the theory of a scientist who lived in Germany, the leader of the hostile powers.

Meanwhile, American astronomers had an earlier opportunity to verify or disprove relativity. In the final months of World War I, a total eclipse passed diagonally across the United States from the state of Washington to Florida. On June 8, 1918, a Lick Observatory team led by William Wallace Campbell and Heber D. Curtis, observing from Goldendale, Washington, pointed their instruments skyward to render a verdict on relativity. The weather was mostly cloudy, but the Sun broke through for 3 minutes during totality. Measuring and interpreting the plates had to wait several months, however, until comparison pictures could be taken of the same region of the sky at night, without the Sun in the way. By this time, Curtis was in Washington, D.C. working on military technology for the government as the war came to a close. Curtis returned to Lick in May 1919 and the results were announced in June. The star images were not as pointlike as desired. Curtis could detect no deflection of starlight. By this time, word was spreading about the findings of the British expeditions, and the Lick paper was never published.[11]

Four months after the armistice, two British scientific teams were poised for departure. Andrew C. D. Crommelin and Charles R. Davidson,

The total solar eclipse of May 29, 1919 featured an enormous prominence. Arthur Eddington's observation of this eclipse from Principe provided the first observational confirmation of Einstein's general theory of relativity. [Royal Astronomical Society]

heading one party, were to set sail for Sobral, about 50 miles (80 kilometers) inland in northeastern Brazil. Eddington, Edwin T. Cottingham, and their team were headed for Principe, a Portuguese island about 120 miles (200 kilometers) off the west coast of Africa in the Gulf of Guinea. On the final day before sailing, the four team leaders met with Dyson for a final briefing. Eddington was extremely enthusiastic and confident that Einstein was right. A deflection of 1.75 arc seconds would confirm relativity. Half that amount—a deflection of starlight by 0.87 arc seconds—would reconfirm Newtonian physics.[12]

"What will it mean," asked Cottingham, "if we get double the Einstein deflection?" "Then," said Dyson, "Eddington will go mad and you will have to come home alone!"

On Principe, worrisome weather conditions greeted Eddington and Cottingham. Every day was cloudy. However, May was the beginning of the dry season and no rain fell—until the morning of the eclipse. The fateful day, May 29, 1919, dawned overcast, and heavy rain poured down. The thunderstorm moved on about noon, but the cloud cover remained. The Sun finally peeped through 18 minutes before the eclipse became total but continued to play peekaboo with the clouds. Said Eddington: "I did not see the eclipse, being too busy changing plates, except for one glance to make sure it had begun and another half-way through to see how much cloud there was."

Arthur S. Eddington. [AIP Emilio Segrè Visual Archives, gift of S. Chandrasekhar]

Eddington had much cause for worry. He was not interested in prominences or the corona. He needed to see clearly the region around the Sun. With great care, the plates were developed one at a time, only two a night. Eddington spent all day measuring the plates. Clouds had interfered with the view, but five stars were visible on two plates. When he had finished measuring the first usable plate, Eddington turned to his colleague and said, "Cottingham, you won't have to go home alone."

Eddington measured the displacement in the stars' positions, extrapolated to the limb of the Sun, to be between 1.44 to 1.94 arc seconds, for a mean value of 1.61 ± 0.30 arc seconds. The deflection agreed closely with Einstein's prediction. In later years, Eddington referred to this occasion as the greatest moment of his life.[13]

The expedition to Sobral had equally threatening weather but there was a clear view of totality except for some thin fleeting clouds in the middle of the event. Crommelin and Davidson stayed in Brazil until July to take reference pictures of the star field without the presence of the Sun and then brought all their photographic plates back to Britain before measuring them. They found that their largest telescope had failed because heat caused a slight change in focus, which spoiled the pinpoint images of

English astronomers photographed the total solar eclipse of 1919 from two sites to test Einstein's general theory of relativity. Andrew Crommelin and Charles Davidson took this photograph from Sobral, Brazil. The white bars mark the location of stars recorded during totality. Careful measurements of their positions showed shifts because the stars' light, passing close to the Sun, was bent by the Sun's gravity, just as Einstein predicted. [Courtesy of the Royal Society of London]

the stars. But the other instrument had worked well, and its plates also supported Einstein's prediction. Their mean value was 1.98 ± 0.12 arc seconds.

It was now September and no news about eclipse results had been released. Einstein was curious and inquired of friends. Hendrik Antoon Lorentz, the Dutch physicist, used his British contacts to gather news and telegraphed Einstein: "Eddington found star displacement at rim of Sun . . . "[14] He also announced the favorable results at a scientific meeting in Amsterdam on October 25, with Einstein in attendance. But no reporters were present, and no word of the discovery was published.

Finally, on November 6, 1919, the Royal Society and the Royal Astronomical Society held a joint meeting to hear the results of the eclipse expeditions. The hall was crowded with observers, aware that an age was ending. From the back wall, a large portrait of Newton looked down on the proceedings. Joseph John Thomson, president of the Royal Society and discoverer of the electron, chaired the meeting, praising Einstein's

Albert Einstein and Arthur S. Eddington in Eddington's garden, 1930. [Royal Astronomical Society]

work as "one of the highest achievements of human thought." He called upon Frank Dyson to summarize the results and introduce the reports of the eclipse team leaders. The astronomer royal concluded his presentation by saying: "After careful study of the plates I am prepared to say that there can be no doubt that they confirm Einstein's prediction. A very definite result has been obtained that light is deflected in accordance with Einstein's law of gravitation."[15]

Einstein awoke in Berlin on the morning of November 7, 1919, to find himself world famous. Hordes of reporters and photographers converged on his house. He genuinely did not like this attention, but he found a way to turn the disturbance to the benefit of others. He told the reporters about the starving children in Vienna. If they wanted to take his picture, they first had to make a contribution to help those children. Suddenly Einstein was a celebrity.

More Tests

For the next half century, scientists from many nations traveled the world to measure and remeasure the bending of starlight around the Sun at every total eclipse. And always the results confirmed Einstein's general theory of relativity.[16]

In the 1970s, a new method emerged for testing relativity by the deflection of radiation around the Sun. Radio astronomers were using widely separated radio telescopes to measure the positions of celestial objects with an accuracy greater than a single optical telescope. Quasars had been discovered and nearly all astronomers interpreted them to be the most distant objects in the universe—so distant that they were essentially fixed markers. Their angular distance from one another provided a new coordinate system against which the positions of all other objects could be referred. Because some of the quasars lay near the ecliptic (the Sun's apparent path through the heavens in the course of a year), the Sun's gravity would annually deflect the quasars' radiation, making them seem to shift slightly in position. Only this time, the radiation would be radio waves rather than visible light. And no eclipse was necessary because radio telescopes do not require darkness. In 1974 and 1975, Edward B. Fomalont and Richard A. Sramek used a 22-mile (35-kilometer) separation between radio telescopes to measure the deflection of light at the limb of the Sun as 1.761 ± 0.016 arc seconds, a result that not only confirmed relativity, but also favored Einsteinian relativity over slightly variant relativity formulations by other scientists.[17]

Solar eclipses may no longer be the only or most accurate way to determine the relativistic deflection of starlight. But when the general theory of relativity predicted that the gravity of the Sun would bend starlight, a phenomenon never before observed, it was an eclipse that provided proof. The 1919 total eclipse of the Sun supplied the first and most dramatic demonstration of Einstein's masterpiece and brought relativity and Einstein to the attention of the entire world.

NOTES AND REFERENCES

1. Epigraph: Arthur S. Eddington as cited in Allie Vibert Douglas: *The Life of Arthur Stanley Eddington* (London: T. Nelson, 1956), page 44.
2. The angular displacement of a star is inversely proportional to the angular distance of that star from the Sun's center.
3. In 1911, Einstein had asked Freundlich to investigate another aspect of his emerging general theory of relativity: the motion of Mercury. Freundlich reviewed the anomaly in Mercury's motion, recognized since the work of Le Verrier in 1859, and reconfirmed that Mercury was indeed deviating slightly from the law of gravity as formulated by Newton. Freundlich's results, which matched Einstein's relativistic recalculation for the precession of Mercury's orbit, were published in 1913 over the objections of his superiors.
4. To make certain that the telescope optics and the camera system introduced no unknown deflection in the positions of stars, it was important to have for

comparison with the eclipse plate a picture of that same region taken with the same equipment when the Sun present was not present. No such plates were available.

5. Charles Dillon Perrine, an American who was directing the Argentine National Observatory, prepared to test the bending of starlight during the October 10, 1912 eclipse in Brazil, but was rained out.
6. Ronald W. Clark: *Einstein, the Life and Times* (London: Hodder and Stoughton, 1973), page 176.
7. Freundlich's team was not alone in the Crimea. An American team from the Lick Observatory also journeyed to Russia to test relativity, but they were clouded out. In 1918, Freundlich left the Royal Observatory to work full time with Einstein at the Kaiser Wilhelm Institute. In 1920, he was appointed observer and then chief observer and professor of astrophysics at the newly created Einstein Institute at the Astrophysical Observatory, Potsdam. Freundlich continued to be plagued by miserable luck on his eclipse expeditions to measure the deflection of starlight. He returned empty-handed in 1922 and 1926 because of bad weather. He finally got to see an eclipse in Sumatra in 1929, although he obtained a deflection (now known to be erroneous) considerably greater than Einstein predicted. When Hitler came to power, Freundlich left Germany and eventually settled in Scotland, where he changed his name to Finlay-Freundlich, based on his mother's maiden name, Finlayson. He built the first Schmidt-Cassegrain telescope, the prototype for almost all large photographic survey telescopes today.
8. Banesh Hoffmann with the collaboration of Helen Dukas: *Albert Einstein, Creator and Rebel* (New York: Viking Press, 1972), page 116.
9. Banesh Hoffmann with the collaboration of Helen Dukas: *Albert Einstein, Creator and Rebel* (New York: Viking Press, 1972), page 125.
10. Robert V. Pound and Glen A. Rebka, Jr.: "Resonant Absorption of the 14.4-kev Gamma Ray from 0.10-microsecond Fe^{57}," *Physical Review Letters*, volume 3, December 15, 1959, pages 554–556. The gravitational redshift was detected previously, based on Einstein's suggestion, in light emitted from extremely dense white dwarf stars, but to nowhere near the same degree of certainty.
11. The Lick team was working with mediocre equipment and improvised mounts. Their excellent regular equipment had stood ready under cloudy skies in Russia in 1914, but it was too cumbersome to transport home after the outbreak of World War I. After the war, it was delayed in shipment home and missed the eclipse of 1918.
12. The "Newtonian prediction" is calculated on the basis that light energy has a mass equivalent (Einstein's $E = mc^2$). That mass is then treated as ordinary matter using Newton's equations for gravity.
13. Allie Vibert Douglas: *The Life of Arthur Stanley Eddington* (London: T. Nelson, 1956), pages 40–41.

14. Ronald W. Clark: *Einstein, the Life and Times* (London: Hodder and Stoughton, 1973), page 226.
15. Frank W. Dyson, Andrew C. D. Crommelin, and Arthur S. Eddington: "Joint Eclipse Meeting of the Royal Society and the Royal Astronomical Society," *The Observatory*, volume 42, November 1919, pages 389–398. The article includes dissenting comments from scientists in audience. The article based on eclipse results was Frank W. Dyson, Arthur S. Eddington, and Charles R. Davidson: "A Determination of the Deflection of Light by the Sun's Gravitational Field, From Observations Made at the Total Eclipse of May 29, 1919," *Philosophical Transactions of the Royal Society of London*, series A, volume 220, 1920, pages 291–333. A summary of the eclipse results had appeared the week after the joint meeting: Andrew C. D. Crommelin: "Results of the Total Solar Eclipse of May 29 and the Relativity Theory," *Nature*, volume 104, November 13, 1919, pages 280–281. Eddington had yet another reason to be proud: "By standing foremost in testing, and ultimately verifying, the 'enemy' theory, our national observatory kept alive the finest traditions of science; and the lesson is perhaps still needed in the world today" (Clark: *Einstein*, page 284). British physicist Robert Lawson made a similar observation: "The fact that a theory formulated by a German has been confirmed by observations on the part of Englishmen has brought the possibility of cooperation between these two scientifically minded nations much closer. Quite apart from the great scientific value of his brilliant theory, Einstein has done mankind an incalculable service" (Clark: *Einstein*, page 297). The world soon forgot this lesson.
16. For a description of other expeditions, see Mark Littmann, Fred Espenak, and Ken Willcox *Totality: Eclipses of the Sun* (New York: Oxford University Press, 3rd edition: 2008; updated 3rd edition: 2009), pages 101–102.
17. Edward B. Fomalont and Richard A. Sramek: "A Confirmation of Einstein's General Theory of Relativity by Measuring the Bending of Microwave Radiation in the Gravitational Field of the Sun," *Astrophysical Journal*, volume 199, August 1, 1975, pages 749–755. The result reported in their article, 1.775 ± 0.019 arc seconds, was later revised to 1.761 ± 0.016 arc seconds. (Personal communication, March 1990.) They used three 85-foot (26-meter) antennas and a distant 45-foot (14-meter) antenna, instruments of the National Radio Astronomy Observatory at Green Bank, West Virginia. The idea of using radio waves and radio telescopes to test the bending of light was first proposed by Irwin I. Shapiro: "New Method for the Detection of Light Deflection by Solar Gravity," *Science*, volume 157, August 18, 1967, pages 806–807.

A MOMENT OF TOTALITY

The Difference Between Partial and Total

Canadian filmmaker and eclipse veteran Jean Marc Larivière is always frustrated when people confuse a partial eclipse of the Sun with a total eclipse. "Oh yeah," they say, "I saw an eclipse of the Sun when I was young. It was no big deal."

A total eclipse of the Sun is utterly different from a partial eclipse.

For his film *Shadow Chasers*,[1] Jean Marc interviewed a busload of French tourists who had traveled to the west coast of Australia in February 1999 to see an *annular* eclipse of the Sun that was almost total. The Moon was just a bit too far from Earth to completely cover the disk of the Sun.

After the eclipse, he interviewed a woman who told him: "It was a wonderful annular eclipse, very tight. But you know, despite so much of the Sun being eclipsed, it was still a partial eclipse. A partial eclipse is like a first kiss. A total eclipse is like a night of passionate love."

[1] The 2000 documentary *Shadow Chasers* was written and directed by Jean Marc Larivière and produced and distributed by the National Film Board of Canada: https://www.nfb.ca/film/shadow_chasers.

9

Observing a Total Eclipse

A total eclipse of the Sun . . . is the most sublime and awe-inspiring sight that nature affords.

Isabel Martin Lewis (1924)[1]

A total eclipse of the Sun. What is this sight that lures people to travel to the ends of the Earth for a brief view at best, and a substantial possibility of no view at all, with no rain check? And how do you get the most out of the experience of a total eclipse?

In these pages, eclipse veterans of today, plus some eclipse seekers from earlier times share their experiences with you. Among them, they have witnessed over 500 total eclipses.

"It's like a religious experience," says Jay Anderson, "the anticipation as the time until totality is counted in days, then hours, then minutes. It's the perfect buildup. Spielberg couldn't do it better. It's an intensely moving event."

Steve Edberg agrees: "It is the intensity of the event. You grab as much as you can. I like action in the heavens and you can't get much better than this."

Mike Simmons starts his countdown to the eclipse four weeks in advance when the new moon becomes visible after sundown in the west. He watches the Moon night by night edging eastward across the sky, increasing from crescent to first quarter to gibbous to full, then waning from gibbous to last quarter to crescent. It seems like an ordinary month—but it isn't. This month the Moon is headed for a perfect alignment between the Sun and Earth.

Roger Tuthill had a ritual he followed the day before an eclipse. He walked around inside the path of totality observing the people who live there. They have no idea what they are about to experience and there are no words to adequately prepare them.

Panel of Eclipse Veterans

Jay Anderson, meteorologist, noted for eclipse weather predictions (Canada)
Paul and Julie (O'Neil) Andrews, information technologist; businesswoman (UK)
Dave Balch, sociologist (California)
John Beattie, eclipse researcher (New York)
Richard Berry, astronomy author and former editor-in-chief, *Astronomy* (Oregon)
Satyendra Bhandari, astronomer (India)
P. M. M. (Ellen) Bruijns, scientist (Netherlands)
Joe Buchman, marketing consultant (Utah)
Kristian Buchman, university student (Utah)
Dennis di Cicco, senior editor (retired), *Sky & Telescope* (Massachusetts)
Stephen J. Edberg, astronomer, NASA Jet Propulsion Laboratory (California)
Alan Fiala, astronomer, US Naval Observatory (Washington, D.C.) (Alan died in 2010)
George Fleenor, planetarium director (retired) (Florida)
Ruth S. Freitag, senior science specialist (retired), Library of Congress (Washington, D.C.)
Thomas Hockey, professor of earth science, Univ. of Northern Iowa
Joseph V. Hollweg, professor of astronomy (retired), Univ. of New Hampshire, Durham
Xavier Jubier, computer scientist; created Google interactive eclipse maps (France)
Jean Marc Larivière, filmmaker (Canada)
Charles Lindsey, astronomer, Southwest Research Institute, Boulder (Colorado)
George Lovi, astronomy author, Lakewood, New Jersey (George died in 1993)
David Makepeace, filmmaker (Canada)
Larry Marschall, professor emeritus of physics, Gettysburg College (Pennsylvania)
Sarah Marwick, physician (United Kingdom)
Frank Orrall, professor of physics & astronomy, Univ. of Hawaii (Frank died in 2000)
Jay M. Pasachoff, Field Memorial professor of astronomy, Williams College (Mass.)
Patrick Poitevin, technology scientist (Belgium, United Kingdom)
Luca Quaglia, physicist and financial engineer (New York)
Joe Rao, senior meteorologist, News 12 (New York)
Leif J. Robinson, editor-in-chief (emeritus), *Sky & Telescope* (Mass.) (Leif died in 2011)
Michael Rogers, writer and futurist (New York)
Gary and Barbara (Schleck) Ropski, attorney; journalist (retired) (Illinois)
Walter Roth, travel agency owner (Florida) (Walter died in 2002)
Glenn Schneider, astronomer, University of Arizona
Mike Simmons, founder, Astronomers Without Borders (California)
Gary Spears, travel agency owner (Oklahoma)
Roger W. Tuthill, electrical engineer; solar filter inventor (New Jersey) (Roger. died in 2000)
Ken Willcox, geologist, Phillips Petroleum Company (Oklahoma) (Ken died in 1999)
Sheridan Williams, rocket scientist (retired) (United Kingdom)
Michael Zeiler, computer scientist, eclipse mapmaker (New Mexico)
Jack B. Zirker, astronomer (retired), National Solar Observatory (New Mexico)
Evan Zucker, software engineer (California)

First Contact

There is a special feeling at the instant when the Moon begins to slide in front of the Sun. In less than a minute, observers with small telescopes see the first tiny "bite" out of the western side of the Sun. That's when the magic starts.

First contact remains a very special moment for Jay Pasachoff. As an astronomer, he can appreciate more readily than most all the factors that go into predicting precisely when an eclipse will occur and exactly where on Earth it will be seen. And when the call "first contact" comes right on schedule, he always finds this accomplishment of mankind astounding.

Famed nature writer Annie Dillard saw the 1979 solar eclipse, the last time totality visited the mainland of the United States. "It began," she said, "with no ado. It was odd that such a well-advertised public event should have no starting gun, no overture, no introductory speaker. I should have known right then that I was out of my depth. Without pause or preamble, silent as orbits, a piece of the sun went away.[2]

Even 175 years ago, the commencement of this rare event was already exerting a powerful effect on its beholders. French astronomer François Arago observed the 1842 eclipse from Perpignan in southern France amid townspeople and farmers who had been educated about the eclipse and who were watching the sky intently. "We had scarcely, though provided with powerful telescopes, begun to perceive a slight indentation in the sun's western limb, when an immense shout, the commingling of twenty thousand different voices, proved that we had only anticipated by a few seconds the naked eye observation of twenty thousand astronomers equipped for the occasion, and exulting in this their first trial."[3]

Yet, as George B. Airy, astronomer royal of England, learned when he saw his first total eclipse in 1842: "No degree of partial eclipse up to the last moment of the sun's appearance gave the least idea of a total eclipse . . . "[4]

The Crescent Sun

The partial phase of a total eclipse has a power all its own, as the Moon steadily encroaches upon the Sun, covering more and more of its face. A partial eclipse, close to total, visited a Russian monastery in medieval times, near sunset on May 1, 1185. A chronicler recorded: "The sun became like a crescent of the moon, from the horns of which a glow similar to that of red-hot charcoals was emanating. It was terrifying to men to see this sign of the Lord."[5]

As the partial phase proceeds, it's a good time to use your cardboard pinhole camera to project the Sun in crescent phase on a sheet of white cardboard. Alternatively, use a strainer or cheese grater and get dozens

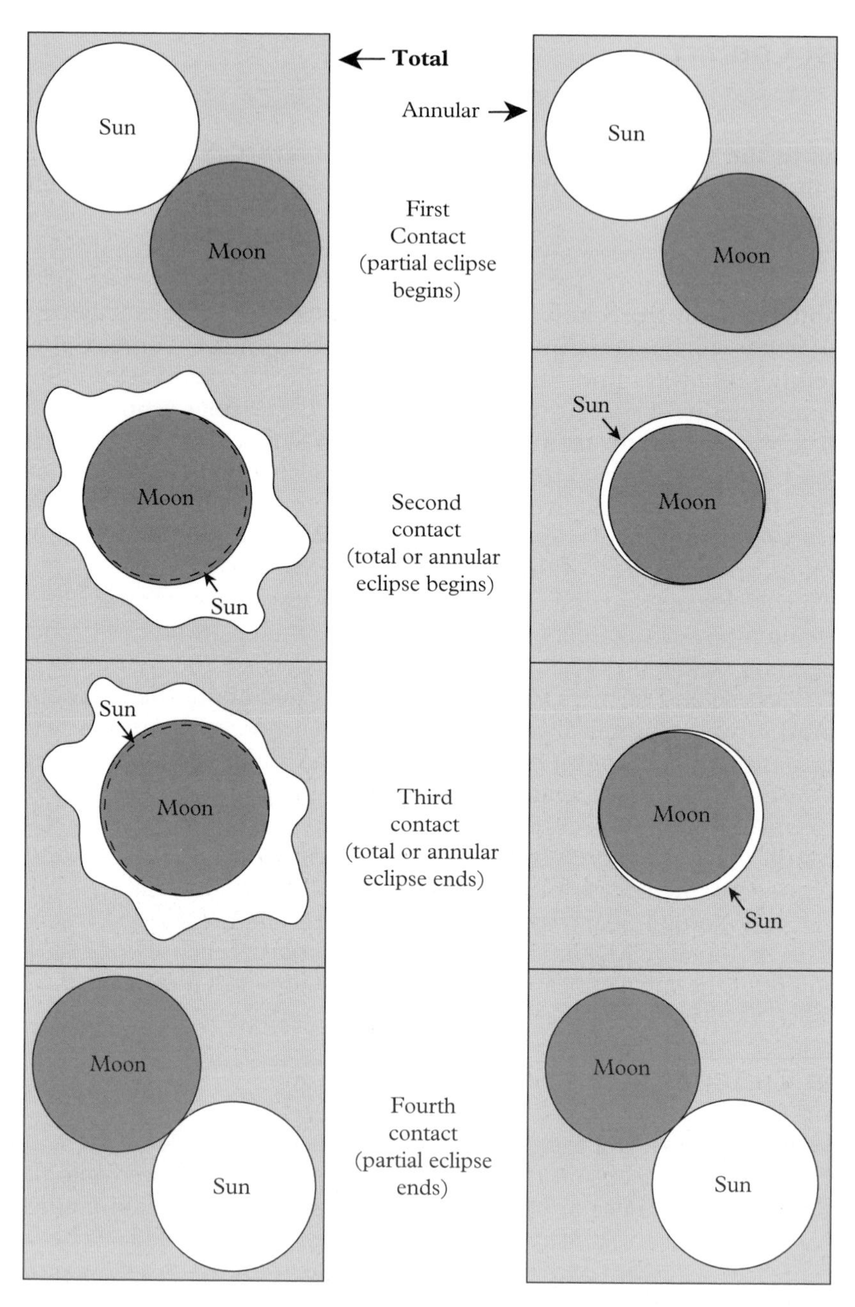

Points of contact in a total solar eclipse *(left)* and an annular eclipse *(right)*. In a partial eclipse, there are only first and fourth (last) contacts.

of crescent Suns. You can also project images of the crescent Sun with just your hands. Lace your fingers and let the sunlight pass between them onto the ground or a piece of cardboard.

Check nearby trees and bushes to see if they are casting a thousand tiny images of the crescent Sun on the ground beneath them. Small gaps between the leaves create natural pinhole cameras to focus the crescent image of the waning Sun.

During the early partial stages of a total eclipse, be sure to notice the shadows cast by the Sun. For now they will be the usual solar shadows: a little fuzzy at the edges. But as the crescent of the Sun narrows, they will be quite different.

If you are observing the partial phases with binoculars or a telescope—with a solar filter covering the front of the binoculars or telescope, of course—make note of any sunspots you see, then watch as the Moon steadily occults them. If there are sunspots visible at the limb of the Sun, remember their positions and, when totality is beginning or ending, see if those positions are marked by vivid reddish prominences—magnetic storms in the Sun's atmosphere associated with the sunspots. When the

Pinhole images of the crescent Sun are formed through the leaves of a tree during the solar eclipse of October 3, 2005, from St. Julian's, Malta. [Canon PowerShot A20, f/2.8, 1/40 second. Wikimedia Commons]

corona flares into view, see if it shows lengthened rays or shock waves near where you saw the sunspots at the edge of the Sun. You may be seeing evidence of eruptions from the Sun—coronal mass ejections—rushing through the corona and out into space at more than a million miles per hour (1.6 million kilometers per hour). But that's too slow for you to notice this motion within the corona during a few minutes of totality.

The Changing Environment

Every eclipse veteran urges newcomers to pause as totality approaches to look around at the landscape and notice the changes in light levels and color. Many people are surprised how little the landscape darkens until the last 15 minutes before the eclipse becomes total. All the better for the drama, because once the light begins to fade, it sinks quite noticeably. Usually, everyone becomes silent. You can feel the tension and rising emotion. A primitive portion of your brain tugs at you to say that something peculiar is going on, that it ought not to be growing dark in the midst of the day.

During the 1976 eclipse in Australia, Dennis di Cicco remembers that the birds responded to the fading light by raising a racket and going to roost. In 1973, he witnessed an annular eclipse in Costa Rica that took place shortly after sunrise. The cows grazing around him ignored the partial phases of the eclipse, but as the eclipse reached maximum, the cows ceased grazing, formed a line, and marched back to their barn.

Walter Roth watched the 1973 total eclipse from a game preserve in Africa. As the light faded in the minutes just before totality, the birds flocked into the trees, complaining madly. Elephants, which had been grazing peacefully, milled around nervous and confused.

"As the light fades, the Sun is a thin crescent and shrinking quickly," says Richard Berry. "The quantity and quality of the light have begun to change noticeably. The temperature is dropping; the air feels still and strange."

The Sharp Edge of Shadows

On ordinary days and in the early partial phase of the eclipse, sunlight falling on an object casts a shadow with fuzzy edges. Now, as the crescent of the Sun narrows and the daylight dims, look again at the shadows cast by trees and leaves and buildings—and you. Spread your fingers and let your hand cast a shadow on the ground. Look how sharp the edges of the shadow are.

On an ordinary day, every part of the disk of the Sun—top, bottom, left side, right side—is casting your shadow from a slightly different angle, creating a fuzzy edge. Now the Sun is reduced to a thin crescent, more like a narrow slit than a disk. Your shadow is fuzzy in one direction but razor sharp in the other. This is no ordinary day.

Color Change

It's now about 15 minutes before totality. What's happening to the colors around you: the sky, the landscape, the buildings, the water, the trees and leaves? The colors are fading. The sky is taking on a steely blue-gray, even purplish cast. You've never seen the sky that color, even with the approach of a storm. Yet today the sky is clear or nearly so. The Sun is still there in the sky. But the light is going out. The shadows are sharp, crisp.

Temperature Drop

As you waited for first contact, for the eclipse to begin, the Sun was warming the Earth. The temperature was rising, as it does on a usual day. But now—what's happening? Is it how weird nature is beginning to look? Is it how weird you feel? The temperature is no longer rising. It's maybe even a little cooler. It will be getting noticeably cooler soon as more and more of the Sun is blocked by the Moon.

If small summertime cumulus clouds have been puffing up, maybe their buildup has stopped. Maybe some clouds have begun to thin and disappear because less warm, moist air is rising from the ground and because the upper atmosphere cools more rapidly than the lower air. The cooling upper air sinks, taking the clouds lower, where the slightly higher atmospheric pressure causes the water droplets that form the clouds to heat up and vaporize, turning the droplets of the clouds into transparent water gas scattered in the atmosphere.

The cooling as the Sun is increasingly blocked from view is not sufficient to break up a developing thunderstorm or a frontal system, but the approach of totality can substantially reduce clouds and improve observing conditions.

The Moon's Shadow Visible

Now look toward the western horizon. Looming there, growing ever larger, is the Sun-cast dark shadow of the Moon. And it is coming toward you. Alan Fiala described its appearance as the granddaddy of thunderstorms, but utterly calm. If you are observing from a hill with a view to the west, the approach of the Moon's shadow can be quite dramatic, even chilling.

In the words of astronomer Mabel Loomis Todd more than a century ago, it is "a tangible darkness advancing almost like a wall, swift as imagination, silent as doom."[6]

Shadow Bands

As the eclipse nears totality, and shortly after it emerges from totality, shadow bands—faint undulations of light rippling across the ground at jogging speed—are sometimes visible. They are one of the most peculiar and least expected phenomena in a total eclipse. Many eclipse veterans have never seen them. Some do not want to take time to try because they

Catching Shadow Bands

by Laurence A. Marschall

Even though shadow bands are only visible for a few fleeting minutes, it is possible to catch them if you prepare in advance. Get a large piece of white cardboard or white-painted plywood to act as a screen—the bands are subtle and can be more easily seen against a clean, white surface. A large white sheet staked to the ground may be more portable and will serve in a pinch, but ripples in the sheet can mask the faint gradations of the shadow bands.

Lay out on the screen one or two sticks marked with half-foot intervals (yardsticks will do nicely). Orient the sticks at right angles to one another so that at the first sign of activity you can move one stick to point in the direction that the shadow bands are moving. Then, using the marks on that stick, make a quick estimate of the spacing between the bands (typically 4–8 inches; 10–20 centimeters). Finally, using a watch, make a quick timing of how long a bright band takes to go a foot or a yard. Jot down the figures or, better yet, dictate your measurements into a small audio recorder. If you practice this procedure before the eclipse, you will be able to see the shadow bands and then swing your attention back to the sky to catch the diamond ring, Baily's beads, and the onset of totality.

The second stick, by the way, is reserved for marking the direction of the shadow bands *after* totality, if they are visible.

After the eclipse, you can take stock of your data. Do the shadow bands seem to move at all? At some eclipses, especially when the air is very still, they just shimmer without going anywhere. If they move, how fast? Typical speeds are about 5–10 miles per hour (8–16 kilometers per hour). Do they change directions after eclipse? Usually they do, unless you happen to be standing directly along the central line of totality.

Dr. Laurence A. Marschall is Professor of Physics (Emeritus) at Gettysburg College and Deputy Press Officer, American Astronomical Society.

occur in the last moments before totality as Baily's beads and other beautiful sights are visible overhead. Other veterans consider shadow bands one of the true highlights of a total eclipse. They resemble the graceful patterns of light that flicker or glide across the bottom of a swimming pool.

Shadow bands occur when the crescent of the remaining Sun becomes very narrow so that only a thin shaft of sunlight enters the atmosphere of

The Personalities of Eclipses

by Stephen J. Edberg

Each total eclipse is different from all others and these differences continue to lure eclipse veterans. They use them in planning their observations.

One factor is the magnitude of the total eclipse, the degree to which the disk of the Moon more than covers the disk of the Sun. When the angular size of the Moon is great, the Moon at mid-totality will mask not only the Sun's photosphere but also its chromosphere and innermost corona. Except at the beginning and end of totality (or near the eclipse path limits), the stunning fluorescent pink of the prominences will be hidden from view, unless an absolutely gigantic prominence happens to be present on the Sun's limb (as there was for the 1991 eclipse).

However, because a large magnitude eclipse blocks from view the relatively bright lower corona, a large magnitude eclipse is the best time to observe the full extent and detail in the corona. Because the Moon appears larger, its shadow is wider. Thus, the total eclipse lasts longer and the darkness is deeper, allowing the full majesty of the corona to shine through, as well as any planets and bright stars that happen to be above the horizon.

The Moon's apparent motion over the Sun's disk is easier to photograph during long eclipses. One photo taken at the onset of totality and a second taken just before the end maximizes the change in the Moon's position with respect to the Sun. When viewing these pictures as a stereo pair, the Moon appears to actually hang in front of the corona and prominences.

By contrast, in a smaller magnitude total eclipse, the Moon's disk is not big enough to mask the lower corona, and prominences may be seen all the way around the disk of the Sun during most of totality. But this eclipse will be comparatively brief and the full extent of the corona may not be evident.

A total eclipse with less obscuration often provides a better display of Baily's beads and the diamond ring effect. Because the Moon's disk is nearly the same apparent size as the Sun's disk, as totality nears, the crescent of the Sun is long and narrow, allowing Baily's beads to glimmer like jewels on a necklace. When the Moon's apparent disk is large, the length of the Sun's crescent is greatly shortened as it thins. There may be only one or two Baily's beads.

As veteran observers plan for upcoming eclipses, they also take into consideration the sunspot cycle. At sunspot maximum, the corona is brighter, rounder, and larger. Prominences also tend to be more numerous. At sunspot minimum, the corona is

Earth overhead. There it encounters currents of warmer and cooler air that have slightly different densities. The different densities act as weak prisms to bend the light passing from one parcel to the next.[7] It is the slight bending and rebending of light by these ever-present air currents in motion that causes stars to twinkle. The Sun would twinkle too if it were a starlike dot in the sky. And so it does, near the total phase of a solar

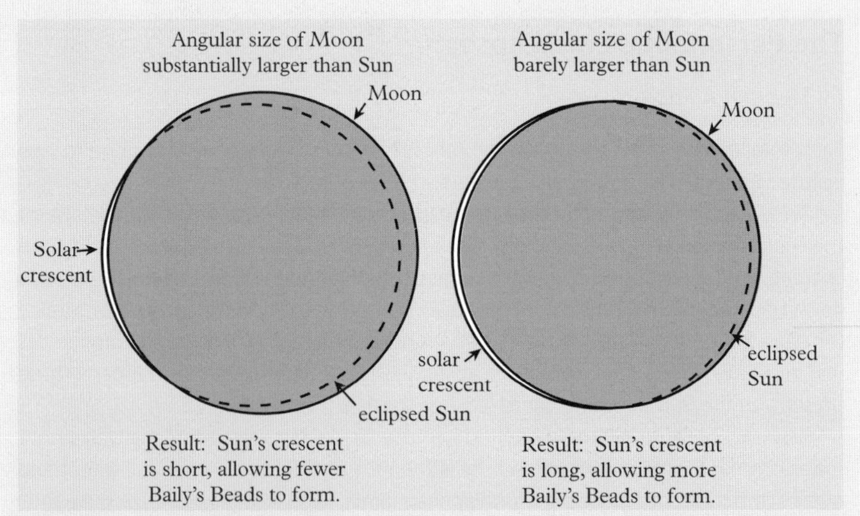

Left: *When the Moon's angular size is large, the Sun's crescent is short, reducing the span of Baily's beads.* Right: *When the Moon just barely covers the Sun, so that the disks appear nearly the same size, the Sun's crescent is much longer and the span of Baily's beads is greater.*

fainter and broader at the equator than at the poles. Brush-like coronal features projecting from the poles are more noticeable.

A third factor used by eclipse followers is the proximity of the shadow path to the Earth's equator, where the rotation of the Earth causes the ground speed of the Moon's shadow to be slowest, making the duration of totality longest. Eclipses with 6 to 7 minutes of totality are almost always found between the Tropic of Cancer and the Tropic of Capricorn. The rotation of the Earth extends not just the period of totality but all aspects of the eclipse, so that, near the equator, the partial phases of the eclipse last longer. The last crescent of the Sun is covered more slowly, so Baily's beads and the diamond ring effect, while still brief, last a little longer, and the view of the prominences at the limb of the Sun is prolonged.

Stephen J. Edberg is a Jet Propulsion Laboratory scientist, astronomy author, and executive director of the RTMC Astronomy Expo, a major annual conference of amateur astronomers.

Shadow bands on an Italian house in 1870. [Mabel Loomis Todd: *Total Eclipses of the Sun*]

eclipse. When the Sun has narrowed from a disk to a sliver, the light from the sliver twinkles in the form of shadow bands rippling across the ground.

In 1842, George B. Airy, the English astronomer royal, saw his first total eclipse of the Sun and recalled shadow bands as one of the highlights: "As the totality approached, a strange fluctuation of light was seen . . . upon the walls and the ground, so striking that in some places children ran after it and tried to catch it with their hands."[8]

The Approach of Totality

During the final 1 to 2 minutes before totality, the Sun is about 99% covered and the light is fading fast. Now everything happens at once. In these brief moments, the Moon's shadow on the horizon rushes at you, while at the rim of the vanishing Sun, the corona looms into view and the sliver of sunlight becomes a brilliant diamond ring, then breaks up into Baily's beads.

This is how astronomer Isabel Martin Lewis in 1924 described the onrush of the Moon's shadow as totality begins: "When the shadow of the moon sweeps over us we are brought into direct contact with a tangible presence from space beyond and we feel the immensity of forces over which we have no control. The effect is awe-inspiring in the extreme."[9]

Glenn Schneider went to Roy, Montana, for the 1979 eclipse. He watched the approaching shadow of the Moon growing on the western horizon dark and vast. Suddenly it flew at him. Instinctively, he ducked.

As the shadow races toward you, it seems to accelerate. Ten seconds before totality, you feel as if you are being swallowed by a gigantic whale.

At this moment, above you—there it is. The soft white glow of the corona begins to silhouette the Moon's dark disk while a last dazzling beacon of sunlight clings to one edge of the Moon—the diamond ring effect.[10] "The remaining light seems to scintillate," says Michael Zeiler. "Your surroundings shimmer—eerie, sharp, electric."

A few seconds before totality begins, the ends of the remaining sliver of the Sun fracture into Baily's beads. The last rays of sunlight are shining through the lunar valleys. Each bead lasts only an instant and flickers out until only one remains. "For one fleeting moment this last bead lingers, like a single jewel set into the arc that is the lunar limb," says John Beattie. And then it is gone.

The eclipse is total.

Diamond ring effect is captured 10 seconds before totality begins during the total solar eclipse of August 1, 2008, from Jinta, China. [Nikon D300, Vixen 90 mm fluorite refractor, fl = 810 mm, f/9, 1/250 second, ISO 320. ©2008 Fred Espenak]

Baily's beads are seen just before and after the total phase of a solar eclipse. They are formed by sunlight passing through deep valleys along the edge of the Moon. This image of Baily's beads was made aboard the *MS Statendam* off the coast of Curaçao on February 26, 1998. [600 mm F/4 Nikon with TC-301 2X converter, fl = 1200 mm, f/11, 1/250 second, ISO 400 film. ©1998 Johnny Horne]

The Corona

It is difficult to find words that do justice to the corona, the central and most surprising of all features of a total eclipse. The Sun has vanished, blocked by the dark body of the Moon. A black disk surrounded by a white glow. "It's the blackest black I've ever seen," says Michael Zeiler. "It is the eye of God," says Jack Zirker.

In the bitter cold of the high plateau in Bolivia, astronomer Frank Orrall was part of a research team studying the total eclipse of 1966. In reviewing his observing notes, he realized that he had written "The heavens declare the glory of God." "My notebooks normally do not say such things," he mused.

The color of the corona is difficult to capture. "I prefer 'pearly,'" says Ruth Freitag, "because it conveys a luminous quality that 'white' lacks." "A silver gossamer glow," suggests Steve Edberg. To Michael Zeiler "it appears to be a living object with delicate filamentary structure." "Remember," said George Lovi, "the human eye is the only instrument that can see the corona in all its splendor."

Take time to examine the corona. Is it symmetrical or bunched up at the Sun's equator? Do you see bristle-like rays extending outward from the poles of the Sun—"polar brushes"? Are there streamers stretching outward from the equator? Do you see loops, arcs, or plumes in the corona? Any swirls or concentrations of material? You might be looking at a coronal mass ejection—the Sun hurling million-degree gases into space. Only during a total eclipse can you see the corona with your unaided eyes.

How far can you see the corona extending away from the Sun? You can estimate it by comparing it to the diameter of the eclipsed Sun, which is conveniently and just barely covered by the Moon. The angular diameter of the Moon—and Sun—is ½°. Does the corona extend away from the black rim of the Moon by two solar diameters? Three? More? The diameter of the Sun is 865,000 miles (1.4 million kilometers)—almost a million miles. If you can see the corona extending outward three solar diameters, that's about 3 million miles (about 4¾ million kilometers) into space.

During Totality

In the midst of totality, it is particularly impressive to look around at the nearby landscape and the distant horizon. Depending on your position within the eclipse shadow, the size of the shadow, and cloud conditions away from the Sun's position, the light level during totality can vary greatly. It also varies from eclipse to eclipse and from place to place within an eclipse. On some occasions, totality brings the equivalent of twilight soon after sunset. On others, it is dark enough to make reading difficult. Yet it is never as black as night.

The color that descends upon everything is hard to specify. Most veteran observers describe it as an eerie bluish gray or slate gray, and then apologize for the inadequacy of the description. The Sun's corona contributes some brightness, but only about as much as a full moon. Primarily, the light that brightens your location within the shadow of the Moon is light reflected from miles away where the Sun is not totally eclipsed.

There in the distance in every direction, it is twilight—painted with sunset oranges and yellows. It surrounds you. The colors are familiar. Yet it's all so strange. This twilight does not tower high into the sky. It is cut off by something dark overhead—a lid that confines the colors to a band around the horizon. You are seeing beyond the Moon's shadow. Out there, far away, the eclipse is not total. Where you stand is very special.

Here, in the midst of totality, looking around at the colors on the horizon and the darkness overhead, you have the fullest impression of standing in the shadow of the Moon. You also have a renewed appreciation for the power of the Sun. There it is, its face completely blocked by the Moon.

How big in the sky is that body whose overwhelming brightness creates the day and banishes the stars from view? Stretch out your arm to full length toward the eclipsed Sun, just as you did to measure the Moon. Squeeze the darkened Sun between your index finger and your thumb until it just fits between them. It is the size of a pea. Yet that little spot in the sky, darkened now, is usually enough to blanket the Earth in light and warmth and completely dazzle the eye. For just a few moments its face is hidden. Daytime has become night and the temperature is falling. Do you feel something of what ancient people must have felt as they watched the Sun, upon which they depended for warmth and light and life itself, disappear?

Leif Robinson emphasized that a total eclipse cannot be experienced vicariously. Even the best photography and video recordings are pale reflections of the event. Taking pictures during an eclipse is fine, but be sure you *look* at what is truly a visual spectacle. Do not miss the ambience of the moment either, he advised. Look at other people to see how they are reacting. Notice the changing colors in the sky. He did not, however,

The glorious corona is visible for just a few short minutes during the total solar eclipse of August 11, 1999, from Németkér, Hungary. [Soligor SR-300MD SLR, MTO 1000a Maksutov, fl = 1084 mm, f/10.5, composite of 22 exposures: 1/4000 to 2 seconds, Fujicolor Superia 800. ©1999 Miloslav Druckmüller]

recommend trying to see planets or stars in the sky during a total eclipse. "Time during totality is too precious to spend straining to see stars when you can see those same objects much better in the nighttime sky." Steve Edberg feels differently: "One of my strongest memories from eclipses is seeing stars during the day."

The transformation in the appearance of the Sun from a bright crescent to a dark disk surrounded by ghostly light was recorded by astronomer François Arago as he watched the eclipse of July 8, 1842, with other astronomers and nearly 20,000 townspeople in southern France.

> When the sun, being reduced to a narrow filament, began to throw only a faint light on our horizon, a sort of uneasiness took possession of every breast; each person felt an urgent desire to communicate his emotions to those around him. Then followed a hollow moan resembling that of the distant sea after a storm, which increased as the slender crescent diminished. At last, the crescent disappeared, darkness instantly followed, and this phase of the eclipse was marked by absolute silence . . . The magnificence of the phenomenon had triumphed over the petulance of youth, over the levity affected by some of the spectators as indicative of mental superiority, over the noisy indifference usually professed by soldiers. A profound calm also reigned throughout the air: the birds had ceased to sing.
>
> After a solemn expectation of 2 minutes, transports of joy, frenzied applauses, spontaneously and unanimously saluted the return of the solar rays. The sadness produced by feelings of an undefinable nature was now succeeded by a lively satisfaction, which no one attempted to moderate or conceal. For the majority of the public the phenomenon had come to a close. The remaining phases of the eclipse had no longer any attentive spectators beyond those devoted to the study of astronomy.[11]

How do you take it all in? It is not possible. There is too much to see—and feel. "Everybody, myself included, tries to do too much," says Steve Edberg. "Save time near the beginning, middle, and end of totality just to stare," urges Jay Anderson. "Make a deliberate effort to store the sights in your mind."

More than a century ago, Mabel Loomis Todd wrote that "When Dr. Peters of Hamilton College was asked what single instrument he would select for observing an eclipse, he replied, 'A pillow.'"[12]

The End of Totality

All too soon, it's over. Totality can never last longer than 7 minutes 32 seconds.[13] Even that is not long enough. The Moon moves on in its orbit, beginning to uncover the Sun. The western rim of the Moon's black disk brightens as the Sun's crimson prominences and chromosphere (lower atmosphere) peek out from behind the Moon. A dazzling dot of white

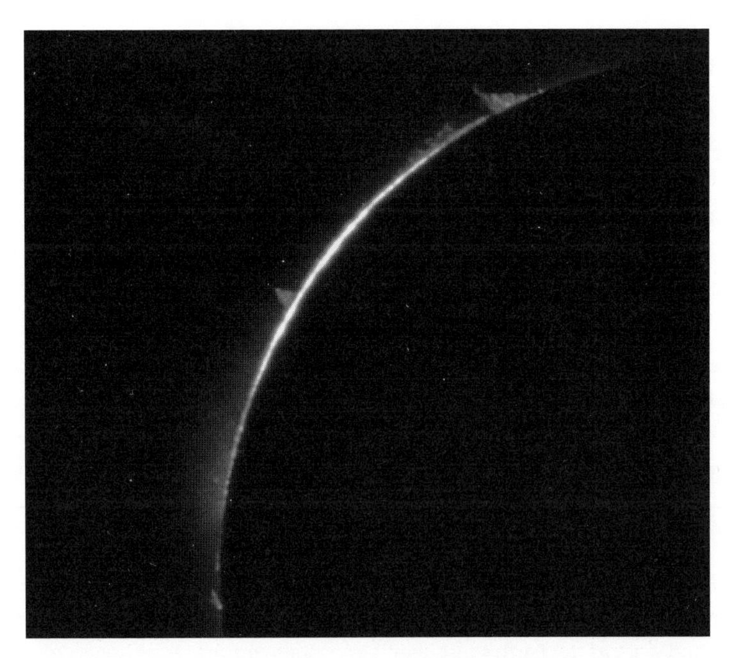

Brilliant prominences and the chromosphere are seen at the total solar eclipse of March 29, 2006, from Jalu, Libya. [35 mm SLR, Borg 100ED f/6.4 refractor, fl = 640 mm, Vixen GP mount, 2x teleconverter, f/12.8, 1/250 to 1 second, Fuji Velvia 50 slide film. ©2006 Dave Kodama]

light appears, then two, three—a diamond necklace, Baily's beads—joining to form the thinnest of crescents, sunlight bursting from the western edge of the Moon. A second diamond ring. Too bright to look at. Filters on. The corona fades as daylight returns. The dark lunar shadow rushes off to the east, bringing totality to observers farther along the path.

Rebecca Joslin and her Smith College astronomy teacher, Mary Emma Byrd, traveled from the United States to Spain for an eclipse in 1905, only to be clouded out. So they shifted their attention to nature around them until light pierced the cloud to tell them that the unseen total phase of the eclipse was over.

> But we hardly had time to draw a breath, when suddenly we were enveloped by a palpable presence, inky black, and clammy cold, that held us paralyzed and breathless in its grasp, then shook us loose, and leaped off over the city and above the bay, and with ever and ever increasing swiftness and incredible speed swept over the Mediterranean and disappeared in the eastern horizon.
>
> Shivering from its icy embrace, and seized with a superstitious terror, we gasped, "WHAT was THAT?"... The look of consternation on M's face lingered for an instant, and then suddenly changed to one of radiant joy as the triumphant reply rang out, "THAT was the SHADOW of the MOON!"[14]

Stages of a Total Eclipse

First Contact The Moon begins to cover the western limb of the Sun.

Crescent Sun Over a period of about an hour, the Moon obscures more and more of the Sun, as if eating away at a cookie. The Sun appears as a narrower and narrower crescent.

Light and Color Changes About 15 minutes before totality, when 80% of the Sun is covered, the light level begins to fall noticeably—and with increasing rapidity. The landscape takes on a metallic gray-blue hue.

Animal, Plant, and Human Behavior As the level of sunlight falls, animals may become anxious or behave as if nightfall has come. Some plants close up. Notice how the people around you are affected.

Gathering Darkness on the Western Horizon About 5 minutes before totality, the shadow cast by the Moon causes the western horizon to darken as if a giant but silent thunderstorm was approaching.

Temperature As the sunlight fades, the temperature may drop perceptibly.

Shadow Bands A minute or so before totality, ripples of light may flow across the ground and walls as the Earth's turbulent atmosphere refracts the last rays of sunlight.

Thin Crescent Sun Only a sliver of the Sun remains, then thinner still until . . .

Corona Perhaps 15 seconds before totality begins, as the Sun becomes the thinnest of crescents, the corona begins to emerge.

Diamond Ring Effect As the corona emerges, the crescent Sun has shrunk to a short, hairline sliver. Together they form a dazzlingly bright diamond ring. Then the brilliant diamond fades into . . .

Baily's Beads About 5 seconds before totality, the remaining crescent of sunlight breaks into a string of beads along the eastern edge of the Moon. These are the last few rays of sunlight passing through deep valleys at the Moon's limb, creating the momentary effect of jewels on a necklace. Quickly, one by one, Baily's beads vanish behind the advancing Moon as totality begins.

Shadow Approaching While all this is happening, the Moon's dark shadow in the west has been growing. Now it rushes forward and envelops you.

Second Contact Totality begins. The Sun's disk (photosphere) is completely covered by the Moon.

Prominences and the Chromosphere For a few seconds after totality begins, the Moon has not yet covered the lower atmosphere of the Sun and a thin strip of the vibrant red chromosphere is visible at the Sun's eastern limb. Stretching above the chromosphere and into the corona are the vivid red prominences. A similar effect occurs along the Sun's western limb seconds before totality ends.

Corona Extent and Shape The corona and prominences vary with each eclipse. How far (in solar diameters) does the corona extend? Is it round or is it broader at the Sun's equator? Does it have the appearance of short bristles at the poles? Look for loops, arcs, and plumes that trace solar magnetic fields.

Planets and Stars Visible Venus and Mercury are often visible near the eclipsed Sun, and other bright planets and stars may also be visible, depending on their positions and the Sun's altitude above the horizon.

Landscape Darkness and Horizon Color Each eclipse creates its own level of darkness, depending mostly on the Moon's angular size. At the far horizon all around you, beyond the Moon's shadow, the Sun is shining and the sky has twilight orange and yellow colors.

Temperature Is it cooler still? A temperature drop of about 10°F (6°C) is typical. The temperature continues to drop until a few minutes after third contact.

Animal, Plant, and Human Reactions What animal noises can you hear? How are other people reacting? How do *you* feel?

End of Totality Approaching The western edge of the Moon begins to brighten and vividly red prominences and the chromosphere appear. Totality will end in seconds.

Third Contact One bright point of the Sun's photosphere appears along the western edge of the Moon. Totality is over. The stages of the eclipse repeat themselves in the reverse order.

Baily's Beads The point of light becomes two, then several beads, which fuse into a thin crescent with a dazzling bright spot emerging, a farewell diamond ring.

Diamond Ring Effect and Corona As the diamond ring brightens, the corona fades from view. Daylight returns.

Shadow Rushes Eastward

Shadow Bands

Crescent Sun

Recovery of Nature

Partial Phase

Fourth Contact The Moon no longer covers any part of the Sun. The eclipse is over.

"I doubt if the effect of witnessing a total eclipse ever quite passes away," wrote Mabel Loomis Todd. "The impression is singularly vivid and quieting for days, and can never be wholly lost. A startling nearness to the gigantic forces of nature and their inconceivable operation seems to have been established. Personalities and towns and cities, and hates and jealousies, and even mundane hopes, grow very small and very far away."[15]

"Beware of post-eclipse depression," warns Steve Edberg. One minute after third contact, with the passage of the Moon's shadow, the disappearance of the shadow bands, and the reappearance of the crescent Sun, you feel exhausted: worn out by totality and the excitement of getting ready for it. Most observers are too tired to watch after third contact, and what follows is anticlimactic anyway. The cure for blue sky blues? "Socialize," says Edberg. "Ask people what they saw. Share war stories."

"The end of totality is a time for celebration," says Jay Anderson. "In the wake of such a powerful shared experience, conversations become animated and casual acquaintances tend to become lifelong friends."

George Fleenor watched the 1991 eclipse from the beach at San Blas, Mexico. Not everyone there had come to see the eclipse. As the Moon covered more and more of the Sun, people continued to play in the ocean. And then the Sun went out. The people in the water stopped where they were and started screaming and cheering. When the Sun emerged from totality, the people in the water started dancing.

As Mike Simmons watched the 2006 total eclipse in Turkey, there was a video camera recording the reactions of the couple near him. The husband was observing with binoculars. His wife was standing behind him. As the eclipse became total, the wife began hammering on the husband's back with both fists and screaming. Afterwards, the wife had no recollection of her reaction to totality. But her husband did. He remembers ducking.

Charles Piazzi Smyth, astronomer royal of Scotland, saw his first total eclipse in 1851 and recognized its ability to overwhelm an observer, distracting him from his carefully planned research. "Although it is not impossible but that some frigid man of metal nerve may be found capable of resisting the temptation," he wrote, "yet certain it is that no man of ordinary feelings and human heart and soul can withstand it."[16]

Confessions of Eclipse Junkies

What if you are clouded out, as happens even with good planning about one out of every six times? Unlike a rocket countdown, which can be "T minus 30 minutes and holding" while the clouds clear, an eclipse countdown is inexorable, and it is not always possible to race to another location where there is a break in the clouds.

"My first eclipse was in Maine in the summer of 1963," Joe Hollweg recalls. Totality was about 1 minute long. "I was working in New York City, but I made a trip to Maine just for the eclipse. The sky was partly cloudy, with lots of small cottony fair weather clouds. What a disappointment when one of those little puffs moved over the Sun just at the start of totality! I got to see Baily's beads, and that was all. A few hundred yards away people were in the clear, and they were cheering with excitement. However, I do remember seeing the Moon's shadow racing across those puffy clouds, both at the start of the eclipse and at the end. The motion of the shadow seemed incredibly rapid, perhaps because the clouds weren't very high. That was the first time I had some sense of how fast 1,000 miles per hour really is."

All the veterans agree: Plan your eclipse trip so that you see things, go places, and meet people that will shine in your memory even if the corona does not. "An eclipse is the perfect excuse to go traveling," says Jay Anderson.

Crucial to the enjoyment of a solar eclipse is eye safety. You wouldn't stare directly at the Sun during a normal day, so you shouldn't stare at the Sun when it is partially eclipsed without proper eye protection (described

The eerie twilight of totality silhouettes astronomers as they quickly make their measurements during the total solar eclipse of February 16, 1980, near Hyderabad, India. [Nikon FE SLR, 24 mm lens, f/2.8, 1 s, ISO 200 film. ©1980 Jay M. Pasachoff]

Equipment Checklist for Eclipse Day

Checklist of your intended activities during eclipse
Solar filters for your eyes
Portable seat or ground covering
Binoculars and/or telescope
Solar filters for binoculars and/or telescope
Camera equipment and tripod
Video camera and tripod
Flashlight with new batteries and a piece of red gel to filter flashlight during totality
Straw hat, kitchen pasta colander, or cooking spoon with small holes to project pinhole images of partially eclipsed Sun on a white piece of cardboard
Suitable clothing and hat (you will be outside for several hours)
Sunglasses (*not* for direct viewing of partial phases)
Bug repellent, sunscreen lotion, basic first aid kit
Snacks and a canteen of water
Audio recorder for your comments and impressions or to capture reactions of people or wildlife near you
Audio recorder with earphones and prerecorded message timed to cue you about what you want to do next (to run from about 5 minutes before totality until 5 minutes after totality)
Pencil and paper to record impressions or to sketch (also to take down the names and addresses of fellow observers)

in the next chapter). However, when the Sun is totally eclipsed—no portion of its disk is showing—it is perfectly safe to look at the Sun without any eye protection. A filter would hide the view.

Jay Pasachoff is irritated by governments and news media in foreign lands—and the United States—that mislead the public by implying that the Sun gives off "special rays" at the time of an eclipse. Instead of teaching simple safety procedures, they frighten people, thereby depriving them of a rare and magnificent sight.

Alan Fiala recalled the 1980 eclipse in India where pregnant women were instructed to remain indoors during the eclipse. For the 1992 eclipse, Roger Tuthill rented a jumbo jet in Brazil to meet and chase the shadow over the Atlantic, using the speed of the aircraft to expand 3 minutes of totality into 6. To serve the 50 passengers aboard the DC-10, the airline assigned 12 flight attendants. There were plenty of windows for them to watch the eclipse, but almost all the stewardesses refused to look. They feared that if they saw the Sun in eclipse, they could never become pregnant.

The pilot, however, was not afraid to look. He was wildly enthusiastic. Even after the Moon's shadow outran the jet and totality was over, the pilot pressed on eastward toward Africa because he was so fascinated by the fast-moving pillar of darkness ahead of him. "It's even more exciting than an engine fire," he explained.

George Lovi, Alan Fiala, and others remember the 1983 eclipse in Indonesia. Before the eclipse, the streets were teeming with people. On eclipse morning, however, there was scarcely anyone outdoors. Fire sirens wailed like an air raid warning for people to take cover. School children, on government orders, were kept indoors, forbidden to see the eclipse, except perhaps on television. A stunning natural event that comes to you, if you are lucky, once in five lifetimes, was denied to them.

On Java for that 1983 eclipse, Jay Anderson recalls soldiers patrolling the streets to discourage unauthorized observers, although the citizens were so frightened by government warnings that almost all stayed indoors. As the Sun was gradually disappearing, the soldiers near him began to be caught up in the excitement. Anderson offered them a view through his telescope, but the warnings they had received overtook them and they refused. Several observers gave the officers a few minutes of careful explanation before the officers nervously took a peek.

> After their first look all apprehension disappeared, and they participated in the event as fully as we did. Ten minutes before totality, the officer in charge was interrupted by a call on his walkie-talkie from his superior at headquarters: "How are things going out there?" "O.K. There are no problems." "It's time to come in now. Collect your men and bring them back to the barracks. We are supposed to have all troops back before the eclipse begins." "There are no problems here; everyone is safe; the tourists are all working on their equipment. Are you sure you want us in?—I see no problems." "Orders are to bring everyone in, and leave the tourists to themselves. Bring the men in." "I'm sorry, I can't hear you. I'm having trouble with the radio. What did you say?"
>
> At this point the officer turned off the radio and put it back in his pocket. The entire troop stayed to watch the eclipse with us.

Even more memorable to Anderson were two dozen Indonesian students from a local college. They were studying English and attached themselves to Anderson's group to practice their speaking skills. As totality neared, the students became extremely nervous. To reassure them, Anderson and his fellow observers held hands with them during totality—the most magical moment in all the eclipses he has seen.

NOTES AND REFERENCES

1. Isabel Martin Lewis: *A Handbook of Solar Eclipses* (New York: Duffield, 1924), page 3.
2. Annie Dillard: "Total Eclipse," in *An Annie Dillard Reader* (New York: Harper Perennial, 1995).
3. François Arago: *Popular Astronomy*, volume 2, translated by W. H. Smyth and Robert Grant (London: Longman, Brown, Green, Longmans, and Roberts, 1858), page 360.
4. George B. Airy: "On the Total Solar Eclipse of 1851, July 28," page 1 in Bernard Lovell, editor: *Astronomy*, volume 1, The Royal Institution Library of Science (Barking, Essex: Elsevier Publishing, 1970).
5. Anton Pannekoek: *A History of Astronomy* (London: G. Allen & Unwin, 1961), page 406.
6. Mabel Loomis Todd: *Total Eclipses of the Sun*, revised edition (Boston: Little, Brown, 1900), page 21.
7. Johana L. Codona; "The Enigma of Shadow Bands," *Sky and Telescope*, 81: 482, 1991.
8. George B. Airy: "On the Total Solar Eclipse of 1851, July 28," page 4 in Bernard Lovell, editor: *Astronomy*, volume 1, The Royal Institution Library of Science (Barking, Essex: Elsevier Publishing, 1970). This speech was given on May 2, 1851, prior to the total eclipse on July 28.
9. Isabel Martin Lewis: *A Handbook of Solar Eclipses* (New York: Duffield, 1924), page 62.
10. On rare occasions, depending on the terrain at the rim of the Moon, a double diamonding is possible—and predictable.
11. François Arago: *Popular Astronomy*, volume 2, translated by W. H. Smyth and Robert Grant (London: Longman, Brown, Green, Longmans, and Roberts, 1858), pages 360–361.
12. Mabel Loomis Todd: *Total Eclipses of the Sun*, revised edition (Boston: Little, Brown, 1900), page 19.
13. Jean Meeus, "The maximum possible duration of a total solar eclipse," *Journal of the British Astronomical Association*, volume 113, number 6 (December 2003), pages 343–348.
14. Rebecca R. Joslin: *Chasing Eclipses: The Total Solar Eclipses of 1905, 1914, 1925* (Boston: Walton Advertising & Printing, 1929), pages 14–15.
15. Mabel Loomis Todd: *Total Eclipses of the Sun*, revised edition (Boston: Little, Brown, 1900), page 25.
16. Mabel Loomis Todd: *Total Eclipses of the Sun*, revised edition (Boston: Little, Brown, 1900), page 174.

A MOMENT OF TOTALITY

The Accidental Eclipse Tourist

The total solar eclipse of 1992 would arc across the South Atlantic Ocean, so Glenn Schneider and his friends chartered a jet to fly from Rio de Janeiro out over the ocean to intercept the eclipse, using Glenn's calculations. The problem was timing. The eclipse wouldn't wait for a late-arriving aircraft and the Rio airport was notorious for ground delays. So the captain was delighted when the Rio control tower gave him clearance a little early to push back from the gate, start his engines, and begin to taxi for takeoff. He accepted. The cabin doors closed, the jetway retracted, the pushback truck eased the aircraft away from the terminal, and the engines began to roar.

Suddenly there was a frantic woman pounding on the cockpit door. She was a member of the ground crew still onboard cleaning the plane. Let me off, she demanded. The plane wasn't scheduled to leave yet. Go back to the gate. Extend the jetway. She had to get off. No, the captain said. The eclipse would wait for no one. He couldn't risk a delay waiting for another clearance. The woman was incensed. You have to let me off, she shouted. She had another plane to clean, and more after that. Her boss would be furious if she wasn't there. She would lose her job. She was being kidnapped.

You are coming with us, the captain said, but I will radio the airline and let your boss know what happened and where you are, that it's not your fault, and that you'll be with us for the next 4 hours.

The flight crew found her a small window to view from, and the eclipse chasers gave her eclipse glasses and instructions on how to

use them. So, without intending to, and even against her will, the cleaning crew woman saw the eclipse—and was flabbergasted. After the eclipse she thanked everyone aboard, grateful that she had been trapped aboard an aircraft bound for a total eclipse.[1]

[1] Glenn Schneider is a University of Arizona astronomer. Interviewed May 5, 2015.

10

Eye Safety During Solar Eclipses

People expect a total solar eclipse to be a curiosity.
They don't expect it to move their souls.

David Makepeace (2015)[1]

Watching the Sun safely during an eclipse is simple. **When the Sun is *partially* eclipsed, you need eye protection. When the Sun is *totally* eclipsed, you need no eye protection at all.**

You would never think of staring at the Sun without eye protection on an ordinary day. You know the disk of the Sun is dazzlingly bright, enough to permanently damage your eyes. Likewise, any time the disk of the Sun is visible—throughout the partial phase of an eclipse—you need proper eye protection. Even when the Sun is nearing total eclipse, when only a thin crescent of the Sun remains, the 1% of the Sun's surface still visible is about 10,000 times brighter than the full moon.

Once the Sun is entirely eclipsed, however, its bright surface is hidden from view and it is completely safe to look directly at the totally eclipsed Sun without any filters. In fact, it is one of the greatest sights in nature.

Here are ways to observe the partial phases of a solar eclipse without damaging your eyes.

Solar Eclipse Glasses

The most convenient way to watch the partial phases of an eclipse is with solar eclipse glasses. These devices consist of solar filters mounted in cardboard frames that can be worn like a pair of eyeglasses. If you normally wear prescription eyeglasses, you place the eclipse glasses right in front of them.

The filters in eclipse glasses are made from one of several types of materials. Black polymer is the most common type of filter and is composed of carbon particles suspended in a stiff plastic. It produces a natural yellow image of the Sun. Here are companies that specialize in black polymer solar eclipse glasses:

A Samburu man wears a pair of eclipse glasses in preparation for an annular eclipse in Kenya. These inexpensive glasses with cardboard frames have become very popular for safe eclipse viewing. [©2010 Fred Espenak]

American Paper Optics: www.eclipseglasses.com/

Rainbow Symphony: www.rainbowsymphony.com/

Thousand Oaks Optical: www.thousandoaksoptical.com/

An older alternative to black polymer is the shiny silver filter made of aluminized polyester (Mylar[2]). A minor problem with this material is that it gives the Sun an unnatural blue color. But more importantly, aluminized polyester may have small pinholes that could allow unfiltered sunlight to reach your eyes and damage them. For that reason, aluminized polyester has largely been replaced by safer materials.

Advertisements for other eclipse glasses may be found in popular astronomy and science magazines and on websites. Check the glasses for a printed message certifying that they are safe for eclipse viewing. And verify the optical density of eclipse glasses by making sure you can look comfortably at the filament of a high-intensity lamp.

When you are using a filter, do not stare for long periods at the Sun. Look through the filter briefly and then look away. In this way, a tiny hole that you miss will not cause you any harm. You know from your ignorant childhood days that it is possible to glance at the Sun and immediately look away without damaging your eyes. Just remember that your eyes can be damaged without you feeling any pain.

Welder's Goggles

Another safe filter for looking directly at the Sun is welder's goggles (or the filters for welder's goggles) with a shade of 13 or 14. They are relatively inexpensive and can be purchased from a welding supply company. The down side is that they cost more than eclipse glasses and give the Sun an unnatural green cast.

The Pinhole Projection Method

If you don't have eclipse glasses or a welder's filter, you can always make your own pinhole projector, which allows you to view a *projected* image of the Sun. There are fancy pinhole cameras you can make out of cardboard boxes, but a perfectly adequate (and portable) version can be made out of two thin but stiff pieces of white cardboard. Punch a small clean pinhole in one piece of cardboard and let the sunlight fall through that hole onto the second piece of cardboard, which serves as a screen, held behind it. An inverted image of the Sun is formed. To make the image larger, move the screen farther from the pinhole. To make the image brighter, move the screen closer to the pinhole.

Do not make the pinhole wide or you will have only a shaft of sunlight rather than an image of the crescent Sun.

Remember, a pinhole projector is used with your back to the Sun. The sunlight passes over your shoulder, through the pinhole, and forms an image on the cardboard screen behind it. Do **not** look through the pinhole at the Sun.

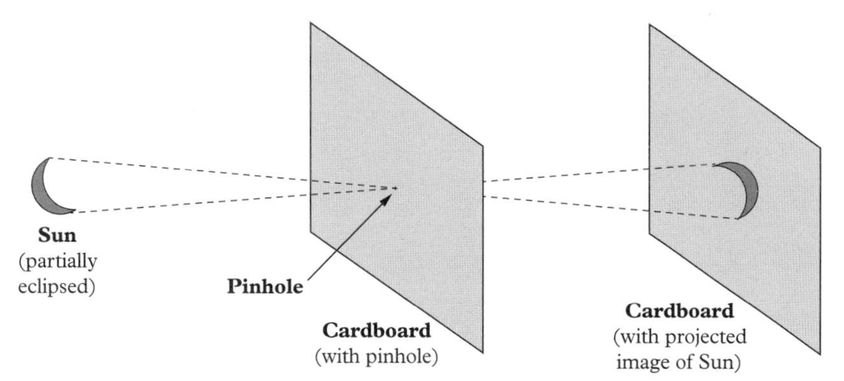

A pinhole projector can be used to safely watch the partial phases of a solar eclipse. It is easily fashioned from two stiff pieces of cardboard. One piece serves as the projection screen. Make a pinhole in the other piece and hold it between the Sun and the first piece. If the two cardboards are held two feet apart the projected image of the Sun will appear about 1/4-inch in size. [Drawing by Fred Espenak]

Solar Filters for Cameras, Binoculars, and Telescopes

Many telescope companies provide special filters that are safe for viewing the Sun. Black polymer filters are economical but some observers prefer the more expensive metal-coated glass filters because they produce

Eye Damage from a Solar Eclipse

by Lucian V. Del Priore, M.D., Ph.D.

The dangers of direct eclipse viewing have not always been appreciated, despite Socrates' early warning that an eclipse should only be viewed indirectly through its reflection on the surface of water. A partial eclipse in 1962 produced 52 cases of eye damage in Hawaii, and a total eclipse along the eastern seaboard of the United States produced 145 cases in 1970. As many as half of those affected never fully recovered their eyesight.

There is nothing mysterious about the optical hazards of eclipse viewing. No evil spirits are released from the Sun during a solar eclipse, and there is no scientific reason for running indoors to avoid "the harmful humors of the Sun." Eye damage from eclipse viewing is simply one form of light-induced ocular damage, and similar damage can be produced by viewing any bright light under the right (or should I say the wrong!) conditions.

Light enters the eye through the cornea and is focused on the retina by the optical system in the front of the eye. Any light that is not absorbed by the retina is absorbed by a black layer of tissue under the retina called the retinal pigment epithelium.

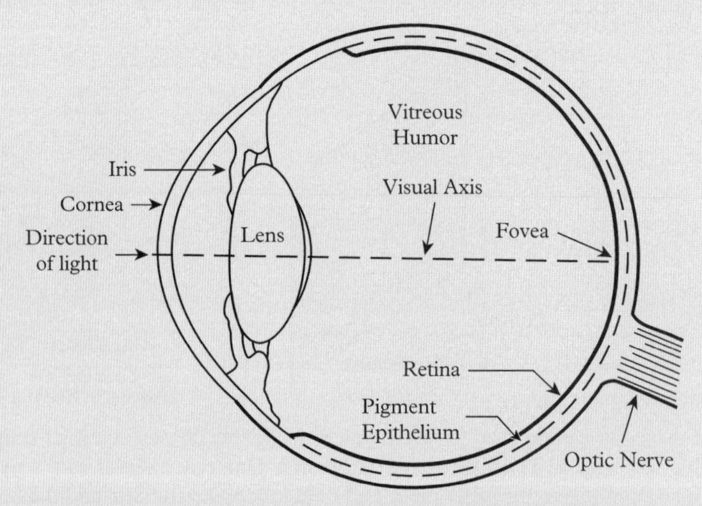

Cross section of the eye. [Drawing by Josie Herr]

sharper images under high magnification. Baader Planetarium AstroSolar Safety Film is another alternative. It's a metal-coated resin with excellent optical quality and high contrast. The company even offers instructions on how to make an inexpensive cardboard cell to mount the filter on your telescope, binoculars, or camera.

The retina is the human body's video camera: it contains nerve cells that detect light and send the electrical signal for vision to the brain. Without it, we cannot see. Most of the retina is devoted to giving us low-resolution side vision. Fine detail reading vision is contained in a small area in the center of the retina called the fovea. People who damage their fovea are unable to read, sew, or drive, even though this small area measures only 1/100th of an inch across, and is less than 1/10,000th of the entire retinal area. Unfortunately, this is the precise area that is damaged if we stare directly at the surface of the Sun. Damage has been reported with less than 1 minute of viewing. The image of the Sun projected onto the retina is about 1/150th of an inch in size, and this is large enough to seriously damage most of the fovea.

Why is sunlight damaging to a structure which is designed to detect light? Bright sunlight focused on the retina is capable of producing a thermal burn, mainly from the absorption of infrared and visible radiation. The absorption of this light raises the temperature and literally fries the delicate ocular tissue. There is no mystery here: every schoolchild knows that sunlight focused through a magnifying glass can cause a piece of paper to burst into flames.

Yet Sun viewing seldom produces a thermal burn. All but the most intoxicated viewer would surely turn away before this occurs. Instead, most cases of eclipse blindness are related to photochemically induced retinal damage, which occurs at modest light levels that produce no burn and no pain. Two types of light lesions are recognized clinically and experimentally, and both are probably responsible for the damage observed after improper eclipse viewing. Blue light (400–500 nanometers) damages the retinal pigment epithelium and leads to secondary changes in the retina, while near ultraviolet light (340–400 nanometers) is absorbed by and directly damages the light-sensitive cells in the outer retina.

Viewing a partial eclipse recklessly is not the only way to produce light-induced retinal damage. Sungazing is a well-known cause of retinal damage even in the absence of an eclipse. Numerous cases of blindness have been reported in sunbathers, in military personnel on anti-aircraft duty, and in religious followers who sungaze during rituals and pilgrimages. The Sun is not even required to produce light damage. Other types of bright lights, including lasers and welders' arcs, will have the same effect.

The common thread here is clear: direct viewing of bright lights can damage the retina regardless of the source. A solar eclipse merely increases the number of potential victims, and brings the problem to public attention.

Dr. Lucian V. Del Priore, M.D., Ph.D. (physics) is a retinal surgeon, researcher, and professor in the Department of Ophthalmology at the Medical University of South Carolina.

Check the Internet or astronomy magazines for dealers offering these filters. Some of the major companies include Celestron, Meade, Orion Telescopes, and Thousand Oaks Optical.

Caution: Do not confuse these filters, which are designed to fit over the *front* of a camera lens or the aperture of a telescope, with a so-called solar *eyepiece* for a telescope. Solar eyepieces are still sometimes sold with small amateur telescopes. They are not safe because they absorb heat and tend to crack, allowing the sunlight concentrated by the telescope's full aperture to enter your eye.

Sources of Safe Solar Filters

Here are some suppliers of filters that are designed for safe solar viewing with or without a telescope. The list is representative, not exhaustive. Additional sources may be found in advertisements in *Astronomy, Sky & Telescope*, and other popular astronomy magazines.

United States
American Paper Optics: *www.3dglassesonline.com/*
Astronomics: *www.astronomics.com/*
Celestron: *www.celestron.com/*
DayStar Filters: *www.daystarfilters.com/*
Lunt Solar Systems: *luntsolarsystems.com/*
Meade Instruments: *www.meade.com/*
OPT Telescopes: *www.optcorp.com/*
Orion Telescopes and Binoculars: *www.telescope.com/*
Rainbow Symphony: *www.rainbowsymphony.com/*
Seymour Solar: *www.seymoursolar.com/*
Spectrum Telescope: *www.spectrumtelescope.com*
Thousand Oaks Optical: *www.thousandoaksoptical.com/*

Canada
Kendrick Astro Instruments: *www.kendrickastro.com/*
Khan Scope Centre: *www.khanscope.com/*
KW Telescope: *www.kwtelescope.com/*

Europe
Baader Planetarium: *www.baader-planetarium.com/*
First Light Optics: *www.firstlightoptics.com/solar-filters.html*
Rother Valley Optics: *www.rothervalleyoptics.co.uk/*
Solar Scope: *www.solarscope.co.uk*

Eye Suicide

Do not use standard or polaroid sunglasses to observe the partial phases of an eclipse. They are *not* solar filters. Standard and polaroid sunglasses cut down on glare and may afford some eye relief if you are outside on a bright day, but you would never think of using them to stare at the Sun. So you must not use sunglasses, even crossed polaroids, to look directly at the Sun during the partial phases of an eclipse.

Also, do not use smoked glass, medical x-ray film with images on them, photographic neutral-density filters, and polarizing filters. All these "filters" offer utterly inadequate eye protection for observing the Sun.

Observing with Binoculars

Binoculars are excellent for observing total eclipses. Any size will do.

Astronomy writer George Lovi's favorite instrument for observing eclipses was 7 x 50 binoculars—magnification of 7 times with 50-millimeter (2-inch) objective lenses. "Even the best photographs do not do justice to the detail and color of the Sun in eclipse," Lovi said, "especially the very fine structure of the corona, with its exceedingly delicate contrasts that no camera can capture the way the eye can." He felt that the people who did the best job of capturing the true appearance of the eclipsed Sun were the 19th century artists who photographed totality with their eyes and minds and developed their memories with paints on canvas.

For people who plan to use binoculars on an eclipse, Lovi cautioned common sense. Totality can and should be observed *without* a filter, whether with the eyes alone or with binoculars or telescopes. But the partial phases of the eclipse, right up through the diamond ring effect, must be observed with filters over the objective (front) lenses of the binoculars. Only when the diamond ring has faded is it safe to remove the filter. And it is crucial to return to filtered viewing as totality is ending and the western edge of the Moon's silhouette brightens with the appearance of the second diamond ring.

After all, binoculars are really two small telescopes mounted side by side. If observing a partially eclipsed Sun without a filter is quickly damaging to the unaided eyes, it is far quicker and even more damaging to look at even a sliver of the uneclipsed Sun with binoculars that lack a filter.

If you don't have solar filters for your binoculars, there is a second way to safely view the partial phases with them. Use the binoculars to project the Sun's image onto a white piece of cardboard. Just hold the binoculars

Binoculars can be used to safely project a magnified image of the Sun onto a piece of white cardboard. Never look at the Sun directly through binoculars unless they are equipped with solar filters. [©2000 Fred Espenak]

two or three feet from the cardboard, using the binoculars' shadow to point them towards the Sun. It takes a little practice but it works great once you get the hang of it. You have two magnified images for the price of one because each half of the binoculars projects a separate image.

If you use binoculars to *project* the image of the Sun, you don't need solar filters. So when totality begins, you have no solar filters to remove. You can quickly bring the binoculars up to your eyes and look through them.

Observing with a Telescope

Some observers, including astronomy historian Ruth Freitag, prefer to watch the progress of the eclipse through a small portable telescope, which offers stability and is much less tiring to use for extended periods than binoculars. A telescope also provides more detail at higher powers, if this is desired.

Obviously, solar filters are required for the partial phases of the eclipse—one for the main telescope and a second for the finder scope. The solar

Inexpensive eclipse glasses are a great way to share the eclipse with others and make new friends. Bring extras to hand out on eclipse day. [©1997 Fred Espenak]

filters are removed at totality. When Freitag wants to see a wider view of the corona, she switches to the finder scope.

Make sure, she says, that your solar filters are easy to remove without bumping the telescope.

A Final Thought

Just remember, George Lovi said, "Don't try to do too much. Look at the eclipse visually. Don't be so busy operating a camera that you don't see the eclipse. And don't set off for the eclipse so burdened down by baggage and equipment that you are tired and stressed and too nervous to enjoy the event."

Astronomer Isabel Martin Lewis also warned of the dangers of too many things to do: "A noted astronomer who had been on a number of eclipse expeditions once remarked that he had never SEEN a total solar eclipse."[3]

NOTES AND REFERENCES

1. Epigraph: David Makepeace, Canadian filmmaker, interviewed, May 1, 2015.
2. Mylar is a registered trademark of DuPont.
3. Isabel Martin Lewis: *A Handbook of Solar Eclipses* (New York: Duffiel, 1924), page 98.

A MOMENT OF TOTALITY

Unexpected Totality

In 1993, when she was 88 years old, Florence Andsager McPherson told this story to her great niece Julie Andsager about growing up with her family on a farm in Reno County, Kansas.

> One day in the summer we were outside in the yard. It became real dark—the chickens all went to roost. Dad was in the field cultivating the corn and came in. Mom got us kids together. She thought the world was coming to an end. I don't think it lasted over 20 minutes. It seemed like hours. In those days, we never had any way to know what was happening. It was an eclipse of the Sun.

The Andsagers were in the path of totality for the eclipse of June 8, 1918.[1]

[1] Florence Andsager McPherson, as told to Julie Andsager: *Florence's Memories—as Remembered at Age 89–90*. Self-published family history, 1998.

11

The Strange Behavior of Man and Beast—Modern Times

Henry Holiday was an artist hired to travel to India for the 1871 eclipse to draw the corona accurately for scientific purposes. Upon seeing the total eclipse, he wrote: "I could with difficulty control myself so as to be fit for making a decent observation." As soon as totality ended "[I] plunged my head into water, for I was in a fever of excitement."[1]

Beasts in Eclipse

In centuries past, eclipse observers were fascinated by the reaction of animals. In modern times, as totality approaches and as it ends, it's equally interesting to watch how animals respond to nightfall in the midst of day.

During the 2001 total eclipse in Zambia, Pat and Fred Espenak were serenaded by thousands of crickets during the 20 minutes surrounding totality.

David Makepeace observed the 1995 eclipse in India, not far from the Taj Mahal. In India, cows are protected animals and roam the streets freely. As the eclipse darkened the morning sky, a half dozen confused cows laid down in the middle of the street and went to sleep. When the Sun returned, it took eight men pulling and pushing with all their might to urge the cows to get back up and get out of the way to allow traffic to move.

At age 13, Kristian Buchman watched the 2008 eclipse from eastern Mongolia. A flock of 20 sheep was grazing contentedly during the partial phases. But as the light began to fade, the sheep gathered together and began a slow trek back to their enclosure.

In 1991, in Baja California, when darkness descended, Mike Simmons remembers nighthawks flying at noon. Mike was in Iran for the 1999 eclipse. As the light faded in the midst of day, ants came out of the ground in huge numbers. When the Sun returned, they vanished.

George Fleenor watched the 1998 eclipse in Aruba near the lighthouse at the north end of the island. He could see the Moon's dark shadow looming in the west and then coming at him across the Caribbean Sea at more than a thousand miles an hour. It made the hair on the back of his neck stand up, he says. He wasn't looking for shadow bands, but they were everywhere, rippling vividly. A herd of goats was grazing on the lighthouse grass, bleating noisily. As the sky darkened, they fell silent. Some laid down and went to sleep.

Glenn Schneider traveled to the shores of the St. Lawrence River northeast of Quebec City, Canada, for the 1972 total solar eclipse. Eclipse day delivered broken clouds, but 30 seconds before totality he and the other eclipse chasers could see the thin crescent Sun. And then it was gone, covered by clouds. The landscape darkened, but there was no view of the Sun in eclipse. In the false twilight, mosquitos awakened—clouds of mosquitos. They fed with delight on the heartbroken eclipse goers.

Australian newspaper writer Stephen O'Baugh reported that during totality for the 1976 eclipse, the wolves and dingos at the Melbourne zoo began to howl, an antelope ran into a fence, and birds went to roost.[2]

Eclipse veteran Patrick Poitevin took his wife Joanne to her first total eclipse in Antigua in 1998, an idyllic site. With the diamond ring approaching, hornets suddenly stormed out of a nearby bush and swarmed around them, perhaps startled by the darkening in the midst of the day. Patrick and Joanne watched totality surrounded by hornets, buzzing furiously. Fortunately they were not stung—but Joanne was badly bitten by the eclipse bug. Ever since, Patrick and Joanne always go to eclipses together.

Joe Buchman went to the Beach of Goats, south of Mazatlán, Mexico, for the 1991 eclipse. There were no goats, but there were chickens roaming the beach near a weather-beaten cinderblock bar and restaurant. As the sky brightness faded to a silvery twilight, the chickens quit clucking and pecking, formed a line, and marched back to their pen.

Glenn Schneider watched the 1976 eclipse from a mountaintop in southeastern Australia after a 14-hour overnight drive from Melbourne to escape bad weather. His observing site overlooked a large valley where a thousand cows were grazing. As the eclipse became total, Glenn and his friends were treated to a bovine symphony—a thousand cows mooing in confusion. The cacophony ended with the return of the Sun.

Humans in Eclipse

In ancient times, lunar and solar eclipses—especially total eclipses of the Sun—caused desperate fear and bizarre behavior in humans. Gradually through the centuries we have come to appreciate the science and the beauty of eclipses and have stripped away our superstitions.

Or have we?

People in southernmost India saw an annular eclipse on January 15, 2010. For the residents in Mangalore, farther up the coast, the eclipse was

A group of people gather on a Paris street corner to watch a solar eclipse in 1912. [Eugène Atget, 1912]

near annular—a very large partial. The *Times of India* reported: "Myths continue to rule. Despite assurances from scientists that the eclipse will have no harmful effects on humans, the majority here preferred to stay indoors during the eclipse hours." Few people walked the streets, few jitney buses ran. "Restaurants and hotels saw a dip in customers as many preferred not to eat during the eclipse hours." Most schools closed because students didn't show up.

But a local amateur astronomy society offered an eclipse observing party at one high school. "Teenagers said they were thrilled at the magic in the sky," the newspaper reported.[3]

Remember Hsi and Ho, the mythical Chinese astronomers who got drunk and bungled an eclipse prediction? Their negligence delayed the drum banging, pot thumping, and vocal screeching needed to scare away the celestial dragon that eats the Sun. But times change, says writer Michael Rogers. When the celestial dragon made another pass at the Sun over Cambodia in 1965 in the form of a near-total annular eclipse, "Lon Nol's troops simply aimed their American automatic rifles straight up in the air and blasted the gluttonous reptile right out of the sky—sustaining, it is reported, only scattered casualties from the bullets that the fleeing reptile spat back to earth."[4]

George Fleenor and his wife Stephanie went to Aruba to see the 1998 eclipse. After the event, they stopped by a bar to hear from native islanders what they had seen. No, they told him, they stayed inside during the eclipse. They would not watch it. "It is forbidden," they said.

Researchers from the Royal Observatory of Belgium went to Baja California for the 1991 eclipse. They set up their equipment in the garden of their hotel. During totality, while the astronomers were concentrating on their measurements, they were surprised "by the weeping and wailing of the hotel staff, who were terrified by the unexpected fall of darkness."[5]

Patrick Poitevin had a similar experience when he watched the 1991 eclipse from San Blas, Mexico. The Mexican government was quite aggressive in its warnings about the potential danger to eyesight from looking at the Sun. As the eclipse neared totality, local citizens screamed at one another: "Don't look, don't look." and they rushed under a shelter to protect themselves from the view.

Sometimes superstitions and misinformation about eclipses can be tragic. In the aftermath of the 1998 eclipse in the Caribbean, an Associated Press story[6] reported:

> They were frightened by an eclipse—and apparently, they were frightened to death. Four members of a Haitian family have been found dead in their homes. Officials say the four may have been killed by an overdose of sleeping pills they took to alleviate their fears of last Thursday's eclipse. They also may have suffocated. They'd plugged all the openings to their home with rags to keep the Sun out.
>
> Radio broadcasts in Haiti say another young girl suffocated Thursday in a home that had been sealed. Thousands of Haitians were afraid that the eclipse would blind them or kill them.
>
> The government declared a national holiday, in hopes of preventing panic. Police ordered pedestrians off the street during the eclipse, yelling, "Go home! It's dangerous to be out!"

Providing Misinformation

Fortunately, superstitions about eclipses seldom result in tragedy. More often, they are colorful—and sometimes humorous. But when they deprive people of an inspiring, once-in-a-lifetime sight, they are truly unfortunate. Far worse though is when modern-day, well-educated government officials and science professionals give out information that robs tens of thousands or millions of people of the chance to see a total eclipse of the Sun.

One such occasion was a total eclipse of the Sun in Australia on October 23, 1976.[7] The headline in the Sydney *Daily Telegraph* on October 23,

1976, blared: "Television is the only safe way to watch today's eclipse of the sun." The first line of the story explained: "Doctors and scientists stress there is NO other safe way to watch the eclipse." Later, the article provided a list of "Don't's" and "Do's" for the eclipse, including: "Do supervise children strictly as they . . . will be tempted to look. Keep them inside during the danger period."

That admonition was mild compared to the warning issued in Tony Murphy's story the day before under this headline: "Lock up children during eclipse—expert." The article began: "Children should be locked indoors tomorrow during the solar eclipse." The advice came from C. Waldron, Queensland President of the Australian Optometrical Association, and was endorsed by the state government. Waldron continued: "Parents still have only themselves to blame if their children's sight is harmed—they are the only ones responsible."

The eclipse was not total in Sydney. It was partial with 90% of the Sun obscured. But the eclipse was total in Melbourne and eastward. And the citizens of Melbourne suffered from misinformation as well.

Notices were posted everywhere on walls and poles throughout Melbourne. In huge letters, they screamed:

WARNING!
SOLAR ECLIPSE TODAY
NEVER look directly at the Sun at any time.
You may damage your eyes . . .
It is DANGEROUS to look at the sun through any kind of filter . . .
It is VERY DANGEROUS to look at the sun through a telescope,
binoculars, a camera lens, or viewfinder.
WATCH THE ECLIPSE ON TELEVISION
this is the only SAFE way to see the eclipse.
Prepared by: Community Services Centre . . . Melbourne.
Authorized by: Solar Eclipse Committee, 1976 . . . Melbourne.

"Eclipse, but watch out" was the eclipse-day headline of John Rentsch's article for *The Age*, a Melbourne daily newspaper. "Authorities again warned that people looking directly at the eclipse risked permanent eye damage," Rentsch reported. As in Sydney, there was no effort by the government to teach people the simple precautions for observing a natural wonder safely—what filters to use and when to use them.

Instead, "Scientific and Government authorities have urged people wanting to observe the moon passing across the sun to watch the magnified view on television." However, for irresponsible people who insist on viewing the eclipse, thereby damaging their eyes, "The Royal Victoria

This road warning was photographed along a highway in France before the total eclipse of August 11, 1999. It means "Speed Limit 70 Km/Hr during eclipse period." [Creative Commons (S. Klüsener)]

Eye and Ear Hospital has set up a special clinic to deal with the expected flood of people seeking treatment after the eclipse."

After such dire warnings, the optometrists and government officials had one concession: "Authorities said yesterday that although people should not look directly at the sun, there was no harm in being outside during the eclipse."

Perhaps the only expression of skepticism and satire for the stories about the danger of the eclipse was sounded by William Ellis Green, a cartoonist for the Melbourne *Herald*. It shows two convicts tied to poles against a cinderblock wall riddled with bullet holes. The convicts' eyes are covered with handkerchiefs tied around their heads. They are about to be executed by firing squad. One is saying to the other: "Luckily we're blindfolded—we won't harm our eyes looking at the eclipse!"

The day after the eclipse the Australian Optometrical Association reported many phone calls and hospital visits, but no confirmed eye injuries. It credited itself for warning the public and convincing people to watch the eclipse on television.[8]

The warnings worked. Most Australians stayed inside and watched the 1976 solar eclipse on television. In Melbourne, where the eclipse was total, a daily newspaper, *The Age*, happily reported that "The eclipse provided one of the biggest television audiences of any event in Australian history.

In Melbourne, two million viewers were glued to their television sets to watch the once-in-a-lifetime phenomenon."[9]

Said another way: two million people could have seen a stunning, unforgettable, life-changing total eclipse of the Sun just by stepping outdoors. But because of professional and governmental misinformation, they missed it.

For more recent total solar eclipses Down Under—2002, 2012—the Australian government conducted effective public education campaigns to help people appreciate the eclipse and view it safely.

Government wrong-headedness has struck in Canada too. Steve Edberg was in Winnipeg for the eclipse of 1979. Students had to have notes from their parents to be allowed outside during the eclipse. All the others were confined to their classrooms where the shades were drawn and the students watched the eclipse on television. Jay Pasachoff recalls that "One school in Winnipeg even asked for permission to ignore fire alarms if any sounded during the eclipse, lest the students rush outside and be blinded."

On a more rational note, Jay Anderson remembers that many Winnipeg parents kept their children home from school on eclipse day and took the day off themselves so that they could witness the eclipse as a family.

American Snippets

Canada? Australia? But surely such misinformation and superstitious behavior could never hold sway in the United States, depriving people of a rare gift of nature. Well . . .

The eclipse of 1970 was the first and most impressive of the many George Lovi was to see. It was visible in the eastern United States and he was watching from Virginia Beach. The news media had trumpeted the event, but laid great emphasis on the dangers. Around him were thousands of people, most of them casual observers, watching as the crescent Sun thinned and vanished. Instantly a cry went up, spreading from group to group: "Look away! Look away!" Many did.

Lovi knew of other people who traveled substantial distances to reach Virginia Beach and then, fearful, stayed in their motel rooms and watched the event on television.

When Californians heard that the Sun would experience an annular eclipse on May 20, 2012, some expressed their fear to the BabyCenter Blog: "I'm pregnant. Today there will be a solar eclipse. Did you know I'm supposed to wear red and some sort of metal to protect the baby?"

Yes, said blogger Katherine Martin. Many Hispanic mothers and grandmothers tell their pregnant daughters and granddaughters to wear red underwear and to attach a safety pin to it over their belly to protect the fetus from birth defects. No safety pin? Any sort of metal will do. No red panties? The mother-to-be can use underwear of any color with a red ribbon attached to it.

Also, a pregnant woman must not go outside during the eclipse.

Martin said that these superstitions date back to the Aztecs who believed that a celestial monster was taking bites out of the face of the Sun or Moon. If an expecting mother watched such an atrocity, the same thing would happen to her baby.[10]

What a Shame

Dennis di Cicco was in a cornfield in North Carolina for the 1970 eclipse and recalls that in the midst of totality, a car came up the road with its lights on and drove right by without stopping, its passengers oblivious or impervious to the wonder above them.

A cruise ship brought Joe Buchman and many eclipse chasers to Aruba for the total eclipse of 1998. He observed from the back patio of a hotel looking out over the Caribbean. It was a hot day and there wasn't any shade, but as the Moon cut off more and more sunlight, the temperature fell and it became quite comfortable for one of nature's most stirring sights. Later that day, back aboard the ship, Buchman found that many passengers had not left the boat. They spent the eclipse in the windowless casino, pulling handles on the slot machines during totality.

The Effect of a Total Eclipse

Sometimes people's behavior during a total eclipse is out of the ordinary, but quite understandable.

The 2013 eclipse in Uganda occurred late in the afternoon. Patrick and Joanne Poitevin noticed that farm laborers continued to work in the fields until the sunlight, increasingly blocked by the Moon, began to diminish. The workers stopped, looked around, and, oblivious to the eclipse, picked up their tools and started walking home from the fields, apparently confused at the early onset of evening.

Total eclipses bring out strong emotions in both animals and humans. Xavier Jubier observed the 2008 total eclipse in northwestern China. Just before totality when the diamond ring effect appeared, a man dropped to one knee and proposed to his girlfriend, asking that she give him an answer when totality was over. (She said yes.)

At the 2017 and 2024 eclipses, and all those in between, make your own observations of animal—and human—behavior. During totality, says Joe Buchman, be sure to spend a moment looking at the people around you. Look at their expressions. It's like a medieval painting—the look of adoration. Eyes wide open, mouth agape. Then, as the Sun emerges from eclipse, cheers for what they have seen.

NOTES AND REFERENCES

1. Epigraph: Henry Holiday: *Reminiscences of My Life* (London: William Heinemann, 1914), pages 209–210. Quoted by Alex SooJung-Kim Pang: *Empire and the Sun: Victorian Solar Eclipse Expeditions* (Stanford, California: Stanford University Press, 2002), pages 72–73 plus photo insert.
2. Stephen O'Baugh: "Eclipse Had Top Rating," *The Age* (Melbourne, Australia), October 24, 1976.
3. "Solar eclipse: Myths continue to rule," *Times of India*, January 15, 2010.
4. Michael Rogers: "Totality—A Report," *Rolling Stone*, October 11, 1973. Reprinted in Robert Gannon, editor: *Best Science Writing: Readings and Insights* (Phoenix: Oryx Press, 1991), pages 168–185. In this article, Rogers concentrates on the 1973 total eclipse he saw in Mauritania, with this Cambodian eclipse mentioned anecdotally. Rogers thought this eclipse happened in 1972, but no total or partial eclipse was visible from Cambodia that year, or in 1971 or 1970. The only significant solar eclipse that was visible from Cambodia close to that date was the annular eclipse of November 23, 1965.
5. Thomas Crump: *Solar Eclipse* (London: Constable, 1999), page 191.
6. Associated Press, March 2, 1998, reporting on the eclipse of February 26.
7. Thanks to Glenn Schneider who provided photocopies of newspaper articles and posters from the 1976 total solar eclipse in Australia.
8. The text of the article provided by Glenn Schneider was complete, but did not include the name of the newspaper, the reporter's name, or the date.
9. Stephen O'Baugh: "Eclipse Had Top Rating," *The Age* (Melbourne, Australia), October 24, 1976.
10. Katherine Martin: "Interesting Pregnancy and Solar Eclipse Superstitions," *BabyCenter Blog*, May 20, 2012. We have not been able to verify that this superstition comes from Aztec or Maya folklore.

A MOMENT OF TOTALITY

Spoiled

On December 14, 2001, George Fleenor drove from Bradenton, Florida to Sarasota to photograph a partial eclipse of the Sun. The maximum of the eclipse, with about half the Sun obscured, would happen at sundown over the Gulf of Mexico. George had a perfect beach site in mind, with a lifeguard stand in the foreground, looking out over the water to the horizon. While George took pictures, people were still lounging on the beach and splashing in the waves, oblivious to the eclipse in progress. But as the Sun reached the horizon, the atmosphere filtered the Sun's brightness enough that people could look at it without eye protection and see that half the Sun was missing. As a couple walked by George on their way off the beach, the woman moaned, "This is our last day in Florida and our sunset is ruined."[1]

[1] George Fleenor, a planetarium consultant and former planetarium director, helped the couple appreciate the rarity and significance of the event they were seeing. Interviewed July 14, 2015.

12

Eclipse Photography

One picture is worth a thousand words.
Anonymous[1]

How do you capture the spectacle of a total eclipse with a camera? Photographing an eclipse isn't difficult. It doesn't take fancy or expensive equipment. You can take a snapshot of an eclipse with a simple camera (even a smartphone) if you can hold the camera steady or place it on a tripod.

The first step in eclipse photography is to decide what kind of pictures you want. Are you partial to scenes with people and trees in the foreground and a small but distinct eclipsed Sun overhead? Or do you prefer a close-up in which the radiant corona or vivid red prominences of the eclipsed Sun fill the frame? Your decision will determine what kind of equipment you need. Look at the photographs and captions throughout this book. They illustrate some of what can be done with a range of cameras, lenses, telescopes, and exposures.

New technologies in cameras and electronics are making eclipse photography easier than ever before. Even beginners can take great eclipse photos with some careful planning. *Planning* is the key. The day of the eclipse is *not* the time to try out a new tripod or lens. You need to be completely familiar with your camera and equipment, and you need to *rehearse* with them weeks before the eclipse. A total eclipse grants you only a few precious minutes and everything must work perfectly. Nature does not provide instant replays.

Digital Photography

In less than a decade, digital cameras have revolutionized the way we shoot pictures. The price of electronic image sensors and memory cards has plunged while the ability to capture fine detail has leapt into many

Smart phones are capable of shooting great eclipse landscapes in either photo or video mode. This image is a single frame from an HD video made with an iPhone 5s during the total solar eclipse of March 20, 2015, from Svalbard. [Apple iPhone 5s, HD video mode. ©2015 Sarah Marwick]

megapixels. The result: Digital cameras have completely replaced film cameras. But perhaps you are too young to remember film cameras!

Back in the old days of film, eclipse photographers had to carefully pace themselves because there were only 36 exposures on a roll of film. To capture the diamond ring effect at the beginning of totality, the corona during totality, and the second diamond ring at the end of totality, you had to take pictures sparingly to make your film last. Today, with a large memory card, you can shoot hundreds of digital eclipse images in a few minutes and see your results immediately after totality ends.

The following sections explain how you can bag some prized eclipse photos no matter what kind of digital camera you own. We'll start with some simple cameras and techniques and progress to the more challenging.

Simple Cameras

Point-and-shoot or compact cameras are the simplest digital cameras you can buy. They usually have autofocus lenses, automatic exposure modes, and a small built-in flash. Although the most basic cameras have a single-focal-length lens, more advanced models include a zoom lens.

Many point-and-shoots are small enough to fit in a pocket, making them perfect for snapshots of vacations, parties, and other events. They are popular with people who do not consider themselves photographers but want an easy-to-use camera.

Smart phones are rapidly replacing point-and-shoot cameras as the most popular type of digital camera. These ingenious devices were the stuff of science fiction only a decade ago. Besides shooting photos, they have downloadable applications or apps for almost everything imaginable: texting, e-mail, web browsing, music, games, and alarm clock, just to mention a few. You can even use them to make phone calls!

Both point-and-shoots and smart phones are great for capturing simple snapshots of an eclipse. You can also use them to shoot a panorama around the horizon during totality. Although you can handhold either one, a tripod is a better choice for support, especially during the subdued twilight of totality. Point-and-shoots have a threaded socket on the bottom for tripod attachment, while smart phones will require a special bracket to securely attach them to a tripod.

After the eclipse, you can quickly review your photos. Of course, a smart phone also lets you share them immediately with friends and family via Facebook, Twitter, Instagram, and Flickr.

A time-lapse of the November 14, 2012, total eclipse appears above a stark Australian landscape. [Nikon D700, 35 mm lens, stacked multiple exposures at 5-minute intervals, solar filter used for partial phases. ©2012 Ben Cooper]

Landscape Eclipse Photography

The easiest way to capture the eclipse is to take a few wide-angle shots of the sky, horizon, and landscape before, during, and after totality. This sequence might show the Moon's dark, fast-moving shadow approaching from the west or racing away over the horizon to the east. Place yourself or some other interesting subjects in the foreground to give the photos some scale. If your camera has a zoom lens, the widest-angle setting will produce the most dramatic images. Your camera's built-in auto-exposure feature should work well. Just be sure to *turn off the automatic flash feature* so you don't annoy observers nearby. If your camera has an exposure-bracketing feature, use it for insurance to get the best exposure of the total eclipse. You'll also capture any bright planets near the Sun during totality. You might even discover a new comet!

Suggested Targets for Solar Eclipse Photographs

Wide-Angle Photos

People setting up telescopes
People watching eclipse while wearing solar eclipse glasses
People watching eclipse through telescope or binoculars
Eclipse crescents on the ground (sunlight falling between tree leaves, holes between fingers, etc.)
The lunar shadow on the western horizon (before totality) or eastern horizon (after totality)
The colors at the horizon during totality
People silhouetted against twilight sky
Wide-angle view of totally eclipsed Sun with interesting foreground (people, trees, buildings, statues, etc.)

Close-Up Photos (with telescope or zoom lens)

Partial phases (through solar filter)
Diamond ring effect during the 15-second period before and after totality—no solar filter
Baily's beads during the 5-second period before and after totality—no solar filter
Chromosphere and prominences during totality—just after totality begins and just before totality ends—no solar filter
Totality—no solar filter—use a range of shutter speeds to get inner, middle, and outer corona

The gaps in the fronds of a palm tree produce thousands of eclipsed Sun images on the sand below during the January 15, 2010, annular eclipse from Elaidhoo, Maldives. [Canon 450D, 24 mm, f/6.3, 1/100 second, ISO 200. ©2010 Stephan Heinsius]

Photographing Pinhole Crescents

Eclipses provide other phenomena that make interesting pictures, such as the crescent images of the partially eclipsed Sun produced by tree foliage. The narrow gaps between leaves act as "pinhole cameras" and each projects its own tiny (and inverted) image of the crescent Sun on the ground. This pinhole camera effect becomes more pronounced as the eclipse progresses.

You can make your own pinhole camera to project the crescent Sun with pinholes punched in thin cardboard, or with a wide-brimmed straw hat. You can even produce the effect in the shadow of your hands by loosely lacing your fingers together. Watch the crescents form when the light passes through the gaps between your fingers. The profusion of crescents on a white wall or on a person's face makes a nice photographic memento. Once again, you'll want to disengage the automatic flash on your camera.

Eclipse Close-Ups

If your goal is to shoot close-ups of the Sun and Moon during the eclipse, a little more effort is required. These kinds of eclipse images are more challenging than wide-angle landscapes and pinhole crescents. You will need a more expensive camera, a powerful telephoto lens or small telescope, a strong tripod and a special solar filter (for partial phases only). Successfully capturing magnified views of the eclipse also requires a basic understanding of how digital cameras work as well as careful planning and preparation before eclipse day.

So if you want more advanced eclipse photography, read on.

Digital "Film"

Digital cameras use solid-state memory cards to store images. This is a type of computer storage made from silicon microchips that record your photos electronically. Most memory cards fall into one of four types: compact flash (CF), secure digital (SD), Memory Stick (Sony), and xD. Depending on the make and model of a digital camera, it is designed to work with one (and possibly two) of these cards. The cards themselves come in a range of capacities: 2GB to 128GB and larger. GB stands for gigabyte (one billion bytes).

When a digital camera records an image onto a memory card, it usually writes a JPEG[2] file. JPEG (pronounced "JAY-peg") is a clever way of compressing the raw image recorded by the image sensor and storing it in less space than it normally would require. JPEGs come in varying levels of compression: the higher the compression, the smaller the file. But there's no such thing as a free lunch and this is especially true of JPEGs. When you save an image as a JPEG, some of the information in the photo is lost forever. JPEGs are called "lossy" because the decompressed image isn't quite the same as the one you started with. The more you compress a JPEG, the *smaller* the file size and the *greater* the loss of quality in the original image.

Images can be saved at several levels of JPEG quality. Typical choices include something like "low, medium, high" or "basic, normal, fine." Although the intermediate level is sufficient for most photography, a total eclipse is an exceptional event warranting the highest-level setting. This means fewer images can be stored on a memory card but they will be the best quality JPEGs the camera can deliver.

How many images can fit on a given memory card? That depends on three factors: the JPEG quality level, the number of pixels in the image and the amount of detail in the image. The storage chart below estimates the capacity for a range of image and memory card sizes assuming a highest JPEG quality ("high" or "fine").

When you put an empty memory card in your camera, the display shows the estimated number of exposures for the camera's current settings. Change the JPEG quality and watch the numbers change. This is just an estimate, but it is still a useful guide.

Storage Chart—Approximate Number of JPEGs per Memory Card

Sensor Size (Megapixels)	File Size (MB)	2 GB	4 GB	8 GB	16 GB	32 GB	64 GB	128 GB
8MP	2.4	715	1,430	2,861	5,722	11,444	22,888	45,776
10MP	3.0	572	1,144	2,288	4,577	9,155	18,310	36,620
12MP	3.6	476	953	1,907	3,814	7,629	15,258	30,516
14MP	4.2	408	817	1,634	3,269	6,539	13,078	26,156
16MP	4.8	357	715	1,430	2,861	5,722	11,444	22,888
24MP	7.2	284	567	1,135	2,269	4,539	9,079	18,157

The size of a JPEG also depends on the amount of detail in the subject matter. An image of an outdoor landscape (with rocks, trees, grass, and clouds) produces a larger JPEG file than a partially eclipsed Sun (dark sky with a relatively featureless yellow crescent).

Professional and higher-end consumer cameras can also store images in something called camera RAW format (RAW for short—it's not an acronym; it just means raw data). Earlier, a JPEG was described as "a clever way of compressing the raw image recorded by the image sensor and storing it in less space than it normally would require." A RAW image is essentially the original image recorded by the camera sensor before any clever processing is applied to compress the image size and make a JPEG. RAW image files are about 10 times larger than high-quality JPEG files, so fewer of them will fit on a memory card. To estimate the number of RAW images on a memory card, divide the number of JPEG images in the previous table by 10.

Why would anyone want to use the RAW format if it takes up more room on a memory card? Because RAW images can store a much wider range of brightness values than JPEGs, and that is what you need to best capture the beautiful corona (see vignette on Dynamic Range: RAW vs. JPEG).

Dynamic Range: RAW vs. JPEG

The sensor in your camera is a two-dimensional array of individual pixels. Each pixel is a tiny light sensor that measures and records the brightness of light falling on it in either 1 of 4096 (12-bits) or 1 of 16,384 (14-bits) values, depending on the camera sensor.

When the camera's processing chip converts a RAW image into a JPEG, it compresses the range of brightness in the original image down to just 256 (8-bits) values.

In other words, some of the brightness information in the original image is thrown away in order to make the JPEG image smaller than the RAW image. When you save an image as a JPEG, this extra brightness information is lost forever.

One of the most remarkable things about the Sun's corona is that it exhibits an enormous range in brightness—something referred to as dynamic range. The inner parts of the corona are thousands of times brighter that the outer parts. The more brightness values your camera can record, the more successful it will be in capturing the dynamic range of brightness in the corona.

If you're a more casual photographer, you can shoot perfectly acceptable photos of the corona using the best JPEG setting in your camera. But if you consider yourself a serious eclipse photographer, and if you plan to process your images in a computer to bring out the maximum amount of detail possible, then you need to shoot your eclipse photos in camera RAW.

Quick and Easy Eclipse Photography

by Patricia Totten Espenak

Is your idea of the perfect eclipse experience—"I just want to watch!"? Do your eyes glaze over when serious eclipse photographers delve into the minutiae of exposure times, f-stops, and cameras that require you to carry the manual with you? *But* do you still have just a tiny desire for some photos of your very own? Then read on.

After 16 total solar eclipses, my methods and gear have evolved to accommodate both. I can watch almost the entire eclipse while taking great photos using the following equipment: a comfortable chair, a sturdy tripod, a cable release, a right-angle finder*, and a solar filter for my telescope or telephoto lens.

I use a Bogen 3001 tripod with a Bogen 3275 (410) compact geared head. The geared head has two large knobs that allow me to make small, precise adjustments in altitude and azimuth every minute or so to easily track the Sun. With a right-angle finder, I can bend over to check the position of the image and then quickly lean back to view the eclipse. I focus manually on a sunspot and the focus ring is then taped down to secure it. After setting the camera in program mode and matrix metering, I'm ready to go. I just sit back with the cable release in one hand and my eclipse glasses in the other.

Thirty seconds before totality begins, I check to see that the Sun is centered in my viewfinder and I remove the solar filter. A dozen or more shots are quickly taken in the seconds before second contact and after third contact. This results in nice diamond ring sequences while the camera automatically adjusts the exposure.

During totality, I might take another dozen shots. I also check the Sun's position in the viewfinder, but mostly I'm gazing up at the spectacle and pressing the cable-release button.

After my third-contact sequence, I replace the solar filter and return to the leisurely pace of the partial phases.

Sensor Sensitivity and ISO

The correct exposure of a digital photograph is determined by three factors: shutter speed, lens aperture, and sensor sensitivity. The shutter speed controls how long the shutter is open during an exposure. The lens aperture controls how much light passes through the camera lens. The sensor sensitivity controls how much light is needed to obtain a well-illuminated exposure of the subject.

The camera setting used to adjust and change the sensor sensitivity is called the ISO value. A powerful feature of digital cameras is the ability to dial in the ISO sensitivity of your choosing. This was not possible with

Eclipse photography doesn't have to be difficult or interfere with your view of totality. Just use a 500 mm lens and set your camera on "program." This image was shot in Libya during the 2006 total solar eclipse. [Nikon D200 DSLR, Sigma 170-500 mm zoom at 500 mm, f/5.0, 0.8 second, ISO 400. ©2006 Patricia Totten Espenak]

Capture the moment with your eyes and your camera. You *can* get the best of both!

* If your digital camera has a fold-out LCD screen, it will work fine in place of a right-angle finder.

Patricia Totten Espenak is a retired chemistry teacher who dotes on her granddaughters Valerie and Maggie—and chases eclipses around the world with her husband Fred.

film cameras because each roll of film had to be exposed and processed for a fixed ISO value, but with digital cameras, you can change the ISO value any time you wish. In bright Sun, you might pick ISO 100, while indoor photography with the available light might call for ISO 800. The downside of higher ISO speeds is the increase in digital "noise," which appears as grainy specks in images shot at higher ISO values. In modern cameras, digital noise usually becomes significant by ISO 1600. You can study the noise in your camera by shooting the same scene using a range of ISO settings and comparing the results.

Fortunately, eclipses are relatively bright, so an ISO value of 400 is generally a good compromise. It's fast enough to minimize blurring from vibrations without sacrificing image quality caused by digital noise.

The greatest threat to image sharpness in eclipse photos, especially among beginners, is camera vibration caused by wind, flimsy tripods, and nervous hands. Other causes of blurry images may include shooting handheld without a tripod, shooting from a rocking ship, and shooting at a very long focal length without adequate support. If any of these conditions are true on eclipse day, you might consider boosting the camera sensitivity to ISO 800 or even ISO 1600. This will let you use faster shutter speeds,

This hand-held shot of the diamond ring effect was made from the deck of rocking ship in the Pacific Ocean during the total solar eclipse of April 9, 2005. [Nikon D100 DSLR, Sigma 170-500 zoom lens, fl = 400 mm, f/6.3, 1/1000 second, ISO 800. ©2005 Patricia Totten Espenak]

which can help minimize blurring from vibrations and camera motion. A little extra noise is a small price to pay for a sharper image.

Digital Single-Lens Reflex Cameras

Close-ups of the eclipsed crescent Sun (through a solar filter) and detailed portraits of the solar corona require the use of a digital single-lens reflex (DSLR) camera. DSLRs also feature interchangeable lenses, from extreme wide-angle to high-power telephoto. You can also remove the lens and hook the camera body up to a telescope. All DSLRs have a continuous shooting mode (or burst mode) that allows you to shoot two or more frames per second. Their electronic shutters have sophisticated metering and exposure modes while lenses autofocus on their target in a split second.

Most consumer DSLRs use something called a crop sensor chip measuring about 16 x 24 mm. The more expensive professional DSLRs use a full-size sensor chip of 24 x 36 mm. So the sensor dimensions of the consumer DSLR (or crop DSLR) are about two-thirds the size of the sensor in the full-frame DSLR and 44% of the area.

Although a camera lens focuses the same size image of the Sun on both DSLRs, it fills a larger fraction of the crop DSLR's imaging area. This so-called "digital magnification factor" means an image shot with a crop DSLR appears to be made with a lens having 1.5 times the focal length compared to the image shot with a full-frame DSLR using the same lens. The increased image scale on crop DSLRs has important implications when shooting eclipses.

Super Telephotos and Telescopes

The size of the Sun's image is determined solely by the focal length of the lens you use. The larger the focal length, the larger is the Sun's image. To shoot close-ups of an eclipse, you'll need a telephoto lens with a focal length of 400 mm or 600 mm or more.

Professional sports and nature photographers typically use these types of lenses. They are big, heavy, and expensive. Fortunately, mirror lenses are an economical alternative. They use a combination of mirrors and lenses to focus the light into the camera. The folded light path allows a long focal length (500 mm to 600 mm) to fit within a short lightweight tube, making small mirror lenses easily portable—ideal for most eclipse photography.[3]

With crop DSLRs, the 1.5x magnification factor (actually 1.6x for Canon crop DSLRs) offers a free ride for eclipse photographers. A 500 mm lens captures the same field of view on a crop DSLR that a 750 mm does on a full-frame DSLR. You can easily determine the size of the

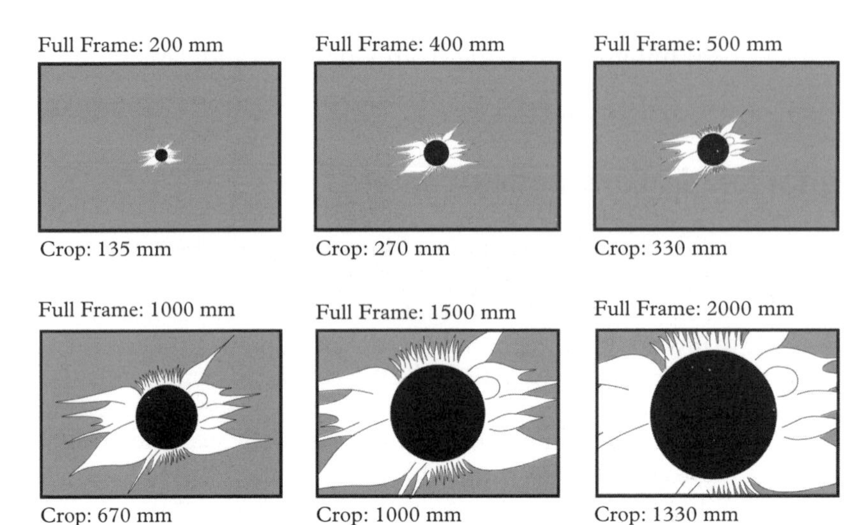

The image sizes of the eclipsed Sun and corona are shown for a range of focal lengths on both professional (full-frame) and consumer (crop-sensor) DSLRs. The same lens produces an image appearing 1.5 times larger on a crop DSLR than on a full-frame DSLR. (Adapted from Fred Espenak and Jay Anderson: *Eclipse Bulletin: Total Solar Eclipse of 2017 August 21* [Astropixels Publishing, 2015]).

Sun's disk in any DSLR, whether crop or full-frame. Take a coin from your pocket and measure its diameter. Now place the coin 110 times its diameter from the lens and shoot a picture of the coin, which now appears the same size as the Sun or Moon. The full moon also makes a great target for evaluating the magnification of your lens for shooting eclipses.

The recommended focal length for close-up photos of solar eclipses ranges from 500 to 2000 mm for full-frame DSLRs (350 to 1350 mm for crop DSLRs), depending on whether you concentrate on the corona or the prominences. Again, allowing one solar radius of corona on either side of the Sun, you can calculate that lenses with focal lengths longer than about 1,400 mm (900 mm for crop DSLRs) may clip the corona, and focal lengths longer than 2,500 mm (1700 mm for crop DSLRs) may not capture the entire solar disk. The longer the lens, the more expensive and less stable a telescope/telephoto system will be, which means that you will need a heavy-duty tripod or mount, adding to your expense and weight.

Modern refracting telescopes are excellent for eclipse photography. The new extra-low-dispersion (or apochromatic) lenses eliminate the color halos around images that handicapped older refractors. Apochromatic refractors have wider objective lenses, yet shorter tubes, making them fast lenses. Apos are expensive and they are not as portable as mirror lenses. Nevertheless, advanced eclipse photographers prize apos for their image quality.[4]

Baily's beads were photographed at the end of totality during the total eclipse of August 1, 2008, from Jinta, China. The beads are formed by sunlight passing through deep valleys along the edge of the Moon. [Nikon D300, Vixen 90 mm fluorite refractor, fl = 810 mm, f/9, 1/1000 second, ISO 320. ©2008 Fred Espenak]

Using a telescope means that you lose the autofocus function (and possibly autoexposure) of your camera. This handicap isn't as serious as it might seem because most cameras don't autofocus well for eclipse close-ups. The rapidly changing light conditions confuse many autofocus cameras, causing them to hunt for the correct focus, so the best solution is to turn autofocus off and focus manually.

Rise of the Superzoom

A less expensive alternative to the DSLR for high-magnification eclipse photography is the superzoom or bridge camera. This digital camera is positioned between the simple point-and-shoot camera and the DSLR. While it does not have an interchangeable lens like the DSLR, it is equipped with a built-in zoom lens with an amazing range of magnifications: 20x, 40x, 60x, and higher. This means you can zoom into the Sun and get good close-ups of the partial phases (through a solar filter). You can also capture the diamond ring and the corona with no telescope or expensive telephoto lens. Of course, a good tripod helps eliminate camera shake.

Superzooms are available from all the major camera manufacturers including Canon, Fujifilm, Nikon, Panasonic, Pentax, and Samsung, and they typically cost hundreds of dollars less than a consumer DSLR.

So why would anyone choose a DSLR instead of a superzoom? DSLRs (even crop DSLRs) have larger sensors than superzoom cameras. A large sensor has less noise and better image quality. And the all-in-one superzoom lens makes some compromises in image quality compared to a good DSLR lens. But it's hard to beat the superzoom for size, convenience, portability, and expense. It's a great choice for first-time or more casual eclipse photographers, and it also makes an excellent vacation camera.

To determine the size of the Sun and Moon in your superzoom, simply take a photo of the full moon. If you don't want to wait for full moon, you can use the coin trick described earlier. Just take a picture of the coin after placing it 110 times its diameter from the superzoom.

Camera Tripods and Cable Releases

Flimsy tripods are the main reason that eclipse photographs come out fuzzy and blurred. Small portable tripods that are convenient for airline travel are not sturdy enough to hold your camera and heavy telephoto lens steady for sharp eclipse pictures. Because you must touch your camera to adjust exposures, your tripod must also dampen any vibrations quickly.[5]

To maximize stability and minimize vibrations, don't extend the tripod's legs more than half way and do not extend the center column. Adjust the tripod height so that you can reach the camera controls while kneeling or sitting on a chair. Test your setup by tapping on the camera or tripod while viewing the Sun (through a solar filter) in your camera. Vibrations should be small and should damp out quickly.

You can help reduce vibrations by suspending some weight under the tripod. Put rocks or sand in a sack or in plastic ziplock bags and hang them from the center of the tripod using string or duct tape. Setting up the tripod on sand or grass is better than concrete or asphalt because the softer surfaces dissipate vibrations faster.

Before you travel to an eclipse, put your camera and lens on the tripod and make sure they can be pointed at the Sun's predicted altitude for the eclipse and that all controls work smoothly. You don't want to discover that your equipment becomes unstable when you try to view the crucial portion of the sky on eclipse day.

You cannot take crisp close-ups of a total solar eclipse without using an electronic cable release. Cable releases have a button at one end and a connection to the camera on the other end. When you press the cable release button, it triggers the camera's shutter without jostling it. Check

Three images of totality and the diamond rings are combined into one using Photoshop. The images were taken during the total solar eclipse of March 29, 2006, in Libya. [Canon 20Da DSLR, William Optics 66 mm triplet Apo refractor, f/6, diamond ring: 1/500 second, totality: 0.3 second, ISO 100. ©2006 Alan Dyer]

with your owner's manual or camera store to choose a cable release that fits your camera.

Solar Filters

When viewing or photographing the partial phases of any solar eclipse, you must always use a solar filter. A solar filter is also needed for observing all phases of an annular eclipse, because the disk of the Moon does not block the entire face of the Sun. Even if 99% of the Sun is covered, the remaining crescent or ring is dangerously bright. Failure to use a solar filter can result in serious eye damage or permanent blindness. *Do not look directly at the Sun without proper eye protection!*

During totality, however, when the Sun's disk (photosphere) is fully covered by the Moon, it is completely safe to look at the eclipse without any solar filter. In fact, you *must* remove the solar filter during totality or you will not be able to see or photograph the exquisite solar corona and prominences.

Solar filters for camera lenses and telescopes come in several different kinds of materials. Metal-coated glass filters are the most expensive but they offer excellent resolution and natural-looking color. Handle glass filters with care since they are fragile and break easily when dropped.

Black polymer is composed of carbon particles suspended in a resin matrix. It produces a natural yellow image of the Sun and is found in both cardboard eclipse glasses and telescope/binocular/camera filters. Black polymer has largely replaced aluminized polyester or Mylar[6] as the inexpensive solar filter of choice. Black polymer filters come mounted in a slip-on cell in any number of diameters to fit most camera lenses and telescopes.

For the do-it-yourselfer, Baader Planetarium AstroSolar Safety Film is a metal-coated resin sold in sheets and has excellent optical quality. The company offers instructions on how to make an inexpensive cardboard cell to mount the filter on your lens or telescope.

All of these solar filters are designed for use on the front end of the lens or telescope (the end pointed at the Sun). They should never be used anywhere else. Materials that should *not* be used for solar filters include exposed color film, stacked neutral density filters, smoked glass, crossed polarizer filters, floppy disks, and CDs (compact disks). These materials are *not safe* even if the image appears dim and no discomfort is felt while viewing the Sun.

If your telescope has a finder scope, be sure to place a small solar filter over its objective lens to protect your eyes and to keep the finder cross hairs from burning. If you don't have a filter for the finder, keep its upper lens covered with a cap or small piece of cardboard taped in place.

Most telescope manufacturers and dealers offer solar filters. Advertisements for them can be found in all the major astronomy magazines (*Astronomy, Astronomy Now, Sky & Telescope, Sky At Night, SkyNews, etc.*). You can also search the web for "solar filter for telescope" (via Google.com, Yahoo.com, etc.).

Photographing the Partial Eclipse

A solar filter *must* be used with your lens to photograph the partial phases of the eclipse. The filter changes the shutter speed and f-ratio of your camera, so you must determine the proper shutter speed and f-ratio for your equipment with the filter in place well in advance of the eclipse.

If your camera has a built-in spot meter that covers a smaller area than the Sun's image, you can simply meter on the Sun's disk through your solar filter and use that exposure throughout the partial phases.

If your camera does not have a spot meter, you need to perform a simple exposure test. Set your equipment up on a sunny day. Carefully center the Sun in your lens or telescope using a solar filter. For telephoto lenses,

The total solar eclipse of June 21, 2001, is captured in a sequence of seven individual images (later combined in Photoshop). The central image of the corona is a composite of several exposures. [Canon Powershot G1 digital camera, Orion 80 mm f/5 refractor, fl = 400 mm, 2x Barlow, 25 mm eyepiece, Orion Solar Filter, f/10, partials: 1/200 s, total: 1/500 s to 1/6 s, ISO 50. ©2001 Fred Bruenjes]

open the aperture to its widest setting. Set the camera in manual exposure mode and shoot one exposure with every shutter speed from 1/15 through 1/2000 of a second. Take notes and use the camera's histogram function to help choose the best exposure. Your camera manual has more information on using the histogram function.

For handy reference, write down the best exposure and tape it to your tripod or solar filter. It should include the ISO, f-number, and shutter speed—for example, "Sun: ISO 400, f/8, 1/125." The exposure doesn't change during the partial phases because the surface brightness of the Sun remains the same throughout the eclipse.[7]

Your best exposure should be determined on a clear, sunny day. If eclipse day has haze or clouds, a longer exposure will be needed to compensate. A thin haze may require an exposure one or two shutter speeds slower than normal, while thicker clouds could call for three or more shutter speeds slower. Use your planned exposure and several longer ones. Memory cards are cheap and eclipses don't happen often.

Photographing the Total Eclipse

The brightness of the solar corona changes tremendously as you move away from the Sun's disk. The inner corona shines as brightly as the full moon, but the outer corona is thousands of times fainter. The challenge is to capture both the brightest and faintest parts of the corona. Unfortunately, this variation in brightness is impossible to record in any one exposure because your camera's sensor just doesn't have the dynamic range of human vision. Only your eyes can see the exquisite detail of this celestial event in all its glory. That's why you should view totality with your eyes and not just with your camera.

Some photographers shoot a series of exposures to capture the wide range of brightness in the corona. This sequence was made with a DSLR during the March 29, 2006, total solar eclipse from Libya. [Nikon D200 DSLR, Vixen 90 mm fluorite refractor, fl = 810 mm, f/9, 9 exposures: 1/125 to 2 seconds, ISO 400. ©2006 Fred Espenak]

The good news is you can photograph some aspect of the corona with almost any exposure you make. There is no one "correct" exposure. Nevertheless, here are some guidelines.

Several factors determine the best shutter speed to use to get a good exposure in your final photograph. The tables accompanying this chapter provide recommended shutter speeds for various eclipse phenomena over a range of ISO speeds and lens f-numbers. Each eclipse phenomenon (partial phases, prominences, inner, middle, and outer corona) has a different brightness value for determining the proper exposure for that aspect of the eclipse. The exposures in the table were determined after photographing more than a dozen solar eclipses, but they are only suggestions. Each eclipse is different and the corona's brightness varies. Weather conditions (haze or clouds) may require longer exposure times. And remember, the partial phases *always* require a solar filter.

During totality, bracket your shots on both sides of the ideal exposure and take several sets of photographs at the same settings to assure success. If you use an ISO speed that is too high, you may discover that your camera does not have a fast enough shutter speed for proper exposure.

Exposure Guide for Solar Eclipse Photography

The following exposure tables are given as guidelines only. The brightness of prominences and the corona can vary. You should bracket your exposures to be safe.

SUN—partial phases and prominences
Use full aperture solar filter for partial phases
DO NOT USE filter for prominences

Film Speed—ISO

f/no.	100	200	400	800
2.8	1/4000	1/8000	–	–
4	1/2000	1/4000	1/8000	–
5.6	1/1000	1/2000	1/4000	1/8000
8	1/500	1/1000	1/2000	1/4000
11	1/250	1/500	1/1000	1/2000
16	1/125	1/250	1/500	1/1000
22	1/60	1/125	1/250	1/500

SUN—total eclipse: inner corona (1° field)
No filter

Film Speed—ISO

f/no.	100	200	400	800
2.8	1/1000	1/2000	1/4000	1/8000
4	1/500	1/1000	1/2000	1/4000
5.6	1/250	1/500	1/1000	1/2000
8	1/125	1/250	1/500	1/1000
11	1/60	1/125	1/250	1/500
16	1/30	1/60	1/125	1/250
22	1/15	1/30	1/60	1/125

SUN—total eclipse: middle corona (2° field
No filter

Film Speed—ISO

f/no.	100	200	400	800
2.8	1/60	1/125	1/250	1/500
4	1/30	1/60	1/125	1/250
5.6	1/15	1/30	1/60	1/125
8	1/8	1/15	1/30	1/60
11	1/4	1/8	1/15	1/30
16	1/2	1/4	1/8	1/15
22	1 sec	1/2	1/4	1/8

SUN—total eclipse: outer corona (4° field)
No filter

Film Speed—ISO

f/no.	100	200	400	800
2.8	1/4	1/8	1/15	1/30
4	1/2	1/4	1/8	1/15
5.6	1 sec	1/2	1/4	1/8
8	2 sec	1 sec	1/2	1/4
11	4 sec	2 sec	1 sec	1/2
16	8 sec	4 sec	2 sec	1 sec
22	15 sec	8 sec	4 sec	2 sec

Determine all of your camera settings before the eclipse so you know in advance what ISO works best with your equipment.[8]

Many eclipse photographers plan their exposures with the following strategy. Using the ISO and f-number, determine the shortest shutter speed for the prominences (bright) and longest shutter speed for the outer corona (dim). After totality begins, shoot a sequence of exposures using every shutter speed, starting with the one for prominences and ending with the one for the outer corona. For instance, at ISO 400 and f/11, the recommended exposure for prominences is 1/1000 second and

Details in the corona are revealed in this Photoshop composite that was processed to emphasize fine structure. The images were obtained during the March 29, 2006, total eclipse from Jalu, Libya. [Nikon D200 DSLR, Vixen 90 mm fluorite refractor, fl = 820 mm, f/9, composite of 22 exposures: 1/1000 second to 1 second, ISO 200. ©2006 Fred Espenak]

for the outer corona 1 second. The shutter speed sequence would then be: 1/1000, 1/500, 1/250, 1/125, 1/60, 1/30, 1/15, 1/8, 1/4, 1/2 and 1. This is a total of 11 exposures. For additional insurance, repeat the sequence in reverse, ending with 1/1000.

Even if your camera's exposure is completely automatic, or if you simply don't want to keep changing exposure settings, you can still get good pictures of the diamond ring and totality using automatic exposure. Set your camera ISO to 400 and grab some shots on auto-exposure. Although the inner corona and prominences may be overexposed, you will still have some fine souvenirs of the event. Best of all, you can devote most of your time to watching the eclipse rather than fiddling with camera settings. Simplicity is especially recommended if you are a novice photographer or have never seen a total solar eclipse (see vignette on Quick and Easy Eclipse Photography).

Eclipse Photography at Sea

For eclipse photography at sea, the pitching and rolling of the ship place certain limits on the focal length and shutter speeds that can be used. In most cases, lenses with focal lengths greater than 500 mm can be ruled out because the ship would need to be virtually motionless during totality. You must also contend with vibration from the engines, wind across the deck, and hundreds of people stomping around, so try to find a location that will minimize these problems.

The ISO choice can be determined on eclipse day by viewing the Sun through the camera lens (with a solar filter) and noting the image motion caused by the ship. In most cases, an ISO of 400 or higher will be needed. As the ship rocks, notice the range of motion and try to snap each picture at one of the extremes.

Some manufacturers have introduced image stabilization (or vibration reduction) in their longer zoom and telephoto lenses. It helps steady the image and lets you use shutter speeds two to three steps slower than possible without this feature. Many photographers relied on image-stabilized lenses to shoot the 2005 hybrid eclipse from ships in the South Pacific because no islands were in the path of totality.

Shooting Video

Total solar eclipses and video are a perfect match. Imagine capturing the excitement of people—their voices and actions—as the Moon's shadow sweeps over the landscape, turning it into an eerie twilight. Virtually all modern point-and-shoot cameras as well as mobile phones are capable of capturing high-definition video.

One of the great things about video is that you can turn it on and let it run. This is especially true if you don't want to be fiddling with a camera during totality. Just start your video recording several minutes before totality begins (no solar filter is needed for wide-angle video). A small tripod is a great asset here because it frees you from holding the camera. Clamps and adapters are available for attaching most mobile phones to a tripod. Just use the same recommendations given previously for landscape eclipse photography.

Video is also the best way of capturing the illusive shadow bands often seen rippling across the ground immediately before and after totality. In this case, the shadow bands show up best against a light-colored surface (snow, sand, a white sheet, even the side of a white car).

If you want close-up video of the eclipse itself, you'll find that super-zoom cameras and newer DSLR models all shoot high-definition video. Check your instruction manual for details. Otherwise, follow the same directions for shooting still images. Just remember that you need a solar filter for the partial phases.

Final Thoughts

If there is one key to successful eclipse photography, it is *preparation*. Set up and test all your equipment at home to ensure that everything works perfectly. Design a photographic plan or schedule and stick to it. Keep things simple. Don't try to do too much. Practice for the eclipse with a full dress rehearsal. Bring extra memory cards, batteries, and other crucial items.

Finally, don't get so overwhelmed with taking photographs that you deprive yourself of viewing the eclipse with your own eyes.[9]

Checklist for Solar Eclipse Photography

Camera Items

Camera (two, if possible, in case one fails)
Lenses (16 mm to 1,000 mm)
Cable release (spare advisable)
Heavy-duty tripod
Extra batteries
Several 8 GB or larger memory cards
Sack or ziplock bags and heavy string (fill with sand or rocks and hang from tripod for stability)
Exposure list for partial phases and totality

Telescope Items

Telescope (consult owner's manual for photographic equipment)
Photo adapters (t-ring, etc., to attach camera to telescope), lens cleaning supplies
Solar filter (full aperture)
Inexpensive solar filter for finder scope

General Items

Eclipse glasses (for viewing partial phases)
Penlight flashlight
Digital voice recorder (to record comments during eclipse)
Pocketknife
Tools for minor repairs: screwdrivers (regular, Phillips, and jeweler's), needle-nose pliers, tweezers, small adjustable wrench, Allen wrench set, etc.
Plastic trash bags (to protect equipment from dust or rain)
Roll of duct tape or masking tape for emergency repairs
Optional: GPS (global positioning system)

NOTES AND REFERENCES

1. Epigraph: Anonymous. Bartlett's *Familiar Quotations* explains that this saying is the creation of Fred R. Barnard, writing in *Printer's Ink*, March 10, 1927. Barnard called it "a Chinese proverb so that people would take it seriously."
2. JPEG stands for Joint Photographic Experts Group. This committee created the JPEG image format in 1992.
3. Mirror lenses are made by the major camera manufacturers (Canon, Minolta, Nikon, Pentax, etc.), as well as independent lens makers (Phoenix, Sigma, Tamron, Tokina, Vivitar, etc.).
4. Manufacturers of apochromatic, fluorite, and extra-low-dispersion refractors include AstroPhysics, Meade, Takahashi, Tele-Vue, Vixen, and William Optics.
5. Bogen/Manfrotto, Gitzo, and Slik all make heavy-duty tripods.
6. Mylar is a registered trademark of DuPont, which does *not* manufacture this material for use as a solar filter. Other aluminized polyester products such as "space blankets" should *not* be used for solar filters.
7. Some photographers like to expose one extra stop during the thin crescent phases because the Sun's limb is a little darker than disk center. This step is important only if you shoot slide film because it has a smaller exposure latitude.
8. Weather conditions (haze or thin clouds) may require longer exposure times. Use the recommended exposures as a starting point and then bracket during the eclipse, especially if the weather is a factor.
9. For tips from Fred Epsenak on still more adventuresome eclipse photography and processing, see Chapter 14: Getting the Most from Your Eclipse Photos in Mark Littmann and Fred Espenak: *Totality: Eclipses of the Sun* (3rd edition or 3rd edition updated) (New York: Oxford University Press, 2008, 2009), pages 201–213.

A MOMENT OF TOTALITY

The Sounds of Totality

Michael Rogers, a young writer for *Rolling Stone* magazine, covered the 1973 total eclipse of the Sun in Mauritania—second longest of the 20th century. His award-winning story, "Totality—A Report,"[1] describes his first total eclipse and the moment when the diamond ring flares brilliantly, then fades, leaving the sky a deep purple and the corona spread out around the blackened Sun.

> The courtyard fills instantly with the simultaneous sound of 200 focal-plane shutters firing constantly and rhythmically—ka-chick- ka-chick-ka-chick—a legion of mechanical crickets. At the same moment, outside the courtyard walls, the high steady chanting of a small group of Mauritanians rises in the sudden darkness: "*Allah-tlag es-shems, Allah-tlag es-shems*"—*Allah*, release the Sun—an ancient prayer that has, thus far, never failed to work.

Rogers describes "the frozen flaring light of the corona" and the black disk of the Moon that blocks the Sun: "a perfect gaping ebony cavity in the fabric of the dusk-dark sky." All too soon, totality nears an end.

> Outside the courtyard the *boubou*-clad Mauritanians are prostrate on the dry dust of the mine road; inside, the astronomers, knowing that time is short, launch another frantic volley of photographs: "*Allah tlag*"—ka-chick-ka-chick—"*es-shems*"—ka-chick-ka — "*Allah Allah*"—ka-chick-ka-chick—"*tlag*"—ka-chick—"*es-shems*." The sounds blend musically in the hot darkness and neither chorus appears particularly aware of the other. One, relentlessly Sun-blasted virtually all their days, prays fervently for the Sun's return. The other, knowing return is inevitable, prays inwardly that it be delayed just a moment longer.

[1] Michael Rogers: "Totality—A Report," *Rolling Stone*, October 11, 1973. These passages are slightly condensed.

13

The All-American Eclipse of 2017

If you ever want to see an eclipse, this is the one.
Paul Andrews, Eclipse veteran (England)[1]

A total solar eclipse is the best show on Earth.
Julie O'Neil, Eclipse veteran (England)[2]

August 21, 2017. At last a total eclipse of the Sun returns to the continental United States. For a landmass its size—third largest country on Earth—the United States has been shortchanged on totality in recent times. On July 11, 1991, a total eclipse passed over Hawaii. But the last total eclipse to visit the continental United States was February 26, 1979—skimming the northwestern tier of states. From 1979 to 2017—38 years without totality.

By comparison, China, almost the same size as the United States, in that same span of time booked total eclipses in 1980, 1997, 2008, and 2009.

The eclipse of 2017 is an all-or-nothing affair. Most total eclipses visit a succession of countries as their paths cross thousands of miles. Not 2017. It starts 2,400 miles (3,800 kilometers) out in the Pacific Ocean and crosses no land before it reaches the United States. It then traverses the country diagonally from northwest to southeast for 2,500 miles (4,000 kilometers). Once it leaves America and sets out over the Atlantic, its totality will not touch land again. The eclipse will travel 3,600 miles (5,800 kilometers) across the Atlantic but vanish from the surface of the Earth before it can reach Africa.

Coming to America

The total solar eclipse of 2017 first touches Earth in the Pacific Ocean 860 miles (1,390 kilometers) south of the western Aleutian Islands of Alaska and 1,500 miles (2,400 kilometers) north-northwest of the Hawaiian Islands. Sailors in the area will see the Sun rise totally eclipsed. Totality reaches the American coast 28 minutes later, having crossed half the Pacific at an average speed of 5,200 miles per hour (8,370 kilometers per hour).

The August 21, 2017 eclipse track runs diagonally across the United States. [Map ©2016 Michael Zeiler, GreatAmericanEclipse.com]

The eclipse races ashore in Oregon at 2,400 miles per hour (3,900 kilometers per hour) near Cape Foulweather (let's hope not). The central line passes over Lincoln Beach, bringing 1 minute 59 seconds of total eclipse. Totality bypasses Portland: the edge of the shadow lies 20 miles (32 kilometers) south of downtown.

It appears that the eclipse of 2017, a presidential inauguration year, has a hankering for politics, for in its campaign swing across the nation, the eclipse pays a visit to five state capitals and winks at two more in passing. The first is Salem, Oregon, and the eclipse delivers on its promises: 1 minute 55 seconds of totality.

Deprivation of Totality

The continental United States has been without a total eclipse of the Sun since 1979. Most of the individual states have suffered without a total eclipse for far longer. For each state through which the central line of the 2017 total eclipse passes, here's a breakdown of the last time that state experienced totality—although seldom in the same area. Many states have been waiting a long time for their opportunity. Total eclipses of the Sun rarely come to you. You must go to them.

State	Most recent total eclipse	Location within state
Oregon	February 26, 1979	Along northern border
Idaho	February 26, 1979	Through the middle of the panhandle
Wyoming	June 8, 1918	Southwestern corner
Grand Teton Peak	January 1, 1889	
	July 29, 1878	
Nebraska	August 7, 1869	Northeastern corner
Kansas	June 8, 1918	Southwestern corner
Missouri	August 7, 1869	Northeastern corner
St. Joseph	June 16, 1806	[211 years between total eclipses]
Illinois	August 7, 1869	Northwest to southeast through center of state, encompassing over half of it
Kentucky	August 7, 1869	All but western end and northeastern tip
Tennessee	August 7, 1869	Northeastern corner
Nashville	July 29, 1478	[539 years between total eclipses]
North Carolina	March 7, 1970	East coast
South Carolina	March 7, 1970	East coast

Now the eclipse ascends the Cascade Mountains, bringing totality to Mt. Jefferson, second highest mountain in Oregon, a dormant volcano towering 10,495 feet (3,199 meters) above sea level. But the limits of total

Eclipse Times* for Cities in the Path of Totality[a]

State	City	Partial Eclipse Begins	Total Eclipse Begins	Total Eclipse Ends	Partial Eclipse Ends	Duration Totality
Idaho	Idaho Falls	10:15:12 am	11:33:01 am	11:34:51 am	12:58:04 pm	01m 49s
	Rexburg	10:15:42 am	11:33:15 am	11:35:32 am	12:58:22 pm	02m 17s
Illinois	Carbondale	11:52:26 am	01:20:06 pm	01:22:43 pm	02:47:28 pm	02m 37s
	Marion	11:53:01 am	01:20:46 pm	01:23:13 pm	02:47:56 pm	02m 27s
Kansas	Atchison	11:40:13 am	01:06:13 pm	01:08:33 pm	02:34:22 pm	02m19s
	Kansas City	11:41:12 am	01:08:33 pm	01:08:59 pm	02:35:54 pm	00m 26s
	Kansas City	11:41:14 am	01:08:35 pm	01:09:02 pm	02:35:56 pm	00m 27s
	Leavenworth	11:40:38 am	01:07:13 pm	01:08:48 pm	02:35:06 pm	01m 34s
Kentucky	Bowling Green	11:58:40 am	01:27:32 pm	01:28:27 pm	02:53:10 pm	00m 55s
	Hopkinsville	11:56:33 am	01:24:42 pm	01:27:22 pm	02:51:44 pm	02m 40s
	Madisonville	11:56:14 am	01:24:32 pm	01:26:17 pm	02:50:55 pm	01m 45s
	Paducah	11:54:02 am	01:22:16 pm	01:24:36 pm	02:49:32 pm	02m 20s
Missouri	Cape Girardeau	11:51:59 am	01:20:21 pm	01:22:08 pm	02:47:37 pm	01m 47s
	Columbia	11:45:40 am	01:12:22 pm	01:14:58 pm	02:40:14 pm	02m 37s
	Independence	11:41:37 am	01:08:37 pm	01:09:51 pm	02:36:19 pm	01m 15s
	Jefferson City	11:46:07 am	01:13:10 pm	01:15:38 pm	02:41:07 pm	02m 28s
	Kansas City	11:41:18 am	01:08:39 pm	01:09:09 pm	02:36:01 pm	00m30s
	St. Joseph	11:40:39 am	01:06:28 pm	01:09:06 pm	02:34:38 pm	02m 38s
Nebraska	Alliance	10:27:08 am	11:49:13 am	11:51:43 am	01:16:43 pm	02m 30s
	Beatrice	11:37:09 am	01:02:13 pm	01:04:48 pm	02:30:25 pm	02m 35s
	Grand Island	11:34:20 am	12:58:34 pm	01:01:09 pm	02:26:36 pm	02m 34s
	Kearney	11:33:03 am	12:57:34 pm	12:59:29 pm	02:25:30 pm	01m 54s
	Lincoln	11:37:15 am	01:02:36 pm	01:03:53 pm	02:29:47 pm	01m 17s
	North Platte	11:30:17 am	12:54:06 pm	12:55:48 pm	02:21:47 pm	01m 43s
	Scottsbluff	10:25:50 am	11:48:11 am	11:49:53 am	01:15:26 pm	01m 42s
Oregon	Albany	09:05:06 am	10:17:06 am	10:18:57 am	11:37:42 am	01m 51s
	Baker City	09:09:52 am	10:24:30 am	10:26:05 am	11:46:54 am	01m 35s
	Corvallis	09:04:54 am	10:16:56 am	10:18:36 am	11:37:24 am	01m 40s
	J. Day Fossil Bed	09:07:57 am	10:21:34 am	10:23:39 am	11:43:42 am	02m 05s
	Lincoln Beach	09:04:35 am	10:16:00 am	10:17:59 am	11:36:09 am	01m 58s
	Madras	09:06:43 am	10:19:36 am	10:21:38 am	11:41:06 am	02m 02s
	Ontario	10:10:07 am	11:25:32 am	11:27:00 am	12:48:38 pm	01m 28s
	Salem	09:05:25 am	10:17:20 am	10:19:15 am	11:37:50 am	01m 55s
South Carolina	Anderson	01:08:59 pm	02:37:51 pm	02:40:25 pm	04:03:10 pm	02m 34s
	Charleston	01:16:59 pm	02:46:28 pm	02:47:56 pm	04:10:02 pm	01m 29s
	Columbia	01:13:06 pm	02:41:49 pm	02:44:19 pm	04:06:19 pm	02m 30s
	Greenville	01:09:13 pm	02:38:02 pm	02:40:13 pm	04:02:56 pm	02m 11s
	Greenwood	01:10:22 pm	02:39:20 pm	02:41:48 pm	04:04:25 pm	02m 28s
	North Charleston	01:16:42 pm	02:45:56 pm	02:47:49 pm	04:09:44 pm	01m 53s
	Sumter	01:14:46 pm	02:43:39 pm	02:45:28 pm	04:07:25 pm	01m 48s
Tennessee	Clarksville	11:57:03 am	01:25:35 pm	01:27:54 pm	02:52:31 pm	02m 19s
	Cleveland	01:03:26 pm	02:33:07 pm	02:34:13 pm	03:58:43 pm	01m 06s
	Cookeville	12:01:17 pm	01:29:45 pm	01:32:17 pm	02:56:02 pm	02m 33s
	Maryville	01:04:57 pm	02:33:54 pm	02:35:26 pm	03:59:02 pm	01m 32s
	Nashville	11:58:30 am	01:27:26 pm	01:29:21 pm	02:54:02 pm	01m 56s
	Oak Ridge	01:04:05 pm	02:33:34 pm	02:33:51 pm	03:58:09 pm	00m 17s
Wyoming	Casper	10:22:17 am	11:42:39 am	11:45:05 am	01:09:26 pm	02m 26s
	Douglas	10:23:34 am	11:44:24 am	11:46:46 am	01:11:17 pm	02m 22s
	Jackson	10:16:42 am	11:34:54 am	11:37:09 am	01:00:28 pm	02m 15s
	Riverton	10:19:33 am	11:39:03 am	11:41:17 am	01:05:20 pm	02m 14s

*All times are in local time (including daylight saving time).

[a]Eclipse times from Fred Espenak and Jay Anderson: *Eclipse Bulletin: Total Solar Eclipse of 2017 August 21* (astropixels.com/pubs/TSE2017.html).

Partial Eclipse Times* for Major Cities in the United States[b]

City and State	Partial Eclipse Begins	Maximum Eclipse	Partial Eclipse Ends	Eclipse Magnitude**
Albuquerque, NM	10:21:19 am	11:45:14 am	01:13:58 pm	0.783
Atlanta, GA	01:05:51 pm	02:36:46 pm	04:01:54 pm	0.971
Baltimore, MD	01:18:32 pm	02:42:57 pm	04:01:18 pm	0.831
Billings, MT	10:21:08 am	11:40:00 am	01:03:00 pm	0.941
Birmingham, AL	12:00:44 pm	01:32:00 pm	02:58:22 pm	0.935
Boise, ID	10:10:34 am	11:27:14 am	12:50:07 pm	0.994
Boston, MA	01:28:27 pm	02:46:50 pm	03:59:29 pm	0.702
Charleston, WV	01:08:12 pm	02:35:28 pm	03:57:26 pm	0.911
Charlotte, NC	01:12:21 pm	02:41:35 pm	04:04:22 pm	0.978
Chattanooga, TN	01:02:34 pm	02:32:57 pm	03:58:15 pm	0.993
Chicago, IL	11:54:18 am	01:19:48 pm	02:42:40 pm	0.889
Cleveland, OH	01:06:31 pm	02:31:01 pm	03:51:07 pm	0.836
Dallas, TX	11:40:23 am	01:09:57 pm	02:39:21 pm	0.800
Denver, CO	10:23:22 am	11:47:07 am	01:14:43 pm	0.933
Des Moines, IA	11:42:50 am	01:08:41 pm	02:33:58 pm	0.954
Detroit, MI	01:03:28 pm	02:27:34 pm	03:47:50 pm	0.831
Dover, DE	01:20:58 pm	02:44:56 pm	04:02:37 pm	0.821
Fairbanks, AK	08:29:39 am	09:21:59 am	10:16:34 am	0.478
Fargo, ND	11:38:46 am	12:59:24 pm	02:20:49 pm	0.839
Hartford, CT	01:25:25 pm	02:45:23 pm	03:59:34 pm	0.732
Honolulu, HI	–	06:35:59 am	07:25:23 am	0.388
Houston, TX	11:46:40 am	01:16:56 pm	02:45:45 pm	0.730
Indianapolis, IN	12:57:53 pm	02:25:03 pm	03:48:39 pm	0.927
Jackson, MS	11:54:34 am	01:26:08 pm	02:53:59 pm	0.862
Knoxville, TN	01:04:54 pm	02:34:29 pm	03:58:45 pm	0.997
Las Vegas, NV	09:09:08 am	10:27:23 am	11:53:00 am	0.771
Little Rock, AR	11:47:56 am	01:18:19 pm	02:46:31 pm	0.905
Los Angeles, CA	09:05:43 am	10:21:13 am	11:44:50 am	0.694
Manchester, NH	01:27:16 pm	02:45:15 pm	03:57:45 pm	0.693
Miami, FL	01:26:56 pm	02:58:25 pm	04:20:48 pm	0.823
Milwaukee, WI	11:53:38 am	01:18:03 pm	02:40:09 pm	0.862
Montpelier, VT	01:24:22 pm	02:41:42 pm	03:54:01 pm	0.678
New Orleans, LA	11:57:42 am	01:29:41 pm	02:57:20 pm	0.800
New York, NY	01:23:15 pm	02:44:59 pm	04:00:45 pm	0.769
Newark, NJ	01:22:53 pm	02:44:42 pm	04:00:34 pm	0.771
Oklahoma City, OK	11:37:07 am	01:05:45 pm	02:34:53 pm	0.872
Omaha, NE	11:38:27 am	01:04:14 pm	02:30:19 pm	0.983
Philadelphia, PA	01:21:14 pm	02:44:19 pm	04:01:19 pm	0.799
Phoenix, AZ	09:13:53 am	10:33:45 am	12:00:26 pm	0.700
Portland, ME	01:29:23 pm	02:45:57 pm	03:57:08 pm	0.667
Portland, OR	09:06:20 am	10:19:12 am	11:38:35 am	0.991
Providence, RI	01:28:05 pm	02:47:13 pm	04:00:28 pm	0.717
Richmond, VA	01:17:57 pm	02:44:11 pm	04:03:58 pm	0.881
St. Louis, MO	11:49:55 am	01:18:16 pm	02:44:17 pm	1.000
St. Paul, MN	11:44:12 am	01:06:59 pm	02:29:20 pm	0.861
Salt Lake City, UT	10:13:58 am	11:33:51 am	12:59:36 pm	0.925
San Francisco, CA	09:01:31 am	10:15:16 am	11:37:08 am	0.802
Seattle, WA	09:08:45 am	10:20:54 am	11:39:00 am	0.931
Topeka, KS	11:39:16 am	01:06:41 pm	02:34:05 pm	0.989
Washington, DC	01:17:59 pm	02:42:53 pm	04:01:41 pm	0.844

*All times are in local time (including daylight saving time)

**Eclipse magnitude is the fraction of the Sun's diameter occulted at maximum eclipse

[b] Eclipse times from Fred Espenak and Jay Anderson: *Eclipse Bulletin: Total Solar Eclipse of 2017 August 21* (astropixels.com/pubs/TSE2017.html).

eclipse pass a little south of even higher, volcanic Mt. Hood and a little north of the slightly shorter Three Sisters cluster of volcanoes.

Then down the eastern slope of the Cascades the eclipse slides and over Warm Springs, Madras, Mitchell, Prairie City, and Unity, the duration of total eclipse now exceeding 2 minutes. It crosses Interstate Highway 84 halfway between Baker City, Oregon and Ontario, on the Snake River, which marks the end of Oregon and the beginning of Idaho.

Backtracking the Oregon Trail

At Baker City, the eclipse begins to trace in reverse much of the Oregon Trail that brought more than a third of a million settlers from Missouri River towns to the Northwest between 1843 and 1869. From Baker City, the immigrants drove their wagons north and west to the Columbia River, then usually floated downstream to the site of Portland and the good farmland of the Willamette Valley to its south.

The eclipse skirts Boise, capital of Idaho, the edge of totality passing 15 miles (25 kilometers) to the north. Sun Valley is more fortunate. How could a solar eclipse resist a name like that? The ski and summer resort receives 1 minute 13 seconds of totality. Stanley, Idaho, an hour's drive northward and closer to the central line, receives an extra minute of total eclipse.

As much as this eclipse seems to target state capitals, it seems even more to prefer high mountains. Darkness at noon (well, 30 minutes short of noon) descends on Borah Peak, tallest mountain in Idaho (12,667 feet; 3,861 meters).

In its trek across Idaho, the eclipse clips a stubby protrusion of the "beard" of Montana. About 10 square miles (26 square kilometers) of the Beaverhead Mountains and its national forest in southwesternmost Montana receive a few seconds of totality.

In eastern Idaho, Idaho Falls and Rexburg luxuriate in eclipse shadow. Rexburg, almost on the central line, gets 2 minutes 17 seconds of totality.

Then the eclipse enters Wyoming, and it's time for more mountain climbing. In a few seconds, the eclipse bounds up and over the Grand Tetons, with the central line passing just 8 miles (13 kilometers) south of spectacular Grand Teton Peak (13,770 feet; 4,199 meters), second highest mountain in Wyoming, 10 miles (16 kilometers) south of Jenny Lake, and just 2 miles (3 kilometers) north of Teton Village. This eclipse does seem to enjoy recreational areas.

The eclipse shadow, now traveling 1,840 miles per hour (2,960 kilometers per hour), touches down at the Jackson Hole airport, landing eastward on the single north-south runway. The town of Jackson, southern gateway to Grand Teton National Park, is 9 miles

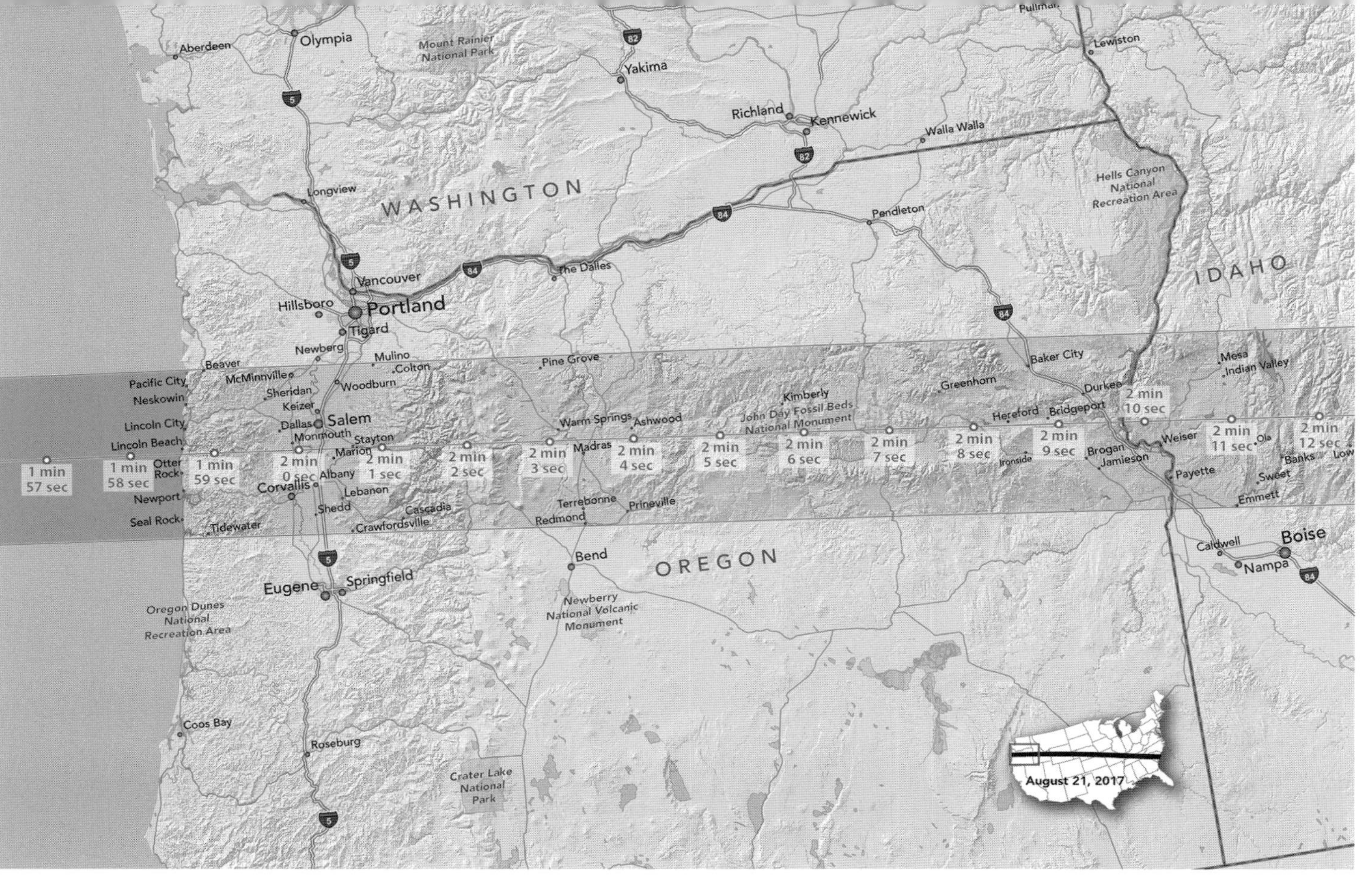

The path of totality as it crosses Oregon and western Idaho. Points along the central line give the duration of totality in minutes and seconds. [Map ©2016 Michael Zeiler, GreatAmericanEclipse.com]

(15 kilometers) to the south and thus very close to the center of the eclipse's path.[3]

The total eclipse of 2017 doesn't visit Yellowstone National Park, just north of Grand Teton. The northern limit of totality barely misses the south entrance to America's first national park.

Crossing the Divide

The path of totality for the 2017 eclipse loves American highways. At Grand Teton National Park in western Wyoming, it hooks up with US Highway 26 and hugs it all the way past Scottsbluff, Nebraska. For more than 450 miles (725 kilometers), the highway never strays from the path of totality.[4]

The eclipse sweeps across Wyoming, angling southeast, rushing up and over the Continental Divide in the Wind River Mountain Range and enveloping Gannett Peak, tallest mountain in Wyoming (13,804 feet; 4,207 meters). Then down the Wind River Basin the shadow slides, with the southern edge of the Owl Creek and Bighorn Mountains to the north and the northern edge of the Rattlesnake Hills to the south. It was across this high plateau that wagon trains rolled westward on the Oregon Trail more than a century and a half ago. Ahead of the settlers, south of the path of totality, lay South Pass, the highest elevation on their journey—7,411 feet (2,259 meters). South Pass was the halfway point—3 months and 1,000 miles (1,600 kilometers) from Independence, Missouri; 3 months and 1,000 miles still to go to the Oregon coast.

Central Wyoming is a wind-carved, wind-burnished landscape with an arctic climate. Homesteaders there used to say: "If summer falls on a weekend, let's have a picnic."[5]

Straight through Casper, Wyoming goes the center of the eclipse, jay-walking Cy Avenue and South Poplar Street without stopping and looking in all directions. Traffic may halt for 2 minutes 26 seconds and look up.

Southwest of Casper, at the edge of eclipse totality, lies massive, mound-shaped Independence Rock on the Oregon Trail. It is a third of a mile long (580 meters) and the height of a 13-story building. If the settlers reached this milestone by July 4, they would celebrate. If they didn't, snow would likely catch them in the mountains ahead.

River Running

At Casper, the eclipse picks up the North Platte River and follows it through Douglas and across one last range of the Rockies, the northern portion of the Laramie Mountains, bathing Laramie Peak (10,274 feet; 3,131 meters) in totality.

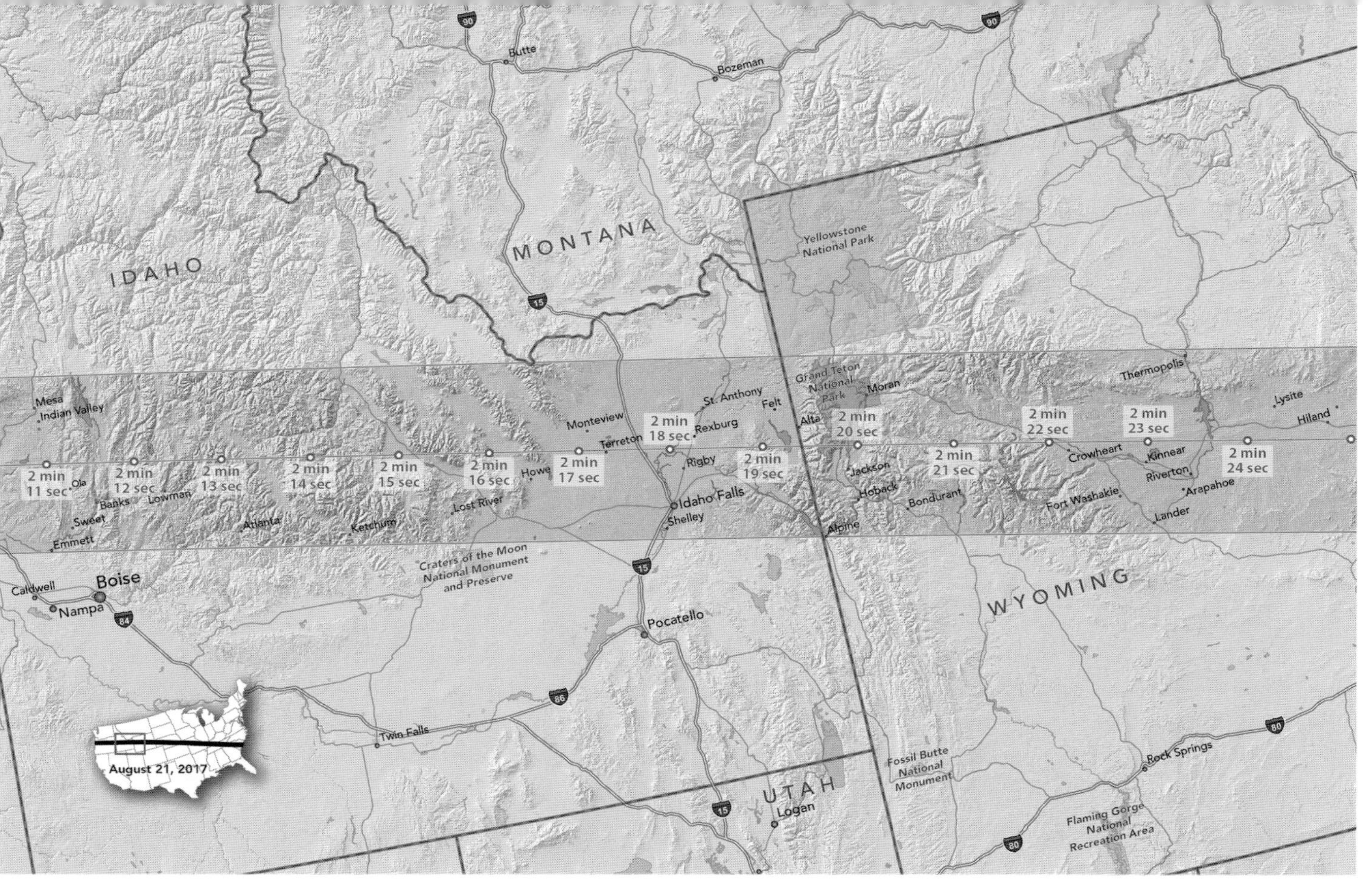

The path of totality as it crosses Idaho and western Wyoming. Points along the central line give the duration of totality in minutes and seconds. [Map ©2016 Michael Zeiler, GreatAmericanEclipse.com]

Then the mountains are gone and the eclipse races off out of Wyoming, into the stubby panhandle of Nebraska, and onto the Great Plains. Still it follows the North Platte River, which meanders along the southern half of the path of totality, through Scottsbluff. It was westward along the North Platte River that the Oregon Trail settlers came, admiring the grandeur of Scotts Bluff, but regretting the detour needed around this obstacle.[6] They averaged 1½ miles per hour (2½ km/hr). Fifteen miles was a good day.

Totality then skims Chimney Rock. Travelers on the Oregon Trail considered this slender stone spire, rising from a conical mound and towering 480 feet (145 meters) above the plain, to be the most spectacular landmark along the Oregon Trail. It was a day-and-a half journey west by wagon from Chimney Rock to Scottsbluff. The eclipse shadow, traveling east, makes the trip in 45 seconds.

The central line runs 3 miles (5 kilometers) south of Alliance, Nebraska, which is surrounded by hundreds of farm fields that are circular rather than rectangular. This land is arid high plain and needs irrigation. The circular fields have a center pivot system with a sprinkler arm that sweeps around the field like the minute hand on a clock. From the Sun's and Moon's perspective looking down, the farmers have drawn more than a thousand giant round images, some light, some dark, as if to honor the creators of eclipses. Alliance is rewarded with 2 minutes 30 seconds of totality.

The North Platte River meanders out of the eclipse path as it approaches Ogallala, but then veers back inside the path of totality as it approaches the town of North Platte. Here, just west of North Platte, the eclipse picks up Interstate Highway 80, which follows the river.

The path of totality shadows I-80 for the next 250 miles (400 kilometers), although the highway usually lies toward one or the other edge of the path of eclipse darkness. To enjoy the longest duration of totality, which in central Nebraska is about 2 minutes 35 seconds, observers want to be on or close to the central line. Reaching the central line is easy using north-south roads off I-80. What I-80 provides is the quickest way of chasing a break in the clouds if local weather disappoints.

Next to receive the eclipse is Kearney. Nearby, once stood Fort Kearny,[7] an unfortified outpost built to protect the pioneers headed west on the Oregon Trail. It was beloved because it was a place where letters could be mailed to those left behind.

Between Kearney, near the southern limit of totality, and Lincoln, near the northern limit, I-80 cuts diagonally across the southeast-headed eclipse path. The central line of the eclipse passes just a little south of downtown Grand Island, straight over the island in the North Platte River that gave the city its name, and right across I-80, 5 miles (8 kilometers) southeast of the island.

On to Lincoln, Nebraska and another state (capital) visit. Lincoln lies near the northern edge of the eclipse. From the steps of its unique,

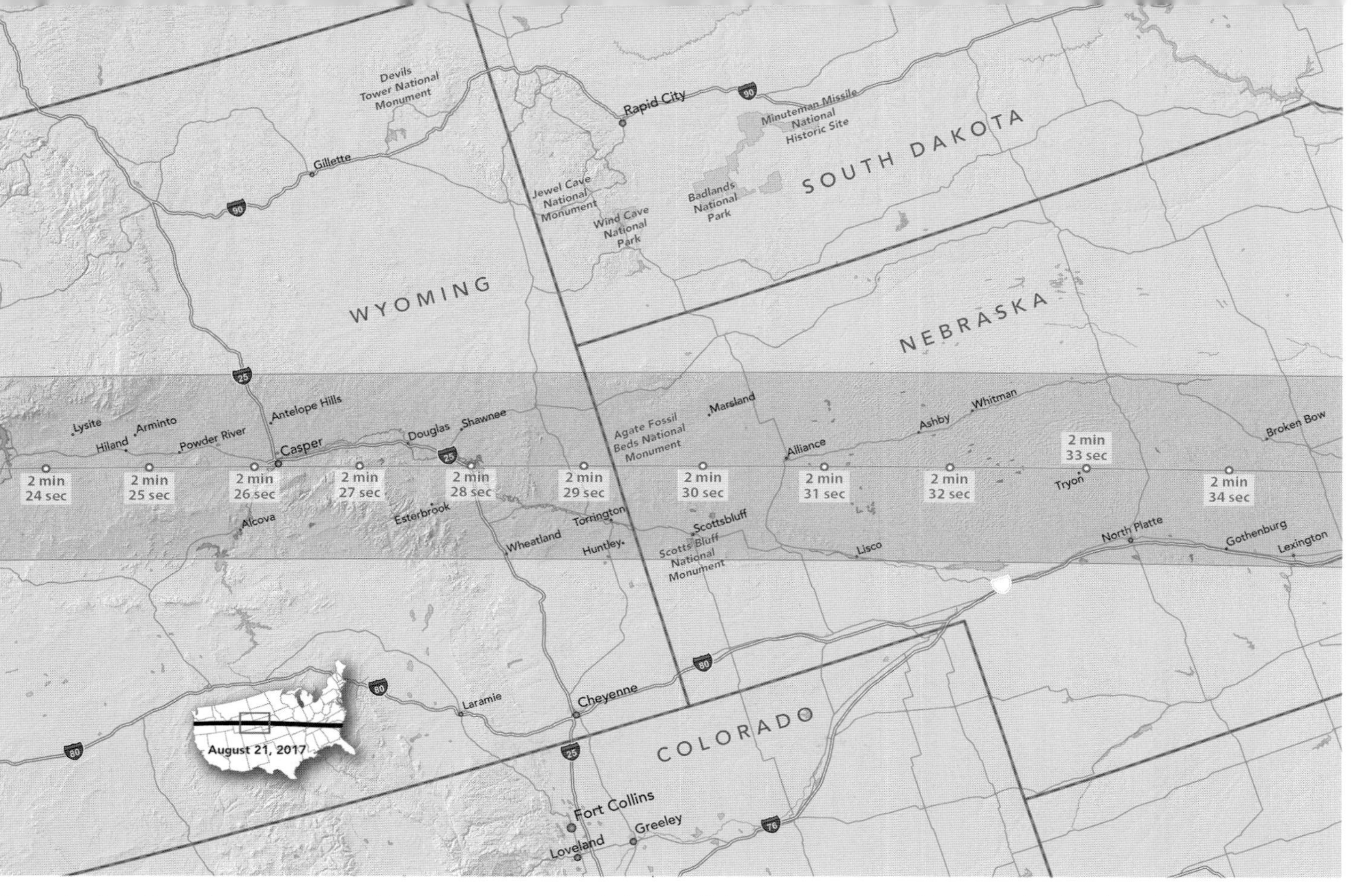

The path of totality as it crosses eastern Wyoming and western Nebraska. Points along the central line give the duration of totality in minutes and seconds. [Map ©2016 Michael Zeiler, GreatAmericanEclipse.com]

high-rise capitol building, the total phase of the eclipse will last 1 minute 26 seconds. The eclipse bypasses Omaha, Nebraska's largest city, which is 35 miles (55 kilometers) too far north.

Cutting Corners

On its way out of Nebraska into Missouri, the total phase of the eclipse clips the tiniest corner of southwesternmost Iowa, with perhaps 2 square miles (5 square kilometers) of farm fields and one farmhouse receiving a few seconds of totality.

The eclipse then skims the northeast corner of Kansas and leaps the Missouri River into St. Joseph, Missouri, one of the great embarkation points for the Oregon Trail. This eclipse takes an interest in modern embarkation points as well, racing through the airport without a security check. It then parades through downtown St. Joe, bringing 2 minutes 38 seconds of totality.

Kansas City, Missouri, 40 miles (65 kilometers) south of St. Joseph, straddles the southern edge of the eclipse path. Just east of Kansas City, within the path of totality, is Independence, where the Oregon Trail began. Two thousand miles (3,200 kilometers) lay ahead of those travelers. Three hundred fifty thousand people began the journey between 1843 and 1869 in search of a new life in a country they had never seen. As many as 35,000 died along the way, most of disease and accidents.

At Independence, the interstate-highway-friendly eclipse hops onto I-70 and brings it Moon-shadow shade all the way to St. Louis. North of I-70, Carrollton and Marshall, Missouri lie very nearly on the central line of totality. Just south of Carrollton, the central line of the eclipse crosses the meandering Missouri River three more times in 4 miles (6 kilometers).

The central line passes right between Columbia and Jefferson City in the center of the state, with plenty of totality for both. On the steps of the state capitol in Jefferson City, totality lasts 2 minutes 28 seconds.

The central line passes 9 miles (14 kilometers) southwest of Fulton, Missouri, where in 1946, ten months after the end of World War II in Europe, Winston Churchill spoke at Westminster College and introduced a new phrase into the world's vocabulary, saying that an iron curtain had descended upon central and eastern Europe. The eclipse provides a dark curtain of a much less ominous and far more aesthetic kind. It lasts 2 minutes 34 seconds, not 44 years.

As it proceeds to and past Columbia and Jeff City and Fulton, the center track crosses the Missouri River three more times.

Like Kansas City, half of St. Louis enjoys the total eclipse. The northern limit of totality passes 2 miles (3 kilometers) south of the Gateway Arch on the banks of the Mississippi River. For the western and southern suburbs

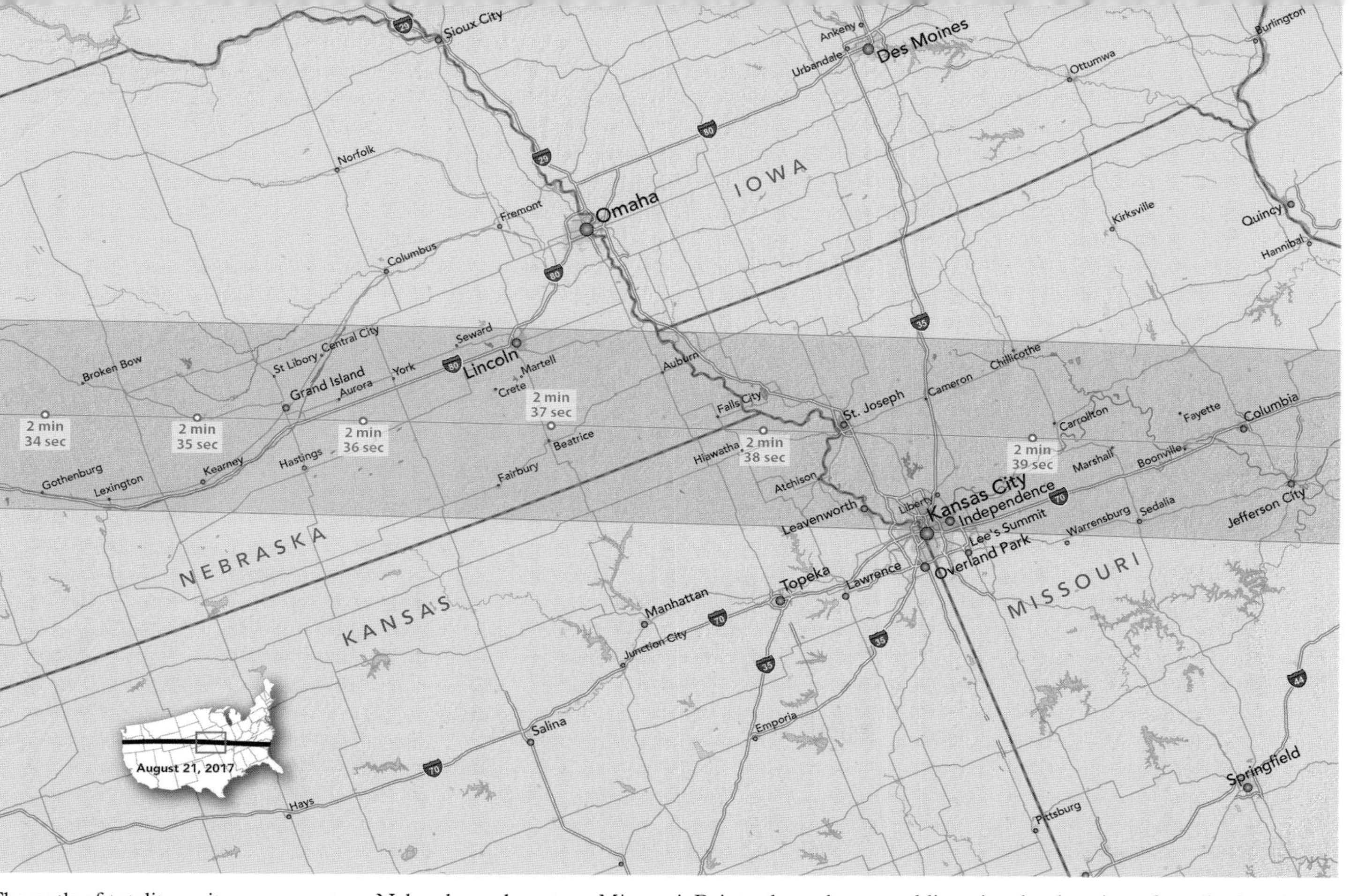

The path of totality as it crosses eastern Nebraska and western Missouri. Points along the central line give the duration of totality in minutes and seconds. [Map ©2016 Michael Zeiler, GreatAmericanEclipse.com]

though, there is a precious minute or so of eclipse darkness. But there are up to 2 minutes 40 seconds of eclipse darkness some 35 miles (60 kilometers) to the southwest or south, near Union or St. Clair, or close to Hillsboro, De Soto, and Festus.

The eclipse reaches the Mississippi River town of Ste. Genevieve, Missouri and angles through downtown. The central line then actually enters Illinois before it crosses the Mississippi River because the boundary between Missouri and Illinois was set before the river changed its course during a spring flood more than a century ago. The central line passes diagonally through Kaskaskia, population 9, the first capital of Illinois. For 4 miles (6 kilometers), the central line, still west of the Mississippi, finds itself back in Missouri, then returns to Illinois at another abandoned oxbow in the river and finally crosses the Mississippi.

Caught in the swath of totality, just north of the central line, are Carbondale and Marion, Illinois. Carbondale owes its name to the coal mined all around it in southern Illinois. Marion is home to a federal penitentiary where some convicted terrorists are serving a good deal more time than 2 minutes 28 seconds in eclipse.

Carbondale and Marion share the honor of being the cities closest to the geographical point where totality for the 2017 eclipse lasts longest: 2 minutes 40.3 seconds. That point is 12 miles (20 kilometers) southeast of Carbondale.[8]

Both Carbondale and Marion hold the even more enviable distinction of lying within the path of America's next total solar eclipse in 2024. For any city, the gap between visits by a total eclipse of the Sun averages 375 years. Carbondale and Marion will experience the magic twice in only 6⅔ years.

Into the South

The shadow of the Moon leaves Illinois by crossing the Ohio River into Kentucky, bringing totality to Paducah, about 18 miles (29 kilometers) south of the central line. Interstate Highway 24 lies within the band of totality for the next 200 miles (300 kilometers) from Goreville, Illinois to Nashville, Tennessee, for those who may need to do some speedy eclipse chasing in case of clouds.

As the eclipse enters Kentucky, the southern half of the path of totality sweeps across the Kentucky portion of the Land Between The Lakes National Recreation Area, where the Tennessee and Cumberland Rivers are dammed side by side a few miles before they flow into the Ohio River.

Twelve miles (20 kilometers) northwest of Hopkinsville, Kentucky, is the point of "greatest eclipse," when the Sun, Moon, and Earth are most precisely aligned. It doesn't coincide with the longest duration of totality because the Moon is not perfectly round. The Moon has more mountains

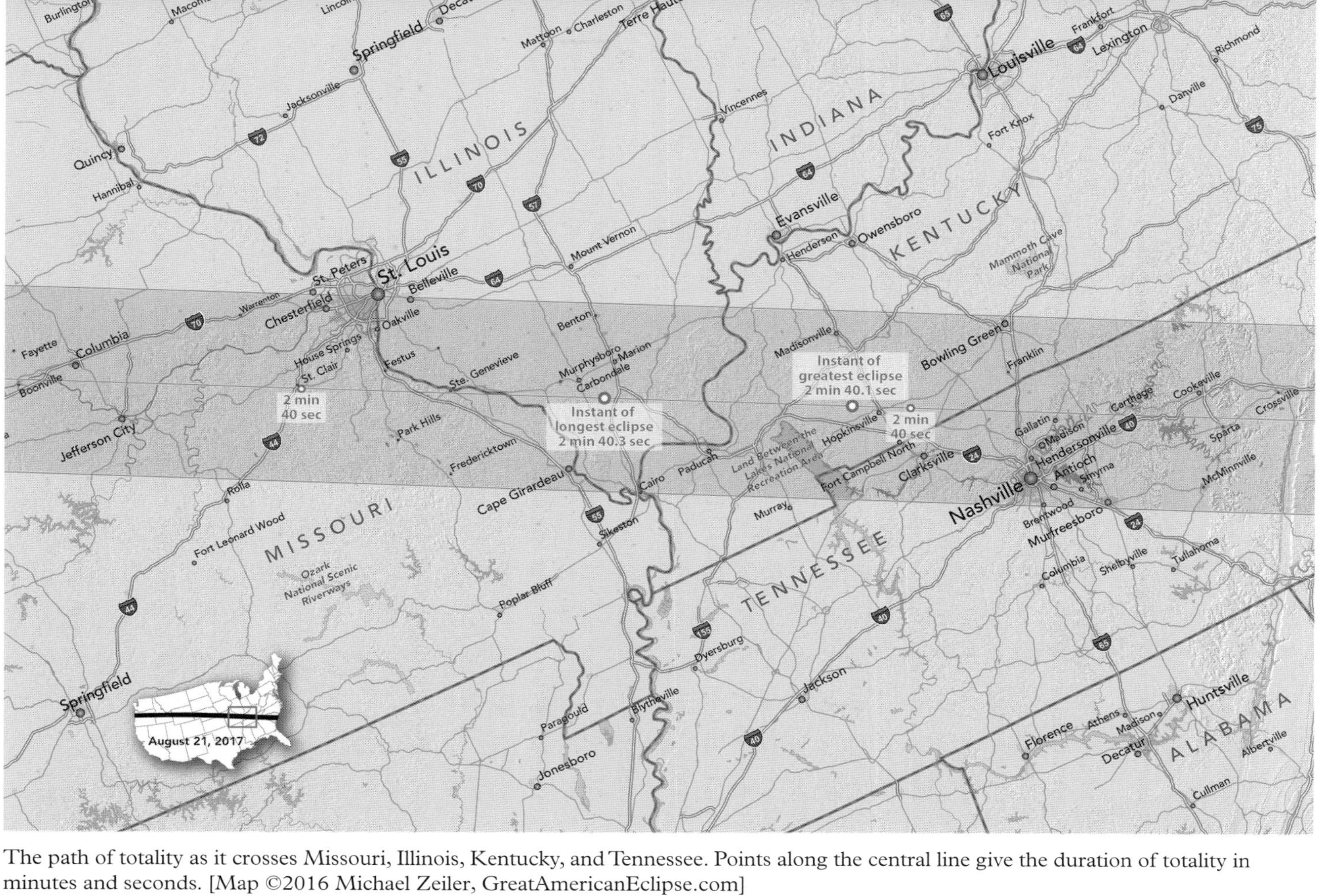

The path of totality as it crosses Missouri, Illinois, Kentucky, and Tennessee. Points along the central line give the duration of totality in minutes and seconds. [Map ©2016 Michael Zeiler, GreatAmericanEclipse.com]

and highlands along its rim in some directions than others. At greatest eclipse in 2017, totality lasts 2 minutes 40.1 seconds, two-tenths of a second less than maximum duration. At that moment, the path of totality is 71 miles (115 kilometers) wide and the Sun is 64° above the horizon.[9]

Two minutes 40 seconds of totality—tops for this coast-to-coast cosmic shadow show. Not exceptionally long for a total eclipse, but this one spreads the riches of its totality quite equally. When the eclipse touches land in Oregon, totality lasts 1 second under 2 minutes. When the eclipse leaves land in Charleston, South Carolina, totality is still 2 minutes 33 seconds in length. From western Missouri to eastern Tennessee—for 600 miles (1,000 kilometers)—the duration of totality along the central line is always within *one second* of 2 minutes 40 seconds, the maximum for the 2017 all-American eclipse. Seems only fitting for an eclipse visiting a country founded on the concept that all men are created equal.

On into Tennessee the eclipse glides, with Clarksville, Springfield, and Westmoreland spread out west to east across the state line and across the eclipse path to welcome it. The eclipse envelops Nashville, yet another state capital. Nashville also boasts the largest metropolitan population of any city fully in the path of totality—1.8 million people. The central line passes 25 miles (40 kilometers) north of downtown.

Closer to the central line is Lebanon, Tennessee, home of the Nashville Superspeedway. The stock cars that race there sometimes turn laps at 200 mph (320 km/hr). The Moon's shadow as it passes over the track will be traveling 1,450 mph (2,330 km/hr).

At Nashville, the eclipse exits Interstate Highway 24 and transfers to I-40. I-40 courteously stays within the lane of totality for 170 miles (275 kilometers), up and over the Cumberland Mountains, from Nashville to the outskirts of Knoxville.

On its climb over the Cumberland Mountains, westernmost chain of the Appalachian Range, the eclipse visits Cookeville and Crossville on I-40, each 5 miles (8 kilometers) north of the central line. Sparta, Tennessee, is not on I-40 but it is on the central line, and receives 2 minutes 39 seconds of totality. The duration of total eclipse is declining, but very slowly.

In Kentucky, totality covered the Tennessee River as it emptied into the Ohio River. Now the eclipse crosses the giant-U-shaped Tennessee River again in eastern Tennessee. Spring City, on Watts Bar Lake, and Watts Bar Dam, creator of that lake on the Tennessee River, are just south of the central line. Also in the path of totality on the Tennessee River and its lakes are Kingston, Loudon, and Lenoir City.

Downtown Knoxville lies outside the path of total eclipse, but some western and southern suburbs, including Oak Ridge and its national laboratory, catch a few seconds of totality. The eclipse is even less kind to Chattanooga, chugging by to the north, venturing no closer than outlying suburb Soddy-Daisy.

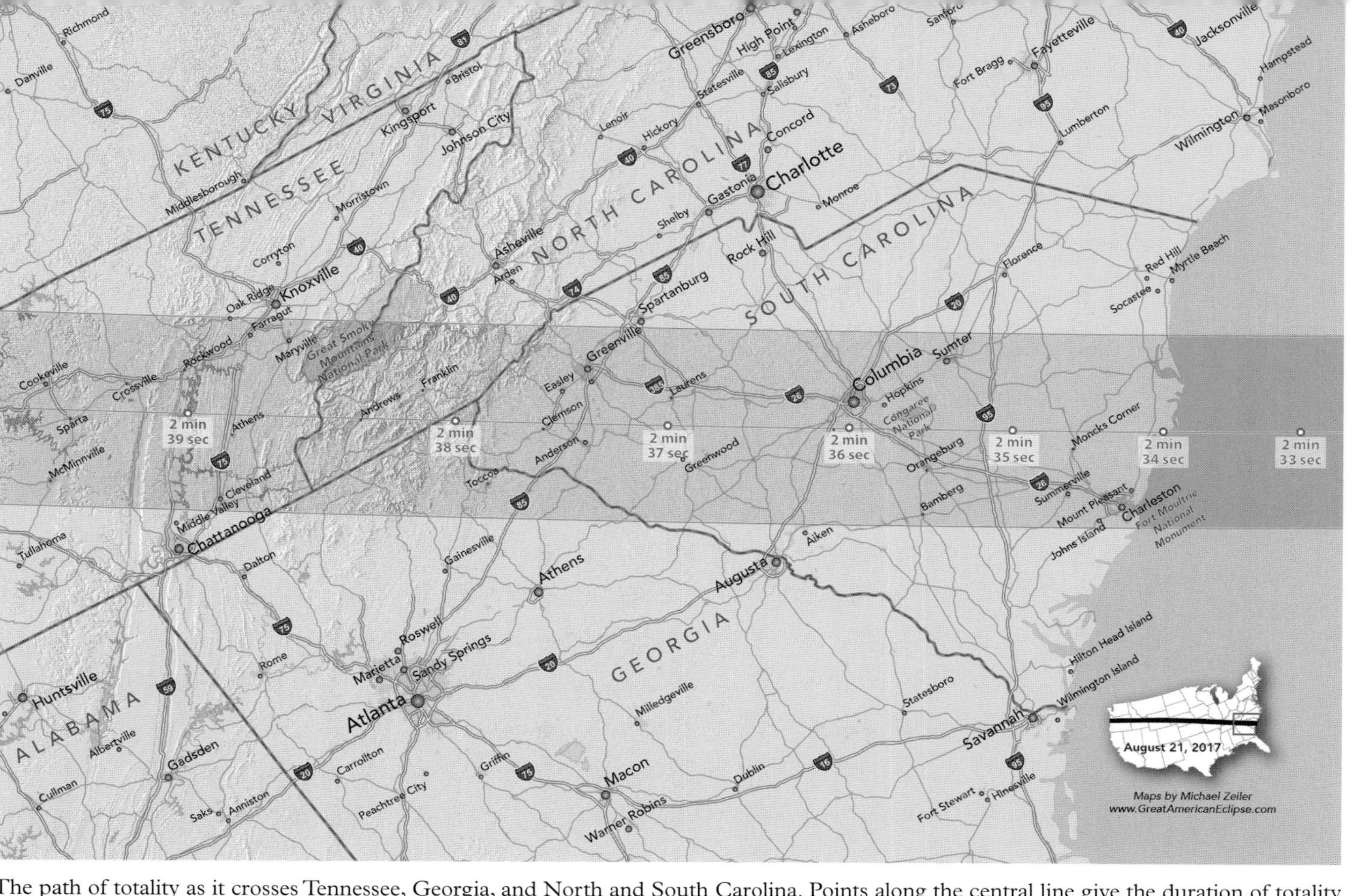

The path of totality as it crosses Tennessee, Georgia, and North and South Carolina. Points along the central line give the duration of totality in minutes and seconds. [Map ©2016 Michael Zeiler, GreatAmericanEclipse.com]

Suddenly this eclipse seems like even more of a world traveler than it is—and with an exceptionally large footprint—as it visits Lebanon, Sparta, Athens, Dayton, Philadelphia, Cleveland, and Louisville, all in the same few moments—all wistfully named towns in the Tennessee Valley. Dayton is where John Scopes was tried in 1925 for teaching evolution. Closer to the central line are Athens, Sweetwater, and Madisonville.

Then it's up and over the main branch of the Appalachian Mountains with more than half of Great Smoky Mountains National Park—the most visited of all the national parks—lying in the zone of totality. Clingmans Dome (6,643 feet; 2,025 meters) is near the northern limit of the eclipse. Cade's Cove is closer to the central line. Gatlinburg and Pigeon Forge, gateways to the Park on the Tennessee side, are excluded from totality. Bryson City, North Carolina, at the east entrance to the Smokies, gets just under 2 minutes of totality. Franklin gets half a minute more, and then the eclipse leaves North Carolina behind. It catches the lightly populated northeastern corner of Georgia and plunges centrally across much smaller South Carolina from Greenville through Charleston, enveloping half the land area of the state.

Last Stops

When the eclipse came ashore in Oregon, it visited the capital of that state. Since then, the eclipse has passed across substantial areas of nine states—and has visited five state capitals: Salem, Oregon; Lincoln, Nebraska; Jefferson City, Missouri; Nashville, Tennessee; and now Columbia, South Carolina, bestowing 2 minutes 30 seconds of totality on the capitol building. The eclipse narrowly missed Boise, Idaho and Topeka, Kansas.

Now the eclipse reaches the South Carolina coast, with totality stretching from Pawley's Island in the north (but not Myrtle Beach) to Sullivan's Island (but not Kiawah Island) in the south. Totally within totality is Charleston, South Carolina, where the American Civil War began, but the city lies toward the southern edge of total eclipse. Fort Sumter in Charleston harbor gets 1 minute 35 seconds of corona time. Charleston's northern suburbs, North Charleston and Mt. Pleasant, are closer to the central line and earn more nearly 2 minutes of totality.

And then the eclipse of August 21, 2017, leaves the United States behind and plunges out across the Atlantic Ocean, never to touch land again. Bermuda is about 400 miles (650 kilometers) too far north. The Bahamas are about that same distance too far south. Totality misses Antigua and Barbuda, at the northeast corner of the Caribbean Sea, by about 350 miles (600 kilometers). The path of totality is narrowing. The shadow of the Moon lifts off the waters of the Atlantic Ocean at sunset about 800 miles (1,300 kilometers) west of the coast of Africa. The eclipse is over.

But the shadow will continue to sweep through space and will find the Earth again, bringing to a most fortunate few a total eclipse of the Sun. In the meantime, you can almost hear it singing as it goes:

> I've been everywhere, man[10]
> I've been everywhere, man
> 'Cross the deserts bare, man
> I've breathed the mountain air, man
> Of travel, I've had my share, man
> I've been everywhere

Been to Broken Bow, Arapahoe, St. Joe, Missouri
Albany, Easley, Kearney, Shoshoni
Greenhorn, Pine Grove, Sweet Home, Antelope
Midvale, Homedale, photograph with telescope
Casper, Culver, Cascade, Idaho
Plattsburg, Orangeburg; no pause, off I go

> I've been everywhere, man
> I've been everywhere, man
> 'Cross the deserts bare, man
> I've breathed the mountain air, man
> Of travel, I've had my share, man
> I've been everywhere

To Seneca, Etowah, Paducah, Sun Valley
Columbia, Aurora, I bring totality
Toledo, Scio, Shaniko, Oregon
Callaway, Norway, Lemay, Jefferson
Shawneetown, Pawnee, Crowheart, Tecumseh
Gasconade, I bring shade; for Merna, corona

> I've been everywhere, man
> I've been everywhere, man
> 'Cross the deserts bare, man
> I've breathed the mountain air, man
> Of travel, I've had my share, man
> I go places you haven't been
> I've been everywhere

Been to Clarksville, Nashville, Crossville, Tennessee
Knoxville, Maryville, Greenville, Liberty
St. Louis, Metolius, Madras, Depoe Bay
Corvallis, Thermopolis: making night of day
Jackson, Atchison, Riverton, Lebanon
Silverton, Torrington, Charleston, I'm gone

The Remarkable Family of the 2017 Eclipse

Every eclipse belongs to a family, a saros series that begins, evolves, and ends. Each saros series has its own personality—its special features and its place in human history. The total eclipse of August 21, 2017, belongs to a distinguished family.

Eclipses in a saros series occur 18 years, 11 days, 8 hours apart. Such eclipses occur at the same node, at the same time of year, and with the Moon at nearly the same distance from Earth. Thus, successive eclipses are very similar in appearance, although they occur about one-third of the way around the world westward. Over the lifetime of a saros series, eclipses start near a pole as very weak partials, move toward the equator and strengthen, then continue toward the opposite pole, weakening as

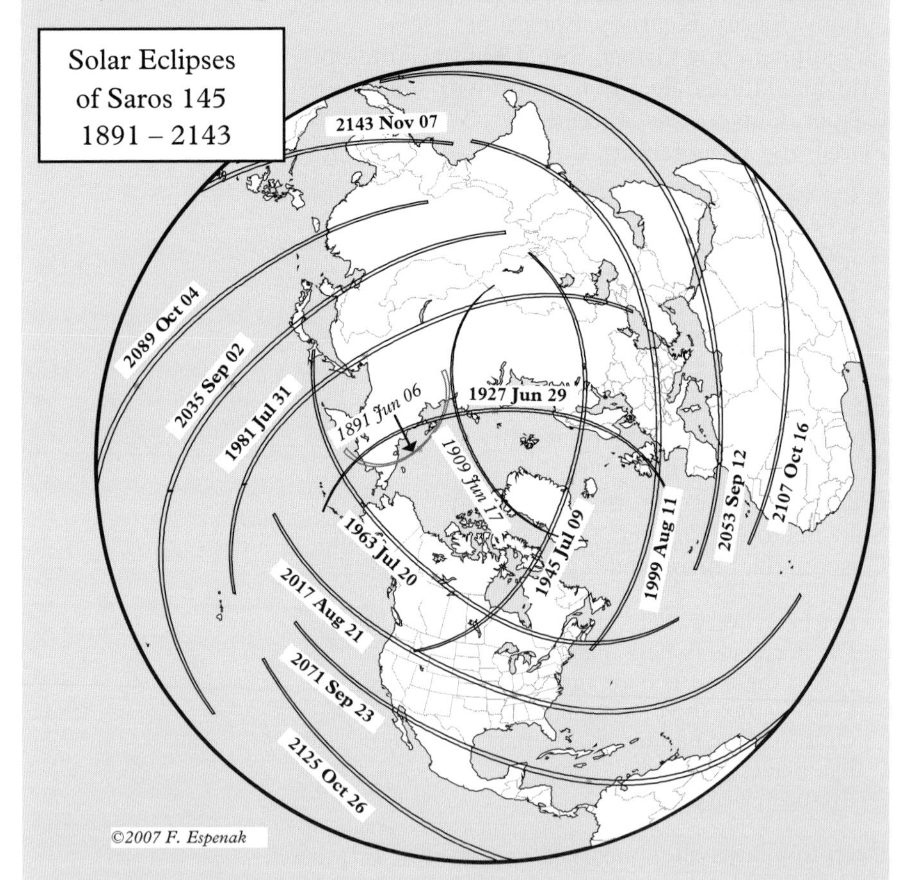

First 15 central eclipses of saros 145. The tracks shift west and south with each succeeding eclipse. The first eclipse (1891) is annular; the second (1909) is hybrid. All the rest from 1927 to 2143 are total eclipses. Saros 145 brings the 2017 eclipse across the United States and provided the 1999 eclipse across Europe and southwestern Asia. [Map and eclipse calculations by Fred Espenak, EclipseWise.com]

they go, until they finally vanish. A saros series typically lasts 12 to 13 centuries, and contains 70 or more eclipses.

The 2017 eclipse cuts diagonally across the United States, with greatest eclipse in Kentucky at latitude 37° north.[c] A total eclipse that peaks at middle latitudes belongs either to a young saros whose central eclipses are just beginning or to an old saros whose flashiest eclipses are dwindling to an end. An eclipse whose greatest occultation occurs near the equator is a saros in its prime.

This All-American eclipse of 2017 belongs to saros series 145. Its odd number tells us that this saros is migrating from the north polar regions to the south. So this saros, peaking in mid-northern latitudes, is still young and strengthening. The 2017 eclipse is number 22 in a sequence that will total 77. It is the sixth *total* eclipse for saros 145. There will be 35 more.

Saros 145 began in 1639 with a tiny partial eclipse near the North Pole. Then followed 13 more partial eclipses, each obscuring a little more of the Sun. Finally, in 1891, the center of the Moon passed across the center of the Sun. But the Moon was a little too far from Earth at that crossing, so it wasn't quite big enough to cover the bright disk of the Sun completely. The eclipse was annular. It would be the only annular eclipse that saros 145 would produce.

In 1909, the next eclipse in the series was a hybrid—starting annular, becoming total, then returning to annular again. After that, saros 145 stopped performing hybrid eclipses as well. One and done.

Instead, saros 145 specializes in total eclipses. It premiered its first on June 29, 1927, for England, Scandinavia, and eastern Russia. Crowds were pleased. It performed encores, with the duration of totality increasing each time:

July 9, 1945	1 min 15 sec	Idaho, Montana, Canada, Greenland, Scandinavia, Russia
July 20, 1963	1 min 40 sec	Alaska, Canada, Maine
July 31, 1981	2 min 02 sec	Russia, Pacific Ocean
August 11, 1999	2 min 23 sec	England, France, Germany, Austria, Hungary, Romania, Bulgaria, Turkey, Iraq, Iran, Pakistan, India

On August 21, 2017, totality lasts as long as 2 minutes 40 seconds. The duration of totality will increase to 2 minutes 54 seconds in 2035 for audiences in China, Korea, and Japan; then 3 minutes 4 seconds in 2053 for Portugal, Spain, North Africa, and Saudi Arabia.

But this eclipse series is just hinting at spectacles to come. Saros 145 is one of those rare eclipse families that creates 7-minute total eclipses, approaching the 7-minute-32-second limit for duration of totality.

June 14, 2504	7 min 10 sec	Australia, Micronesia
June 25, 2522	7 min 12 sec	Southern Africa
July 5, 2540	7 min 04 sec	South America

After these, as saros 145 slides toward the South Pole, eclipse times decline. The final total eclipse for saros 145 takes place in 2648.

For the next 361 years, saros 145 will fade away with weaker and weaker partial eclipses, until the last one on April 17, 3009.[d e] But what a run!

Summary for Saros 145

Total eclipses	41
Annular eclipses	1
Hybrid (annular-total) eclipses	1
Partial eclipses	34
Number of eclipses	77
Duration of saros 145	1,370.3 years

[c] Greatest eclipse is the moment and position in a solar eclipse when the shadow of the Moon passes closest to the center of the Earth. It is also the instant when the greatest fraction of the Sun's disk is obscured. It does not quite correspond to the place on Earth experiencing the longest duration of totality—which for the 2017 eclipse, is in Illinois.

[d] Saros data from Fred Espenak and Jay Anderson: *Eclipse Bulletin: Total Solar Eclipse of 2017 August 21* (Portal, Arizona: Astropixels Publishing, 2015), pages 41–43, astropixels.com/pubs/TSE2017.html.

[e] A complete list and maps of all eclipses in saros 145 can be found in the EclipseWise.com catalog of solar eclipse saros series: <eclipsewise.com/solar/SEsaros/SEsaroscat.html>

Saros Series Statistics

	Range	Average
Number of solar eclipses in a saros series	70–85	73
Timespan for a series	1,244–1,514 years	1,315 years
At any time, 42 saros series are running simultaneously.		

Total Solar Eclipses in the United States Through the 21st Century

The year 2017 marks the first of 8 total solar eclipses visible from parts of the contiguous United States during the 21st century. The eclipses occur in 2017, 2024, 2044, 2045, 2052, 2078, 2079, and 2099. Two maps below show the paths of these eclipses. Although the United States' mainland experienced a drought of total eclipses from 1979 to 2017, the period from 2017 through 2099 is much richer.

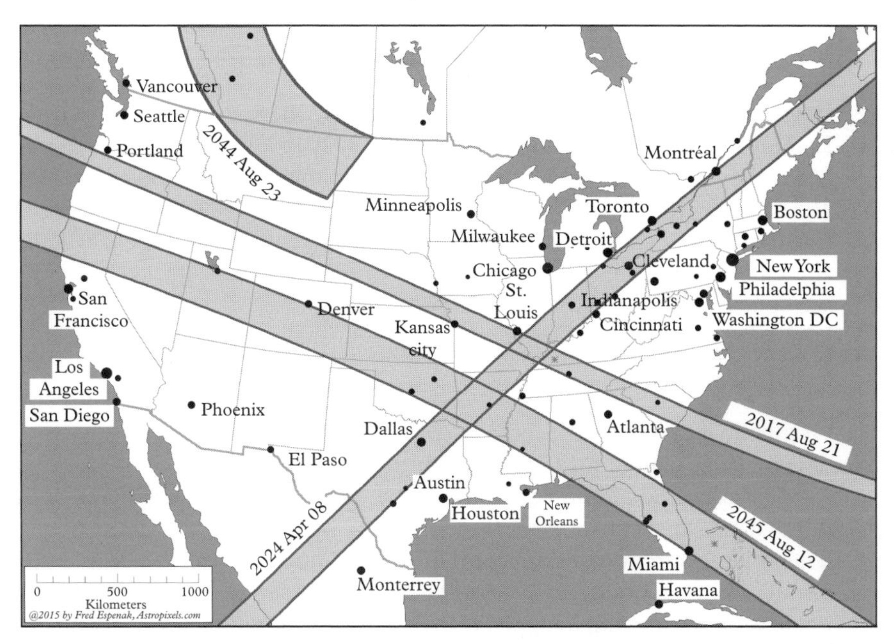

Beyond 2017, the next total solar eclipses through the continental USA are in 2024, 2044, 2045, and 2052. [Map and eclipse predictions by Fred Espenak, EclipseWise.com]

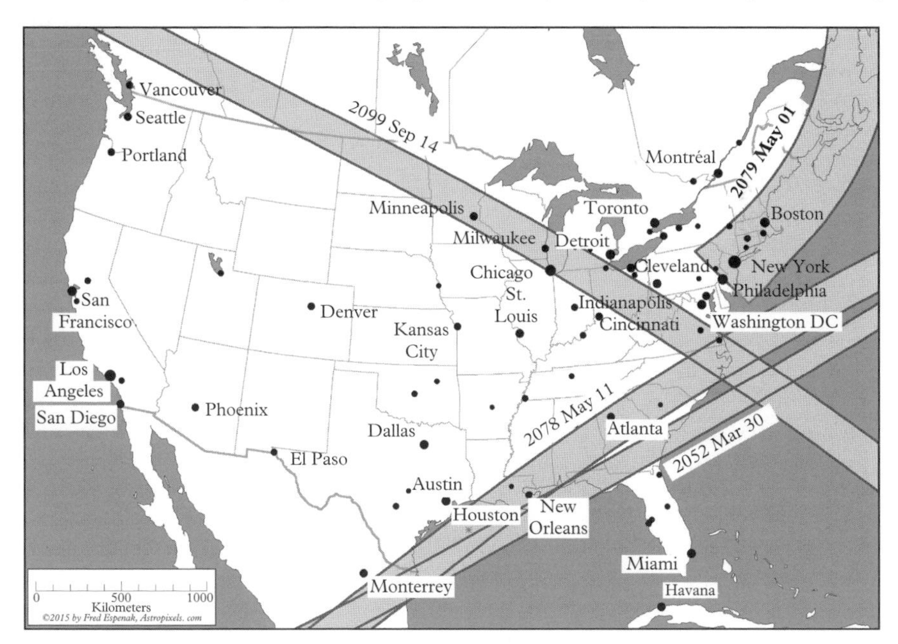

During the last half of the 21st century, total solar eclipses through the continental USA occur in 2052, 2078, 2079, and 2099. [Map and eclipse predictions by Fred Espenak, EclipseWise.com]

NOTES AND REFERENCES

1. Epigraph 1: Paul Andrews, interview, April 28, 2015.
2. Epigraph 2: Julie O'Neil, interview, April 28, 2015.
3. The valley region, bounded on three sides by mountains, is called Jackson Hole. The town is Jackson.
4. For road maps along the entire path of totality, see Fred Espenak: *Road Atlas for the Total Solar Eclipse of 2017* (Portal, Arizona: Astropixels, 2015); astropixels.com/pubs/Atlas2017.html.
5. John McPhee: *Rising from the Plains* (New York: Farrar Straus Giroux, 1986), page 15.
6. The steep rock formation is Scotts Bluff, now a national monument. Originally it was spelled Scott's Bluff. The city is Scottsbluff.
7. The fort was named for General Stephen Watts Kearny. A later mistake by the post office added an "e" to the name of the town that has never been removed.
8. The longest duration of totality on the central line for the 2017 eclipse is 2 minutes 40.3 seconds and is based on a round Moon. When a correction is made for a Moon with a ragged rim, the greatest duration of totality on the *central line* expands to 2 minutes 41.5 seconds. But the mountains in the Moon's southern hemisphere contribute to a still slightly longer duration—2 minutes 41.6 seconds—that occurs 3/8 mile (0.6 kilometers) south of the central line. Fred Espenak and Jay Anderson: *Eclipse Bulletin: Total Solar Eclipse 2017 August 21* (Portal, Arizona: Astropixels, 2015); astropixels.com/pubs/TSE2017.html.
   At greatest eclipse on August 21, 2017, the axis of the shadow cone passes 1,731 miles (2,785 kilometers) north of the center of the Earth.
9. A distance of 90 miles (145 kilometers) separates the point where the longest duration of totality occurs—near Carbondale, Illinois—and the point where greatest eclipse occurs—near Hopkinsville, Kentucky.
10. The song "I've Been Everywhere" was written by Australian composer Geoff Mack in 1959. It gained fame when recorded by Australian singer Lucky Starr in 1962. That same year, American singer Hank Snow adapted the lyrics for American place names and made the song popular in the United States. Johnny Cash recorded "I've Been Everywhere" in 1996 and it became one of his signature pieces.

○◑●◐

A MOMENT OF TOTALITY

The Eclipse Trip from Hell

Science writer Dawn Levy and three friends set off in a 21-year-old Volkswagen bus from San Francisco to Mexico to see the almost-7-minute-long 1991 total solar eclipse. The van's engine, recently rebuilt, began smoking outside Los Angeles. They replaced a seal and pushed on. Across the Mexican border, nearing Guaymas, the engine began spewing smoke. At the VW dealership, the repairman said the engine was irreparable and offered a rebuilt engine.

While they waited for the installation, the eclipse chasers went for a swim in the ocean. Within minutes Dawn had excruciating jellyfish stings on her face, neck, and hands. Then things got worse. Soaked in sweat, her stomach cramped, her hands went numb, and she couldn't breathe—a severe allergic reaction. An ambulance rushed her to the hospital where she received antihistamine injections.

The next day, when she could breathe normally again, Dawn and her friends were informed that the rebuilt engine was bad. The owner abandoned his van and the four travelers waited 8 hours for a train to Mazatlán. But the arriving train was full. Still determined to see the eclipse, they took a bus, arriving just in time for eclipse day.

The morning was overcast and, as the hours passed, the clouds refused to part. The sky turned peach as if it were dusk—then black. "Darkness at noon," Dawn says. People cheered, although they could not see the Moon covering the Sun. A cool breeze replaced the oppressive heat and humidity. Then, like the beam of a giant flashlight, brightness swept back across the sky. After a journey of 1,700 miles (2,700 kilometers), Dawn and her friends returned to their motel and watched a replay of the eclipse on television.[1]

Twenty-six years later, Dawn, now a resident of Tennessee, can't wait for 2017. The eclipse is coming to her. The path of totality includes her house.

[1] Dawn Levy: "An Eclipsed Vacation," *Los Altos* [California] *Town Crier*, August 7, 1991, pages 22–23. Also, interview, May 23, 2015.

14

The Weather Outlook

If you've got a weather forecast and a day to travel, there's no reason to miss the eclipse of 2017.

Jay Anderson, Canadian meteorologist and eclipse weather climatologist[1]

Total solar eclipses may be wondrous, but they are not always easy to reach. The 2015 eclipse passed over a few very cold, frequently overcast islands in the Arctic Ocean. The 2010 eclipse swept across the southern Pacific Ocean, missing almost every inhabited plot of land except Easter Island, and ended in the mountains near the southern tip of South America. The 2003 eclipse visited only Antarctica.

The three rules for seeing a solar eclipse are—just like buying real estate—location, location, location. First, of course, make sure you choose a site located inside the zone of totality. You can use the maps in this book (and others) or the Google interactive eclipse map by Xavier Jubier[2] to position yourself perfectly. Second, check climate conditions to find an observing location with a good probability of clear skies. Third, make sure you have a means of transportation and study the location of roads leading from your site to other good observing sites in case you need to move to a more promising location a day or two in advance because of incoming clouds.

For total eclipses of the Sun, the saying actually should be "location, location, location, location." Many people travel a substantial distance to see a total solar eclipse. Why not take advantage of the travel to enjoy the location—to see some cultural and natural wonders as part of your trip?

That's why eclipse veterans are very excited about the 2017 American eclipse. It's easy to get to the path of totality. It crosses the continental United States diagonally from coast to coast—2,500 miles (4,000 kilometers). Plenty of available land to observe from—and lots of good roads in case you need to change your location. In fact, the 2017 eclipse seems to love American highways. Much of the path of totality runs along or near interstate expressways and US routes.

And what a place for a total eclipse: so many other sights to see—national and state parks, historic cities and towns.

The path of the 1999 total eclipse through Europe and the Middle East brought totality to or within easy reach of more than 100 million people—more than any previous eclipse. Alas, cloudy weather spoiled the view for most of them in Europe.

The path of the 2009 total eclipse through India and China again brought totality to or within easy access of more than 100 million people—probably a new world record. But again, most people found their view clouded out.

The 2017 total eclipse in America occurs in late August. Almost everywhere along the track, National Weather Service maps show August sky conditions to be as dependably sunny or sunnier than at any other time of the year.[3]

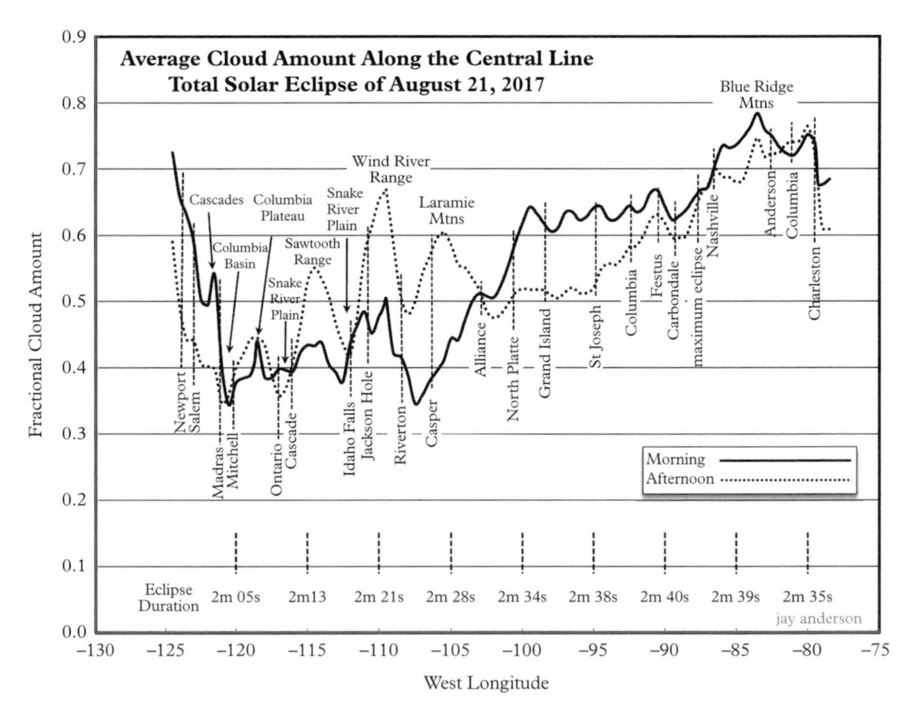

The average morning and afternoon cloud cover along the central line of the 2017 eclipse path has been derived from 20 years of satellite imagery. The locations of cities and towns are indicated by dashed vertical lines above their names. Prominent topographical features are named above the graphs. The ups and downs are due to the change in cloudiness on the western (cloudier) and eastern (less cloudy) sides of mountain ranges. [Source: Jay Anderson and Patmos-X: CIMMS/SSEC].

Might 2017 be the year in which more people than ever before actually *see* a total eclipse of the Sun?

Probably 100,000 foreign tourists will pour into the United States specifically for this stellar event. About 14 million Americans lie in the path of totality or so close they can drive to it in half an hour.[4] And many more will join them. The population of the United States is about 325 million. More than 90% of those people live within a day's drive of the long, narrow zone where the 2017 eclipse is total.

August 21, 2017, will mark the biggest outdoor spectator event in American history—a transcontinental tailgate party to watch the heavenly performance of the Moon and Sun.

Be Your Own Weatherman

You've picked an observing site with reasonably good chances of sunny weather. You have a means of moving to another site if bad weather threatens as eclipse day approaches. Now here's how you can follow weather forecasts and, if necessary, make the best possible decision a day or two in advance about moving to a site with better weather.

The National Weather Service is directly or indirectly the source for almost all the forecasts Americans hear on television and radio and read in newspapers and on the internet. It's government information, so it's all free. You can get forecasts directly from the National Weather Service website. Commercial weather services provide their own forecasts using NWS information and their own tools—also available free on the internet. All offer radar maps showing clouds in motion. Among the easiest to use are Weather Channel, AccuWeather, Weather Underground, Forecast.io, WeatherSpark, and SkippySky.[5]

The National Weather Service and the commercial providers offer weather predictions for 10 days into the future. These 10-day forecasts are useful as alerts about what weather systems are developing, but the NWS urges people not to put much faith in forecasts for a particular spot more than a week in the future. Clouds could arrive earlier or later than expected.

Forecasts for a week in advance are much more accurate. Five days in advance: more accurate still. Keep track of predictions for your chosen site and alternate sites. Once forecasts give the same prediction for eclipse day several days in a row, you can begin to trust it, says Jay Anderson, who teaches at the University of Manitoba after a career as a meteorologist with Environment Canada. He is known around the world for the weather information he offers for every total and annular eclipse and for his many observations of eclipses everywhere on Earth.[6]

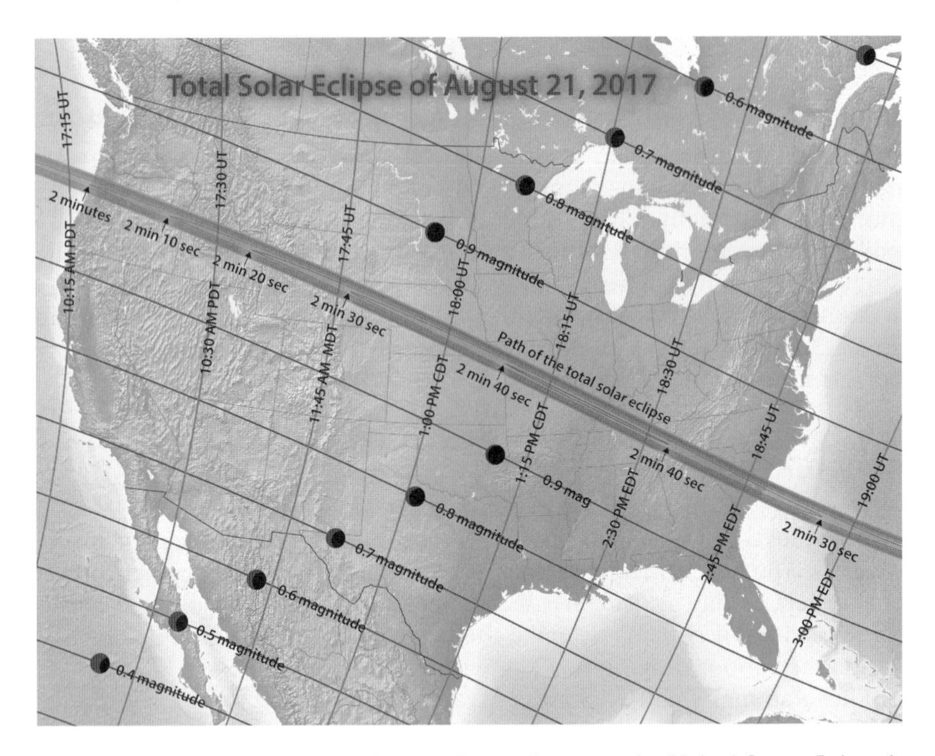

The August 21, 2017, eclipse track runs diagonally across the United States. It is only within this 68-mile-wide "path of totality" that the total phase of the eclipse is visible. The rest of the United States (and North America) will see a partial eclipse as shown on this map. The eclipse magnitude (fraction of the Sun's diameter eclipsed) is given as a decimal number less than 1.0. [Map ©2016 Michael Zeiler, GreatAmericanEclipse.com]

By two days before eclipse day, National Weather Service and commercial weather forecasts are quite dependable. A day or two ahead of time is when you should make your decision to move if your site is likely to be cloudy and the forecast for another site is quite favorable. Allow more time than usual for driving. Other folks may be altering their eclipse-observing plans too.

Fearsome Clouds

Clouds are the enemy of eclipse observing. No place along the eclipse path across the United States is immune to the threat of clouds. Some clouds pose a great threat to eclipse watching. Other clouds pose almost no threat at all; they will vanish as the eclipse approaches.

Clouded out. Glum observers at the August 9, 1896, eclipse in Lapland see only the twilight glow on the horizon in this painting by Lord Hampton. [Annie S. D. Maunder and E. Walter Maunder: *The Heavens and Their Story*]

The clouds you need to escape are frontal systems. Cold fronts and warm fronts bring overcast skies for many hours, or a day or more. Even if you catch a glimpse of the Sun through broken clouds, the clouds may quickly cover the Sun again. Weather systems in the continental United States push across the country generally from west to east. Cold fronts move northwest to southeast. Warm fronts move southwest to northeast.

A weather forecast two days in advance of eclipse day that calls for a cold front or warm front to be over you on that special day begs you to move to a better location. Remember, clouds build up even before the front officially arrives. Check the professional weather forecasts for your alternative sites. Just be sure your alternate sites are within the zone of totality.

Another threat to eclipse visibility is forest and grass fires. They are particularly common in the western United States in the summer when the weather is hot and dry. Clouds of smoke from these fires can block the corona from view. Too bad: the corona is the emblem of a total eclipse. If there is a forest or grass fire in your vicinity, be sure to be upwind of it. The smoke can spread hundreds of miles downwind.

Because this eclipse occurs in August, there is a threat of thunderstorms. Thunderstorms can spring up anywhere, but this hazard to watching the 2017 eclipse is especially menacing east of the Mississippi River where the eclipse occurs in the afternoon, peak time for thunderstorm cloud development.

"Pop-up" thunderstorms develop because the Sun steadily warms the Earth's surface and heat and moisture rise from the ground. The ground warms unevenly—more heat rises from dark forests and asphalt roads than from lighter-colored meadows and farm fields, Anderson notes.[7] So clouds billow up here and there—separate from a frontal system. Winds—also created by the Sun's uneven heating—carry the clouds generally eastward. The probability of thunderstorms is part of every forecast. That probability differs from county to county. Even more, it differs by terrain.

The cumulonimbus clouds of a thunderstorm in America may tower to a height of 10 miles (52,000 feet; 16 kilometers). Thunderclouds in the tropics can soar even higher—14 miles (75,000 feet; 22 kilometers). An average thunderstorm is about 15 miles (24 kilometers) in diameter, says Anderson. An isolated thunderstorm moving about 30 miles per hour (50 kilometers per hour) will drench you and move on in about 30 minutes. Trouble is, not all thunderstorms are isolated. There may be another heading your way.

You know from driving into a thunderstorm that you usually come out the other side in a matter of minutes. If you drive down a highway going the same direction as a thunderstorm, you can outrun it. But you have to get well ahead of it or it will catch up with you and eclipse the eclipse. It's usually better to go westward through a thunderstorm or go north or south of it to find clearer skies. If you need to dodge a thunderstorm, just be sure that the sunny spot you find is in the path of totality.

The Eclipse as Cloud Killer

Your chance of seeing totality is greatly increased by . . . the eclipse itself. Clouds are caused by transparent water vapor rising from the ground to high in the sky. The higher the water vapor goes, the lower the air pressure, and the more the gas expands—which cools it. As the water gas cools, it changes to tiny droplets of water (and ice) suspended in the air—an all-too-visible, non-transparent cloud.

Clouds are always changing, either growing or dissipating. More water vapor may be arriving from below and condensing into droplets. Water droplets elsewhere in the cloud are constantly vaporizing—changing back to transparent water gas. The pretty little cumulus clouds of a summer day will last only 5 to 10 minutes if the updrafts of heat and moisture are shut off. Unless thunderstorms are constantly fed a new supply of hot air and water vapor from below, they will die out in 20 to 30 minutes.

So if there are scattered small cumulus clouds as the partial phase of the total eclipse begins, don't worry, says Jay Anderson. They will disappear. As the Moon blocks more and more of the Sun's light from hitting the land around you, the heating of the ground slows. Less heat and moisture rise into the sky. The clouds stop building. But the water in the cloud continues to change from tiny liquid droplets back to transparent water vapor. The cloud steadily shrinks and vanishes.

A solar eclipse eats clouds. The partial phase of an eclipse can't destroy a thunderstorm, especially if it has begun to rain. And it can't erase a frontal system. But as the Sun becomes an ever thinner crescent and the atmosphere and ground cool, the vanishing Sun helps enormously to clear the sky for the show that's about to happen.

Sunniest Spots Along the Eclipse Route

Late summer weather means that almost everywhere along the eclipse path has a good chance for successful viewing. And excellent highways allow a quick move to a luckier site in case of bad weather.

Here, region by region, are the places most likely to be sunny along the path of totality across the United States, based on years of weather reports for August.[a]

- East of the Cascade Mountains in Oregon, near Madras
- The Columbia Plateau near Ontario in eastern Oregon, near the Idaho border
- The Snake River Plain in western Idaho near the Oregon border
- The Snake River Plain in eastern Idaho, near Idaho Falls and Rexburg
- Central Wyoming east of the Continental Divide, from Riverton to Casper
- Western Nebraska near Alliance
- The confluence of the Mississippi and Ohio Rivers near Carbondale, Illinois and Paducah, Kentucky
- The South Carolina coast just north of Charleston

[a] Based on climatology research by Jay Anderson. See Fred Espenak and Jay Anderson: *Eclipse Bulletin: Total Solar Eclipse of 2015 August 21*.

Clouds and Mountains

You can also use the terrain to influence the weather you experience. Winds generally carry weather systems in the United States to the east. When air encounters a mountain, it flows up the slope, expands, cools, and condenses into clouds. So clouds tend to build up on the windward (usually west-facing) sides of mountains.

If the weather is clear, mountaintops are great places to watch a total eclipse. It's easier to see the Moon's shadow approaching from the west and rushing toward you. After totality, you can see it rushing away to the east. But mountaintops can be risky observation sites—they are cloudier than the valleys below them.

Anderson explains: As air flows up a mountainside, it cools. After crossing the mountain peak, the cool air sinks. As it descends to a lower altitude where the air pressure is greater, the sinking air warms by compression. The tiny water droplets of the cloud vaporize back into transparent water gas as the air warms and dries, clearing or partially clearing the skies on the downwind (leeward) side of mountains, usually the east. So the valleys east of mountains tend to have clearer skies than mountaintops and the western slopes of mountains.

This "rain shadow effect" is most pronounced if the mountains are tall, but even modest mountains produce some cloud clearing on their downwind sides.

Clouds and Beaches

The interior valleys of Oregon offer some of the most promising sky conditions along the eclipse route, but the Oregon beaches are more iffy. The cool Pacific Ocean tends to generate morning fog over the water and beaches, which may or may not have burned off when totality arrives at 10:16 a.m.

Summertime in the American Southeast generates frequent afternoon thunderstorms, although by late August, they are tapering off a bit. Yet the beaches close to the eclipse central line north of Charleston, South

Additional Eclipse-Planning Resources

Fred Espenak and Jay Anderson's *Eclipse Bulletin: Total Solar Eclipse of 2017 August 21* provides additional tables, charts, maps, weather data, and eclipse circumstances for more than 1,000 cities in the United States and elsewhere. Go to astropixels.com/pubs/TSE2017.html for more information.

Espenak has also published *Road Atlas for the Total Solar Eclipse of 2017* with detailed road maps covering the entire path from Oregon to South Carolina. The duration of totality is plotted in 20-second steps, making it easy to estimate the length of the total eclipse from any location in the eclipse path. Visit astropixels.com/pubs/Atlas2017.html for details.

Eclipse chasers aboard the *MS Paul Gauguin* experienced a real cliff-hanger. Hidden behind thick clouds, the Sun emerged into view just as the diamond ring formed at second contact. The total solar eclipse of April 9, 2005 was observed from the South Pacific. [Canon EOS 20D DSLR, 10–20 mm lens at 10 mm, f/4, 1/4 s, ISO 100.] [©2005 Alan Dyer]

Carolina are some of the more favorable observing sites in the region. During the sunlight hours, the land warms faster than the ocean. The rising heat generates clouds over the land. The heat rising from the land also creates lower pressure. The lower pressure sucks in air from the cooler ocean. This sea breeze often pushes the clouds off the beach and a mile or two inland. So, for instance, on a hot summer afternoon, western Charleston can be cloudy, but the area near the shore can be clear.

Clear skies over the beaches north of Charleston would allow observers to provide a fitting send-off for the great eclipse of 2017 as it leaves America and sets sail across the Atlantic, never to touch land again.

NOTES AND REFERENCES

1. Epigraph: Jay Anderson, interview, February 10, 2015.
2. Also available at EclipseWise.com. The path of totality shown on maps in this book and on the Google interactive maps use eclipse calculations by Fred Espenak.

3. El Dorado Weather: map of United States showing "Mean Number of Cloudy Days Sunrise to Sunset."
4. Authors' estimate by adding the populations of towns and cities in and very close to the path of totality.
5. SkippySky excels at graphically mapping cloud forecasts. When you first go to the website, click on the North America region of the map. Near the top of the next page is a list of geographic areas of the USA (North West USA, North Central USA, North East USA, South West USA, South Central USA, and South East USA [*sic*]). Click on the region you are interested in. Then click on "total cloud" on the next line. You will now see a forecast map displaying predicted cloud cover for a particular time in that region. Above the map are a series of times in the future expressed as hours (+6, +9, +12, +15, etc.). Click on these to see cloud cover forecasts for different times in the future. All the times given are in Universal Time (Greenwich Meridian Time) so you have to convert to local time (minus 4 hours for eastern daylight time; minus 7 hours for Pacific daylight time). SkippySky gives you an excellent idea of where clouds may be and where it is clear, especially 24 to 48 hours before the event.
6. For people more familiar with meteorology, Anderson recommends his eclipse weather website Eclipser: https://home.cc.umanitoba.ca/~jander/tot2017/tse17intro.htm—or the instructions he provides at the end of section 5 of *Eclipse Bulletin: Total Solar Eclipse of 2015 August 21* by Fred Espenak and Jay Anderson. Anderson particularly likes to gather information from the College of DuPage weather website: http://weather.cod.edu. Click on "Weather Analysis Tools" and select "Numerical Models." The most useful is "NAM." A menu that drops down on the left side then allows you to select "Precipitation Products," which will give you cloud-cover forecasts.
7. White or silver surfaces (concrete, mirrors) *reflect* more visible light than dark surfaces. Dark surfaces (forests, asphalt) *absorb* more visible sunlight, warm up more quickly than white surfaces, and re-emit the energy they absorbed as infrared radiation—heat—rather than visible light. Heat rising from forests is a major source of storms that provide fresh water for the land, although they challenge eclipse seekers.

A MOMENT OF TOTALITY

Paris to Zambia for 38 Hours to See an Eclipse

by Luca Quaglia

I traveled from Italy, where I grew up, to France to see the great total eclipse of 1999 – and was clouded out. Even though I didn't see the Sun in eclipse, I was stunned by the passage of the Moon's shadow. I promised myself: Wherever the next total eclipse is, I will go!

I did not know the 2001 eclipse would be visible only in southern Africa. At the time I was a student with very little money. Every travel option I found was well out of my price range. But I did not give up. One day I saw an ad for a flight from Paris to Zambia and back in less than 60 hours just to see the eclipse. It was still quite expensive but it only took me 5 minutes to decide. I withdrew almost half the money I had in my bank account and bought that ticket. It was the best money I ever spent.

On June 19, 2001, we left Paris on a chartered overnight flight to Lusaka, capital of Zambia. We arrived on June 20 and headed for a specially built campsite in the midst of the bush north of Lusaka. Night fell and gave me my first view of southern hemisphere stars.

June 21 dawned: eclipse day. It was the middle of the dry season, so the sky was cloudless—perfect. We were told not to leave the campsite, but I wanted a different kind of eclipse experience. I left our group of more than 300 people and walked down a path into the bush.

I reached a small humble village with a dozen tiny mud huts with thatched roofs. It was almost deserted. Only a young boy came my way. I gave him eclipse glasses and we stayed together to watch the eclipse. We found a leafy tree and observed beneath it the images of a hundred crescent Suns. About 15 minutes before totality, as the light was dimming, some cows laid down and started bellowing. A hen and a row of small chicks quietly walked past us, heading for the box where they would roost at night.

Some women with a couple of young children joined us. I gave them eclipse glasses and we all looked at the now very slim crescent Sun.

One minute to go. I did not know what to expect. I had read books and articles. I had stood in the lunar shadow once. But I had never seen the corona.

All of a sudden moving bands of shimmering light wavered across the ground and the walls of the huts—a striking display of shadow bands. The sunlight was fading fast. The lunar shadow on the horizon was rushing toward us. A last ray of sunlight gleamed brightly for one last moment on the limb of the Moon.

We all screamed—the kids, the women, and me. The black disk of the Moon was surrounded by the pearly corona with beautiful tendrils. The pinkish red prominences of the Sun shone above the dark edge of the Moon. Our screaming ended. We fell into rapt silence.

Frogs started croaking in the bush around the village. The boy who had joined me first was the best observer among us. "Stars!" he said suddenly. And there they were—easy to see. The sky around the eclipse was a deep blue with tinges of indigo. The horizon in every direction was bathed in orange.

As abruptly as it started, totality ended. I was so overwhelmed that I cried.

In the evening we flew back to Paris. We had been in Africa only 38 hours.

That was my first time to see totality—an experience etched forever in my mind. Now, many years later, I still think back to that day in that humble African village. I was very lucky to have shared that experience with those kids and those women. It taught me a great lesson: Go see a total eclipse of the Sun. Who would have thought that I would end up in Zambia? Be open to whatever experiences come your way. I've travelled to five more total eclipses since 2001, each of them awesome.

Luca Quaglia has a Ph.D. in physics, conducts research at total solar eclipses, and is completing his master of financial engineering degree at Baruch College, City University of New York.

15

When Is the Next One? Total Eclipses: 2018–2023

Now eclipses are elusive and provoking things . . . visiting the same locality only once in centuries. Consequently, it will not do to sit down quietly at home and wait for one to come, but a person must be up and doing and on the chase.

Rebecca R. Joslin (1929)[1]

There are no total eclipses of the Sun in 2018. Or annulars either. Just three partial eclipses. Perhaps 2018 is our chance to catch our breath—and save up some money to fly off to the next experience of totality. Between 2019 and 2023, four total eclipses beckon us to travel to places on our "maybe someday" list. Or even to places we never thought to go. These foreign lands offer us a chance to enjoy other cultures and unexpected scenery. But now they have an added lure. While there, we have a chance to see again a total eclipse of the Sun, one of nature's most breathtaking, if briefest, sights—a sight that is always different.

July 2, 2019—Chile and Argentina

The shadow of the 2019 total solar eclipse splashes down in the South Pacific Ocean over a thousand miles east of New Zealand and heads for South America 5,000 miles (8,000 kilometers) away. En route it encounters only one speck of land—Oeno, a tiny uninhabited atoll in the Pitcairn Islands. The eclipse path arcs across the Pacific, always south of the equator, and, mid-ocean, offers totality lasting 4 minutes 33 seconds—to sailors and fish.

At last the eclipse shadow hits the shoreline of Chile, with its alternating beaches and cliffs, providing 2 minutes 33 seconds of totality. La Serena (population 400,000) is 25 miles (40 kilometers) south of the central line. As soon as the shadow comes ashore, it begins its climb up the Andes Mountains, second highest mountain chain on Earth. It's also the *longest*

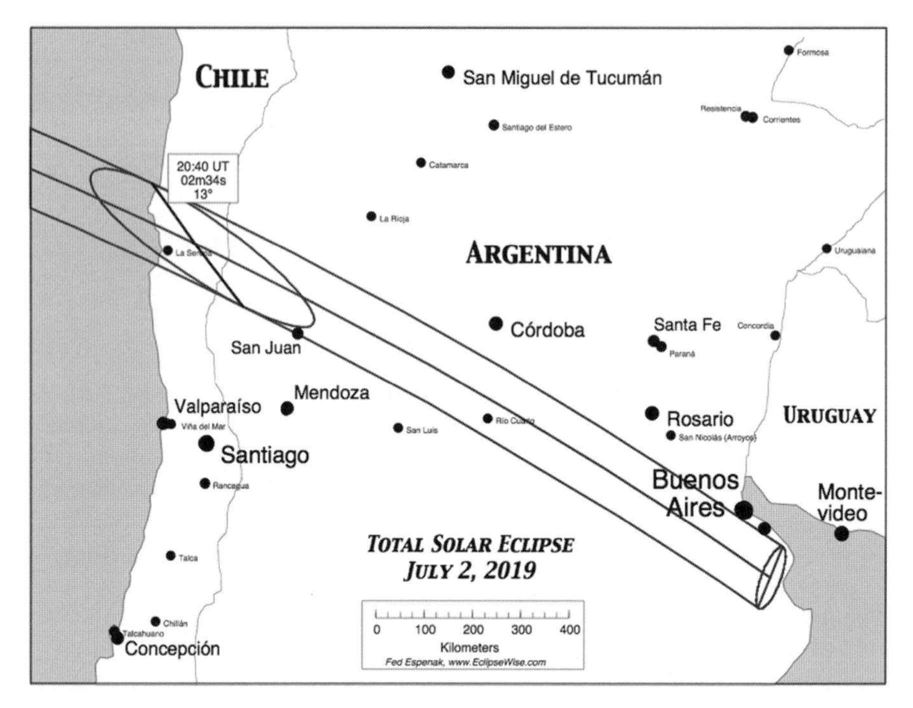

The July 2, 2019, total eclipse track runs diagonally across Chile and Argentina. [Map by Fred Espenak, EclipseWise.com]

mountain chain on Earth that's not buried under water. Just 50 miles (80 kilometers) from the coast, the path of totality collides with Cerro Las Tórtolas, reaching up 20,210 feet (6,160 meters) to greet it. Cerro Las Tórtolas is only 27 feet shorter than Denali (Mount McKinley), tallest mountain in North America, yet it ranks only 52nd among peaks in the Andes. Cerro Las Tórtolas receives 2 minutes 26 seconds of totality.

The eclipse cuts across the southern limits of the Atacama Desert, a place so dry for millions of years that now it plays the role of Mars in films and is used by NASA to test instruments sent to Mars to detect life.[2]

Dry? Dependably clear skies? High altitude? Might you place an astronomical observatory here? Chile is home to seven huge international observatories. The 2019 eclipse visits three of them, dispensing totality of about 2 minutes to each—Cerro Tololo Inter-American Observatory (with a 157-inch [4-meter] telescope), Gemini Observatory (with 319-inch [8.1-meter] and 161-inch [4.1-meter] telescopes), and La Silla Observatory (with two 142-inch [3.6-meter] telescopes). Another observatory, Las Campanas, is just north of the zone of totality.[3]

Chile is a long, skinny country. It stretches 2,700 miles (4,300 kilometers) north to south—almost exactly the distance across the United

States. But Chile averages only 110 miles (175 kilometers) wide. The 2019 eclipse dashes across Chile in less than a minute, enters Argentina, and skids down the Andes onto the Dry Pampas (plains). The central line streaks 80 miles (130 kilometers) southwest of Córdoba (population 1.5 million) and 90 miles (150 kilometers) southwest of Rosario (1.2 million).

2019 might be a truly good-weather eclipse. July 2 is the wintertime south of the equator, but the low-altitude parts of Chile and Argentina visited by this eclipse have overwhelmingly mild, sunny days, with rarely a drop of rain.

Last stop for the 2019 eclipse is Buenos Aires, home to 13 million people. As the Sun begins to set, the northern edge of the eclipse path covers the capital's western and southern suburbs, including the airport. The central line passes 45 miles (75 kilometers) southwest of downtown. Just shy of the Atlantic Ocean, the eclipse shadow slides off the edge of the Earth.

It will return to Chile and Argentina in less than a year and a half.

December 14, 2020—Chile and Argentina Again

Did you enjoy your visit to Chile or Argentina for the 2019 eclipse? Here's a perfect excuse to return to see your new friends: another total eclipse of the Sun, just 17 months later. The United States waited from 1979 to 2017—38 years—for a total solar eclipse to visit its mainland. Chile and Argentina wait less than a year and a half for a return engagement.

But the 2020 eclipse is not a repeat. It's not like a movie you saw last year and now see again. Each eclipse has its own personality—shape of the corona, number and shape and size of prominences, the display of the diamond ring, and Baily's beads. And more: the 2020 eclipse dips farther south, touring different parts of Chile and Argentina. Different regions—and a different season. December is summer in South America.

Like 2019, the 2020 eclipse starts in the middle of the Pacific Ocean and stays south of the equator. But the 2020 eclipse encounters no land at all on its way to South America. It sweeps ashore in Chile 650 miles (1,050 kilometers) south of where the 2019 eclipse reached the coast.

There are no beach resorts here, just farms, farmhouses, and tiny villages. Seventy miles (110 kilometers) inland, the central line of the 2020 eclipse runs close to Villarrica and Pucón, resort communities situated on Lake Villarrica with looming Mount Villarrica in the background—a volcano. It last erupted on March 3, 2015, and previously in 1985, 1971, and 1964. No injuries were reported. Villarrica is one of the few volcanoes to have a lake of lava within the crater at its peak.[4] The eclipsed Sun and Moon will look down into the mouth of the volcano: the central line of

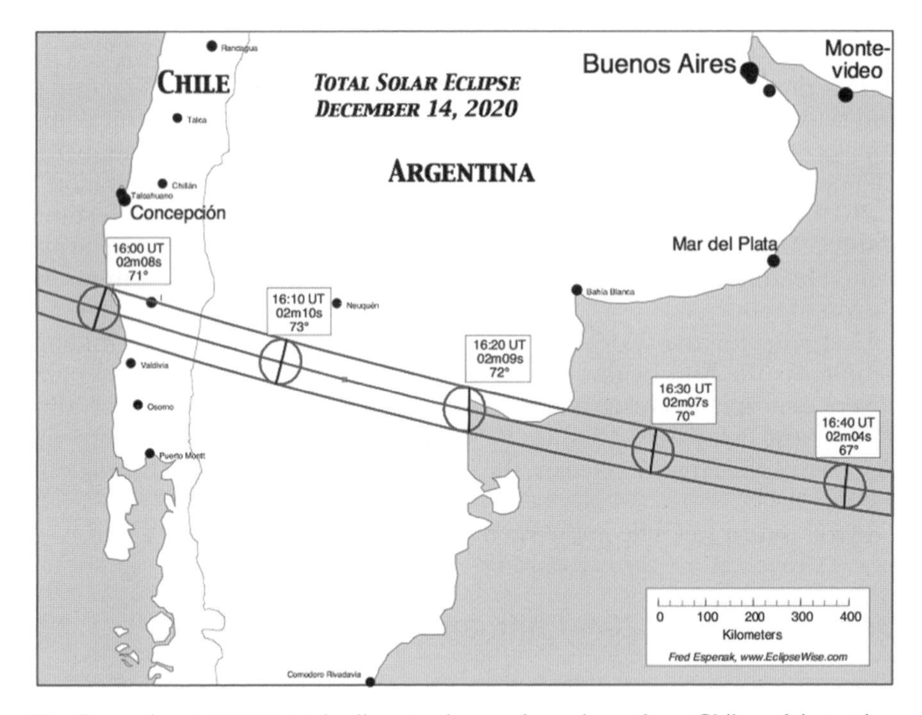

The December 14, 2020, total eclipse track runs through southern Chile and Argentina. The boxed labels along the track give the time (Universal Time or Greenwich Mean Time), the duration of totality, and the altitude of the Sun. [Map by Fred Espenak, EclipseWise.com]

the eclipse passes only 5 miles (8 kilometers) to the north. The lava boiling there can reach a temperature of nearly 2,200°F (1,200°C)—about one-fifth the surface temperature of the Sun. The volcano (a national park) and the two resort towns that admire it from a distance all receive 2 minutes 8 seconds of totality.

December is part of the dry season in this part of Chile, suggesting clear skies for this spectacle.

The eclipse speeds on southeastward, the central line passing just north of Villarrica's two companion volcanoes—Quetrupillán and Lanín. The volcanic trio are about 15 miles (25 kilometers) from one another and are part of a fault system that roughly parallels the track of the eclipse.

These three volcanoes remind us that Chile lies on the Pacific Ring of Fire, where tectonic plates, vast blocks of the Earth's crust and upper mantle, are colliding, pushing up mountains and creating earthquakes and volcanoes. Wherever there is seismic activity, there is gorgeous landscape.[5]

At Lanín, tallest of these three volcanoes, the eclipse crosses the border into Argentina. Hello, Patagonia.

Patagonia is a knee-length sock on the southern foot of South America. It's a dry, sparsely populated region in Argentina and Chile. In December, this northern part of Patagonia is warm and sunny.

One hundred ten miles (180 kilometers) into Argentina, the eclipse crosses the Limay River, offering 2 minutes 9 seconds of totality. The river runs a braided course where the central line wades across. But to the north and south, within the path of totality, are two canyon lakes formed by dams on the Limay. Lake Limay, to the south, has the higher and more colorful canyon walls.

The eclipse rushes on toward the coast of Argentina. About half-way between its Limay River crossing and the Atlantic Ocean, the eclipse reaches the point where its totality lasts longest—2 minutes 10 seconds.

At the coast, two small cities stand ready to bid the eclipse buen viaje (bon voyage) as it sets off across the ocean. The port town of San Antonio Oeste (San Antonio West; population 17,000) receives 1 minute 55 seconds of totality. The central line lies 28 miles (45 kilometers) by highway to the south, with 2 minutes 9 seconds of totality.

The eclipse then skims across a corner of the San Matias Gulf before its central line touches land one last time 40 miles (65 kilometers) southwest of Viedma. Viedma (population 55,000) stands at the northern boundary of the path of totality. The city lies near the mouth of the Black River (Rio Negro), the continuation of the Limay River that the eclipse crossed at the other side of Patagonia.

The eclipse of 2020 will not touch land again. And before it reaches Africa, it will lose touch with Earth.

December 4, 2021—Antarctica

You've been to North America, South America, Europe, Africa, Asia, and Australia—or you are intending to travel there in the future. A world traveler: all the continents! Except for one—Antarctica. A total eclipse of the Sun is a perfect excuse to go there. And you can. Eclipse tour companies offer cruises into the path of totality along the Antarctic coast and some even offer expeditions ashore in Antarctica.

The bottom of the world is staging this eclipse on December 4, summertime on the ice continent. Chilly, yes. Unbearable, no. Think of the lifetime of stories you will gather. You will have been to a place where fewer than one in a million people on Earth have ever ventured—and even fewer have seen a total eclipse there in the southern land of the midnight Sun. Back home, might you write an article for a local newspaper or magazine about what you have experienced, perhaps using photography you shot, thereby slightly defraying the cost of your travels? Probably local television and radio stations and local magazines and newspapers will be eager to interview you. "You did *what*?"

The continent of Antarctica is almost entirely buried under ice. The ice cap averages 7,000 feet in depth—1⅓ miles (2.1 kilometers). Here in Antarctica you are standing on 90% of the world's ice—and the ice is melting faster than falling snow can replace it. Within that ice is 70% of all the fresh water on Earth.

The total eclipse of 2021 starts in the South Atlantic Ocean, arcs across Antarctica south of the tip of South America, and ends in the South Pacific Ocean. It ventures inland from the coast as much as 205 miles (330 kilometers), but never gets closer to the South Pole than about 600 miles (975 (kilometers)).

If you are in Antarctica near the coast, the average daytime high temperature in early December is about 1° below freezing. The average nighttime low temperature is . . . well, there is no nighttime in Antarctica in the summer.[6] But the Sun does drop lower in the sky and the temperature cools on average to 19°F (−7°C).

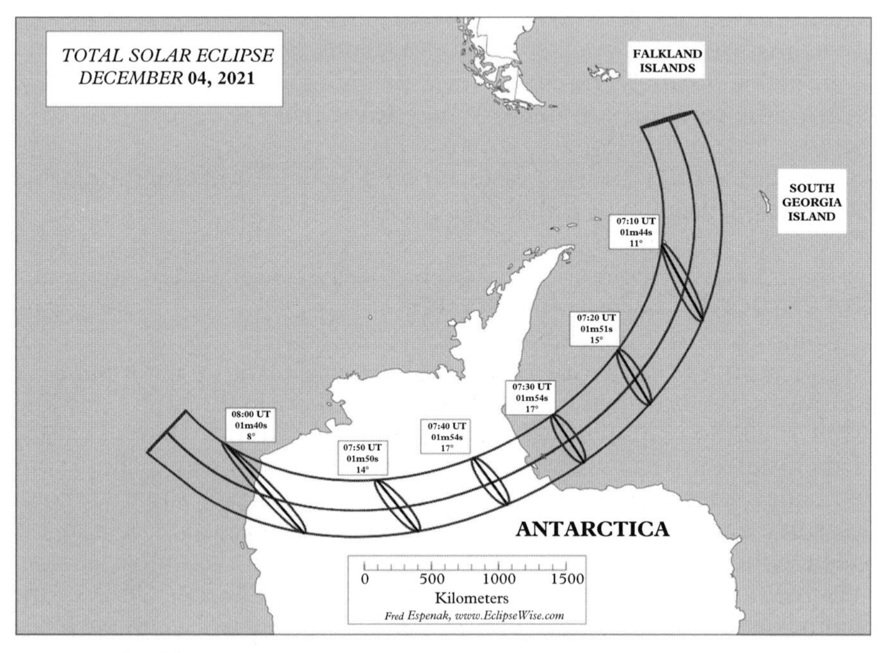

The December 4, 2021, total eclipse track runs through Antarctica and the Southern Ocean. The boxed labels along the track give the time (Universal Time or Greenwich Mean Time), the duration of totality, and the altitude of the Sun. [Map by Fred Espenak, EclipseWise.com] This map should be somewhere within or at the end of the section on the 2021 eclipse in Antarctica.

The chances of seeing an eclipse in Antarctica or from the ocean nearby is good. Antarctica is a desert—the driest place on Earth.

April 20, 2023—Australia and Indonesia

Antarctic eclipses are a little expensive, so the Sun and Moon give you a year off to replenish your eclipse-questing budget. No total—or annular—eclipses in 2022.

2023, though, brings a hybrid eclipse—the rarest kind. Over a 5,000-year period, from 2000 BCE to 3000 CE, only 4.8% of eclipses are hybrid.[7]

Hybrid eclipses usually start off annular, become total, and then revert to annular. The April 20, 2023 eclipse begins in the South Indian Ocean as an annular eclipse. The Moon is a little too far away to completely cover the disk of the Sun. But as the shadow travels northeastward, it rides up the roundness of the Earth, which brings the surface of the Earth closer to the Moon, making the Moon a little larger in the sky, just enough to completely hide the surface of the Sun. Less than 1 minute after the 2023 eclipse touches down in the Pacific and rushes northeastward, the surface of the Earth is close enough to the Moon that the eclipse becomes total—for a fraction of a second. Mile by mile as the shadow sweeps on, totality lasts longer. By the time it reaches Australia, the total phase of the eclipse lasts 1 minute 1 second.

The eclipse of 2023 barely nips Australia, slicing across an outstretched, upraised finger of land, the North West Cape. It travels on land only 37 miles (60 kilometers). But what a peninsula. It's as if the Sun has chosen to reward a permanent admirer. Close to the central line is Learmonth Solar Observatory, a collaboration between Australia and the United States. The site for this observatory of the Sun was chosen for its abundant sunshine. In April the weather is warm and dry.

At Learmonth, scientists use a broad array of instruments to monitor sunspots and solar storms, providing warnings of coronal mass ejections likely to strike the Earth and disrupt communications (while providing lovely displays of the aurora—the northern and southern lights). Here too, researchers detect shock waves (vibrations) coming from the Sun's interior—solar quakes—that reveal the Sun's interior just as earthquakes reveal the inner structure of the Earth.

Learmonth Solar Observatory stands on the beach on the eastern side of the peninsula, right next to a Royal Australian Air Force base that doubles as the airport for the town of Exmouth just up the coast. Both Exmouth and the observatory lie close to the central line of the eclipse and experience 56 seconds of totality.

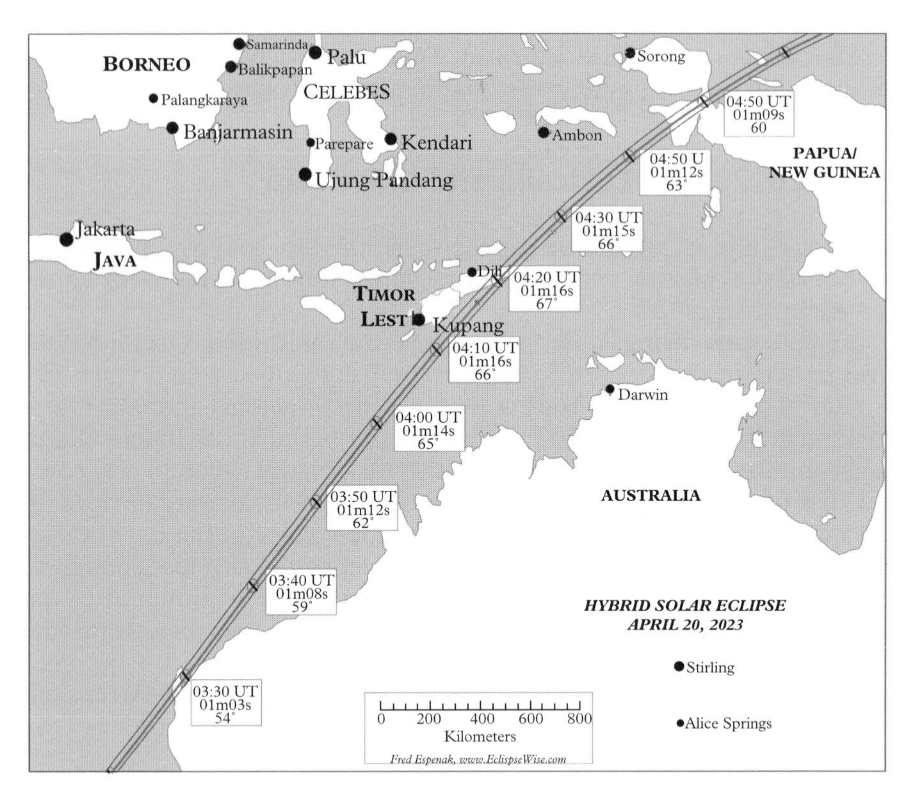

The April 20, 2023, hybrid eclipse track provides totality that runs through Australia and Indonesia. The boxed labels along the track give the time (Universal Time or Greenwich Mean Time), the duration of totality, and the altitude of the Sun. [Map by Fred Espenak, EclipseWise.com]

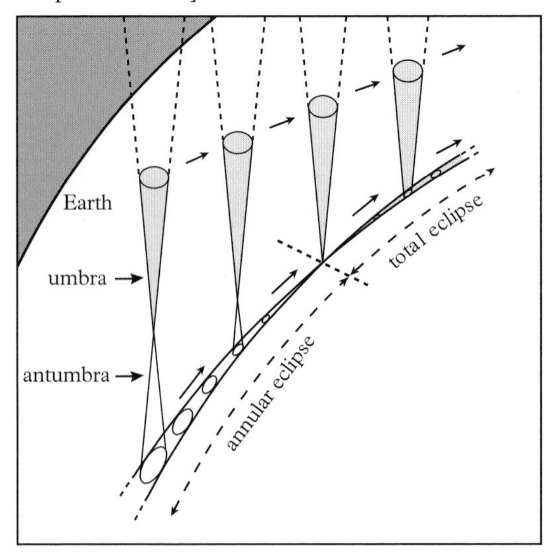

An occasional solar eclipse may start off annular but become total as the roundness of the Earth reaches up to intercept the shadow. The eclipse then returns to annular as the curvature of the Earth causes its surface to fall away from the shadow. These unusual eclipses are called hybrid or annular-total eclipses.

Just a few miles to the west, across the peninsula, where the eclipse reaches land for the first time, lies Cape Range National Park and, just offshore, Ningaloo Reef, a World Heritage Site.

Past Australia now, the eclipse of 2023 treads across the Indian Ocean 1,400 miles (2,200 kilometers) from Australia to Timor-Leste (East Timor), a country that occupies half of the island of Timor. Timor-Leste was a Portuguese colony that became part of Indonesia before gaining independence in 2002. The western half of the island remained with Indonesia.

Just before the eclipse reaches Timor-Leste, its duration of totality is longest: 1 minute 16 seconds. This is a short eclipse. There is less time to stare at the corona, but a little more time to admire the diamond ring effect and Baily's beads and the Sun's red chromosphere and prominences—they stay in view longer.

The path of total eclipse glides across the eastern end of this island country. The largest city in Timor-Leste is Dili (population 250,000), on the northern coast, far west of the path of totality. About 75 miles east of Dili, near the northern coast, is Baucau, second largest city in Timor-Leste (population 16,000). From there, it's 50 miles (80 kilometers) farther east by coastal road to intercept the central line of the eclipse. Or, if the weather is better on the south coast, it's a drive of 45 miles (75 kilometers) from Baucau over the mountains to greet the eclipse just after it comes

Annular Eclipses 2018–2023

2019 December 26 — 3 minutes 39 seconds

Saudi Arabia, Qatar, United Arab Emirates, Oman, India, Sri Lanka, Malaysia, Indonesia, Philippines, Guam

2020 June 21 — 0 minutes 39 seconds

DR Congo, Central African Republic, South Sudan, Ethiopia, Eritrea, Yemen, Saudi Arabia, Oman, Pakistan, India, China, Taiwan

2021 June 10 — 3 minutes 51 seconds

Canada, Greenland, North Pole, Russia (Siberia)

2023 October 14 — 5 minutes 17 seconds

Pacific Ocean, United States (Oregon, California, Nevada, Utah, Arizona, Colorado, New Mexico, Texas), Mexico, Belize, Honduras, Nicaragua, Costa Rica, Panama, Colombia, Brazil

The 2023 Annular Eclipse Visits America

On October 14, 2023, an annular eclipse will be visible in the western United States, continuing through Mexico, Central America, Colombia, and Brazil. Here is the track for the portion of the annular eclipse that crosses the United States. The event provides a chance to compare an "almost total" with a true total eclipse of the Sun that occurs less than 6 months later, on April 8, 2024, for parts of Mexico, the United States, and Canada.

An annular eclipse occurs because the Moon is a little too far from Earth to completely cover the Sun, leaving some of the Sun's bright disk visible around the rim of the Moon—a "ring of fire." About 91% of the Sun's disk will be occulted by the Moon.

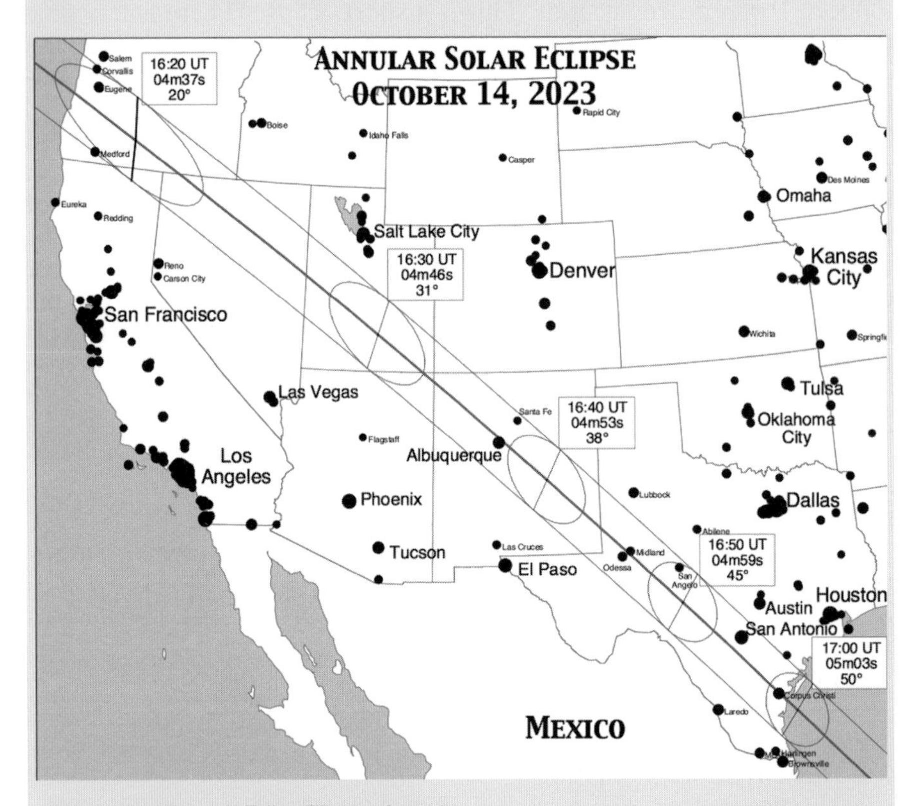

[Map by Fred Espenak, EclipseWise.com]

Annular eclipses are a special, more dramatic form of partial eclipse. But they are still *partial eclipses*. No corona is visible. Solar filters must be used throughout the eclipse.

The boxed labels along the track give the time (Universal Time or Greenwich Mean Time), the duration of totality, and the altitude of the Sun. Depending on your exact location, the "ring of fire" annular phase will last from 4½ to over 5 minutes.[1]

For more information, see http://eclipsewise.com/solar/SEprime/2001–2100/SE2023Oct14Aprime.html

ashore. Either way, totality on the central line lasts the maximum—1 minute 16 seconds.

Timor-Leste is a tropical country with hot and humid coastlines. April marks the transition from the wet season to the dry.

The path of 2023 totality then traverses the Bardar Sea (part of the Pacific) for the next 500 miles (800 kilometers) until it reaches the island of New Guinea. New Guinea is the second largest island in the world. Greenland is the biggest. The eastern half of New Guinea is the nation of Papua New Guinea. The western half is part of Indonesia.

On a map, the island of New Guinea looks like a turkey seen from the side, with its head and long neck (wattle included) stretched out to the west. This part of Indonesian New Guinea is called the Bird's Head Peninsula—and the total eclipse of 2023 slices across the bird's neck (and wattle).

The path of totality crosses 140 miles (225 kilometers) of land but touches no cities. Almost all of New Guinea is covered by dense jungle. Roads are few. This eclipse is hard to reach. It may also be hard to see in this part of Indonesian New Guinea. As in Timor-Leste, the hot, tropical climate has a wet season and a dry season, with April as a month of transition. But rain is common year round.

This eclipse hurries on northeastward, touching a few small Indonesian islands on its way to the Central Pacific Ocean, declining to a total eclipse of 0 seconds, then becoming an annular eclipse a few seconds long before it departs from Earth.

The eclipse of 2023 is rare in being a hybrid—annular at beginning and end; total in the middle. It is also quite shy. It avoids people. The eclipse plants its footprint of totality in three countries—Australia, Timor-Leste, and Indonesia (Western New Guinea), but visits only the most remote, least inhabited portions, bringing the spectacle of totality to the homes of fewer than a quarter million people. The eclipse of 2021 in Antarctica is even shyer. No one lives there permanently.

And now, the total eclipse of 2024. It will be different . . .

NOTES AND REFERENCES

1. Epigraph: Rebecca R. Joslin: *Chasing Eclipses: The Total Solar Eclipses of 1905, 1914, 1925* (Boston: Walton Advertising and Printing, 1929), pages 1–2.
2. Jim Farber: "Can't Get to Mars? Try Chile's Atacama Desert, the driest place on earth and the most otherworldly," *New York Daily News*, January 16, 2012. Kenneth Chang: "Life on Mars? Could Be, But How Will They Tell?" *New York Times*, March 29, 2005. "*Space Odyssey* (TV Series)," Wikipedia (online encyclopedia).

3. In 1991, a total solar eclipse enveloped the observatory complex atop Mauna Kea in Hawaii—seven optical telescopes with mirrors ranging in diameter from 120 inches (3 meters) to 400 inches (10 meters).
4. The lava lake is listed at intermittent.
5. Ninety percent of the world's earthquakes and 75% of the world's active volcanoes lie along the Pacific Ring of Fire. This belt of seismic activity runs along the Pacific coast of South America, Central America, Mexico, the 48 contiguous United States, Canada, Alaska, Japan, the Philippines, Indonesia, and New Zealand.
6. On December 4, the Sun does not set anywhere along the path of totality traced by the 2021 eclipse. Only the Antarctic Peninsula, the stub of Antarctica jutting toward South America, pokes out above the Antarctic Circle. North of the Antarctic Circle, the Sun will set or rise at least briefly every day of the year.
7. Fred Espenak and Jean Meeus: *Five Millennium Canon of Solar Eclipses: −1999 to +3000 (2000 BCE to 3000 CE)* (Greenbelt, Maryland: NASA [NASA/TP-2006-214141], 2006), page 19.

○◐●●◑

A MOMENT OF TOTALITY

Vacation Planning

"If you see one total solar eclipse, you know where you'll be going for vacations for the rest of your life," says Sheridan Williams, retired British rocket scientist.

"In 2028, I'll be in Australia, looking at the eclipse with the Sydney Harbor Bridge in the foreground.

"For the eclipse of 2037, I'll be on the North Island of New Zealand. And I think I'll stay there a year and a half to see the eclipse of 2038 that crosses the path of 2037. Total eclipses in the same place in consecutive years. I'll be 90 years old at the time."

Interview, May 23, 2015.

16

Coming Back to America: The Total Eclipse of 2024

Few can imagine how much I longed for another minute, for what I had witnessed seemed very much like a dream.

Edwin Dunkin (1851)[1]

Party animal. That's the total solar eclipse of April 8, 2024. It visits many big cities—seems to be genuinely fond of people. And it's generous. As a party favor, it hands out more than 4 minutes of totality along its central line throughout Mexico and much of the United States and never less than 3 minutes until the party ends in Newfoundland. And as it sweeps across North America, it spreads the party widely. The swath of totality is as wide as 126 miles (202 kilometers) in Mexico and never less than 101 miles (163 kilometers) through the United States and Canada—a much wider band of totality than the 2017 eclipse. Party, party. More people are invited.

In Mexico, Durango lies within the zone of totality. In the United States, San Antonio, Austin, Dallas, Fort Worth, Indianapolis, Cincinnati, Cleveland, Buffalo, and Rochester have the honor of lying wholly or partly within the path where the eclipse is total. In Canada, Hamilton and Montreal. Many middle-size and small cities are equally blessed. Other major cities lie just outside of the path of totality. If people are shy about going to a total eclipse, this one comes to them.

The totality of the 2024 eclipse first touches Earth south of the equator about halfway between Australia and Mexico. Its darkness flies northeastward across the Pacific, visiting only one island, Socorro, an active volcano that is part of Mexico.

It sweeps ashore in mainland Mexico at Mazaltlán. This is Mazatlán's second total eclipse in 33 years. It played host to the 1991 eclipse. For cities that sit still and wait, a total eclipse usually comes along once in 375 years.

The April 8, 2024 eclipse track runs diagonally across the United States. [Map ©2016 Michael Zeiler, GreatAmericanEclipse.com]

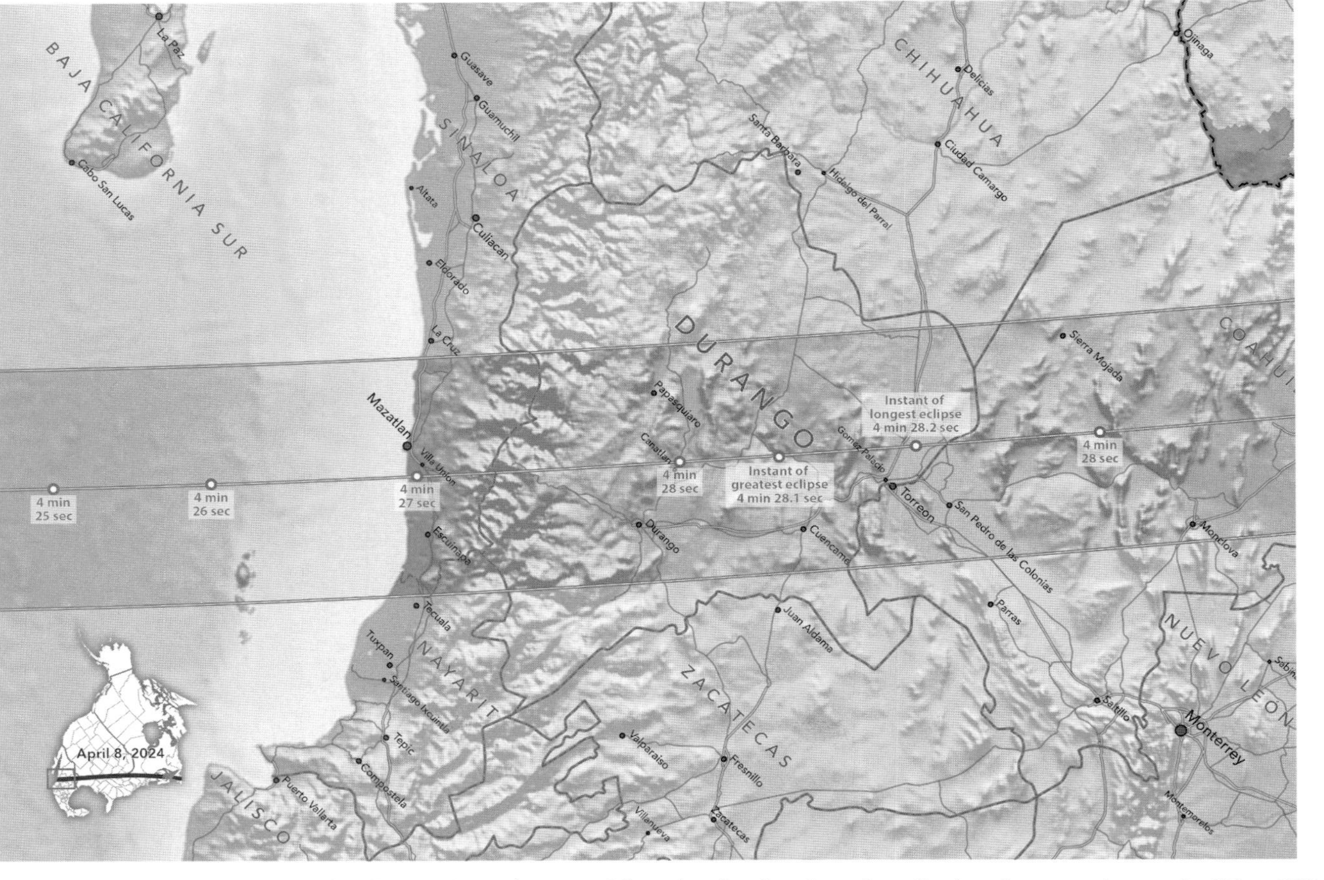

The path of totality as it crosses Mexico. Points along the central line give the duration of totality in minutes and seconds. [Map ©2016 Michael Zeiler, GreatAmericanEclipse.com]

The Moon and Sun, both fond of bright lights, grant El Faro, the historic lighthouse on a peak at the south end of downtown Mazatlán, 4 minutes 19 seconds of totality. The central line of the eclipse passes 18 miles (30 kilometers) south of downtown, offering an extra 8 seconds in the shadow of the Moon. April is a comparatively dry month in Mazatlán.

The 2024 eclipse heads northeast, up into the Sierra Madre Occidental, a southward continuation of the Rocky Mountains. From Mazatlán to Durango, the eclipse embraces scenic Mexican Highway 40D. In 143 miles (230 kilometers) of sensational canyons, the expressway passes through 61 tunnels and across 115 bridges. Most spectacular is the leap over the Baluarte River gorge on a cable-stayed bridge suspended 1,280 feet (520 meters) above the river. The Baluarte Bridge is the highest roadway bridge in North America.

Next, the eclipse arrives in Durango, largest of the Mexican cities on its itinerary, with a metropolitan population of 1.7 million. Durango lies 30 miles (50 kilometers) southeast of the central line. In the western part of downtown, amid colonial buildings and plazas, is an aerial tramway that carries passengers to the top of a hill 270 feet (82 meters) high with a fine view of the city—and the approach of the eclipse shadow from the west that will provide a total eclipse of 3 minutes 51 seconds.

Highway 40D continues on northeastward from Durango to Torreón, always within the zone of totality, but well south of the central line. Along the central line, about two-thirds of the way between Durango and Torreón, the 2024 eclipse reaches its point of longest duration—4 minutes 28 seconds. Durango and Torreón lie between the western and eastern ranges of the Sierra Madre. April is the second driest month of the year for this region.

The eclipse bounds over the Sierra Madre Orientale to reach the Rio Grande, which provides a watery boundary between Mexico and the United States for over a thousand meandering river miles. Straddling the border, almost on the central line of the eclipse, are Piedras Negras and much smaller Eagle Pass. On the International Bridge connecting them, totality lasts 4 minutes 25 seconds.

The Eyes of Texas Are Upon Totality

The 2024 eclipse gives a big Texas hug to four of its five largest cities. It starts by sideswiping San Antonio. The western and northern suburbs of the city fall within the southern limit of the eclipse. The central line crosses Interstate Highway 10 about 75 miles (120 kilometers) northwest of the San Antonio, just west of Kerrville, with 4 minutes 26 seconds of totality.

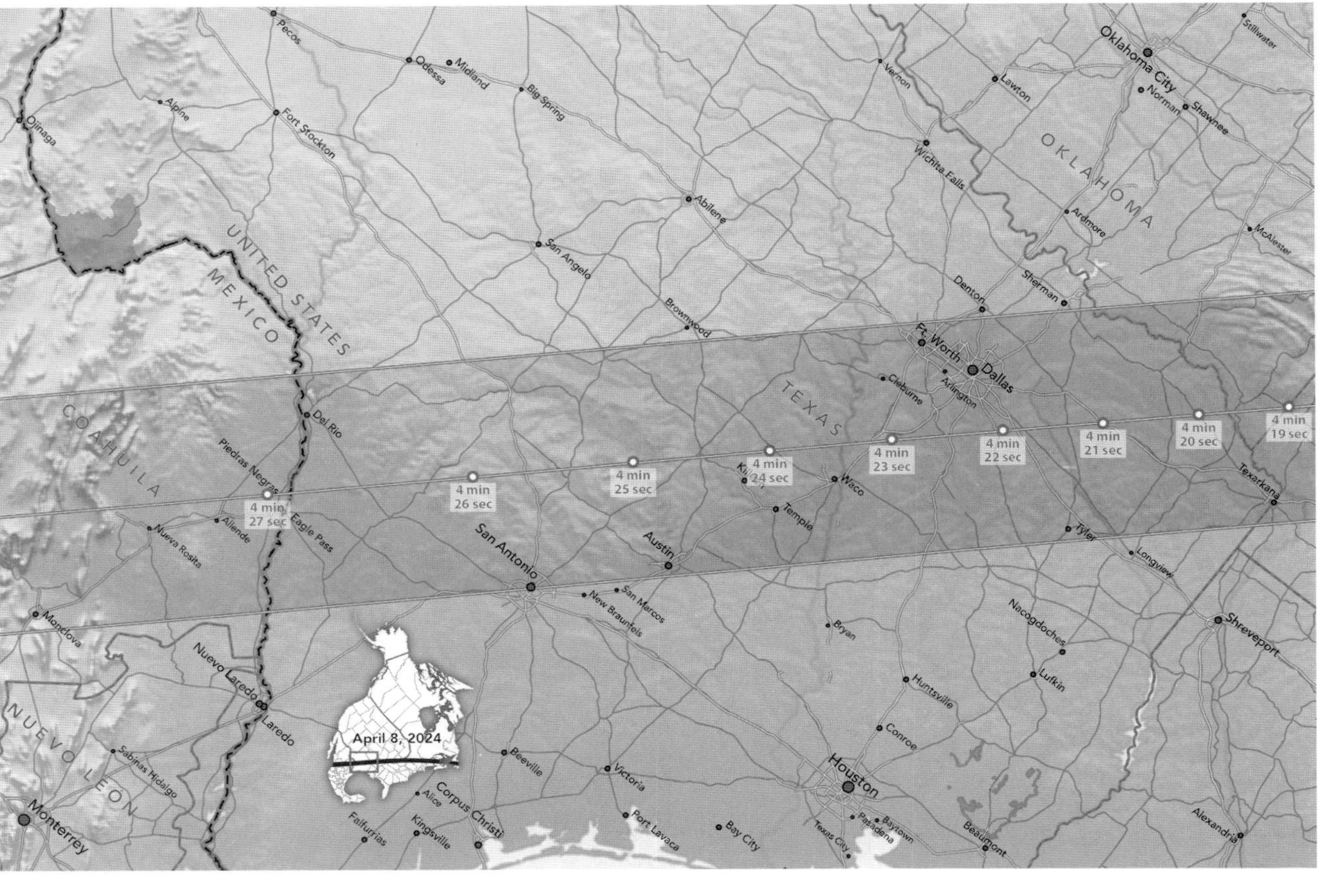

The path of totality as it crosses Texas. Points along the central line give the duration of totality in minutes and seconds. [Map ©2016 Michael Zeiler, GreatAmericanEclipse.com]

Next along the eclipse route is Austin, a second American city with a metropolitan population over 2 million. All of Austin's city limits lie within the path of totality, although at the southern edge of the zone. The Texas State Capitol Building receives a tax-free 1 minute 52 seconds of celestial umbrage.

Flying across Texas, the central line of the eclipse passes just north of Killeen and Waco, bringing times of total eclipse of 4 minutes 24 seconds for those who walk the line.

For 200 miles from Austin to Dallas and Fort Worth, Interstate Highway 35 runs continuously within the path of totality. Within the Dallas-Fort Worth metropolitan area—fourth largest in the United States—live 7 million people who will be completely submerged in the Sun-cast shadow of the Moon. The central line passes about 35 miles (60 kilometers) to the southeast and east of downtown Dallas, bringing 4 minutes 22 seconds of totality.

Interstate Highway 30 escorts the eclipse 320 miles from Dallas to Little Rock. Almost halfway between the two cities, the eclipse crosses the meandering Red River where Oklahoma and Arkansas border Texas. The Red River got its name from the soil it carries downstream from its headwaters in the Texas panhandle.

On its diagonal path across Arkansas and the southeastern corner of Oklahoma, the total eclipse envelops all of Ouachita National Forest as its stretches from the middle of Arkansas into Oklahoma. It is the largest national forest in the eastern half of the United States.

Toward the southern edge of the zone of totality is a second American state capital for the 2024 eclipse: Little Rock, Arkansas. From the capitol building, the total eclipse lasts 2 minutes 31 seconds. Interstate Highway 40 runs northwestward from Little Rock toward Fort Smith. Almost halfway in between, the central line of the eclipse dashes across the highway and totality lingers for 4 minutes 16 seconds. The length of totality is slowly declining, but 2024 is still a long total eclipse.

As the eclipse path leaves Arkansas, its southern edge nips a tiny northwestern corner of Tennessee at a tight horseshoe bend in the Mississippi River. At the top of that bend is New Madrid, Missouri, close to the epicenter of four of the most powerful earthquakes in United States history. These earthquakes, all close to magnitude 8, occurred between December 1811 and February 1812. The last, on February 7, 1812, created ground upheaval and shock waves so violent that the Mississippi River in that region flowed backward for several hours. The seismic jolt damaged many homes in St. Louis, 146 miles (235 kilometers) away, and shook church steeples in New York and Boston enough to set church bells clanging—at a distance of 1,000 miles (1,600 kilometers) and more. The US Geological Survey expects major earthquakes in the future in the New Madrid area.

Déjà Vu All Over Again

As it prepares to cross the junction of the Mississippi and Ohio Rivers, the total solar eclipse of 2024 crosses the path of the total solar eclipse of 2017, bringing totality to a region twice in 6⅔ years, a rare gift. For most of the larger communities within this region of almost 8,840 square miles (22,900 square kilometers), the eclipse of 2024 brings even longer totality.

Paducah, Kentucky is a twice-blessed city although it lies near the southern edge of totality for 2024. Paducah is home to the National Quilt Museum and late each April hosts an international quilt festival.

Across the Ohio River from Paducah and to the west is Metropolis, Illinois, home of Superman—that is, a 15-foot-tall statue of him. It stands in Superman Square. Two blocks away there is a statue of Lois Lane, less than half as tall. Metropolis was founded a century before the comic-strip character appeared on magazine racks. Superman, disguised as a newspaper reporter for the *Daily Planet*, constantly saved his great city of Metropolis. The newspaper of the real city is the *Metropolis Planet*.[2]

Across the Ohio River from Paducah and to the east is Kincaid Mounds State Historic Park.[3] Here, almost a thousand years ago, Native Americans of the Mississippian Culture built a trading and ceremonial center with at least 11 mounds. One mound was 30 feet tall with a flat top larger than a football field. Another, 20 feet tall, had a flat top closer to two football fields in size. At the top of each mound was a building—some

Sampling of Cities that Share Totality in 2017 and 2024

Length of Totality in Minutes and Seconds		
	2017	**2024**
Missouri		
Cape Girardeau	1m 46s	4m 5s
Ste. Genevieve	2m 40s	2m 40s
Illinois		
Chester	2m 40s	3m 23s
Carbondale	2m 37s	4m 8s
Marion	2m 31s	4m 7s
Metropolis	2m 23s	2m 37s
Kentucky		
Paducah	2m 20s	1m 41s

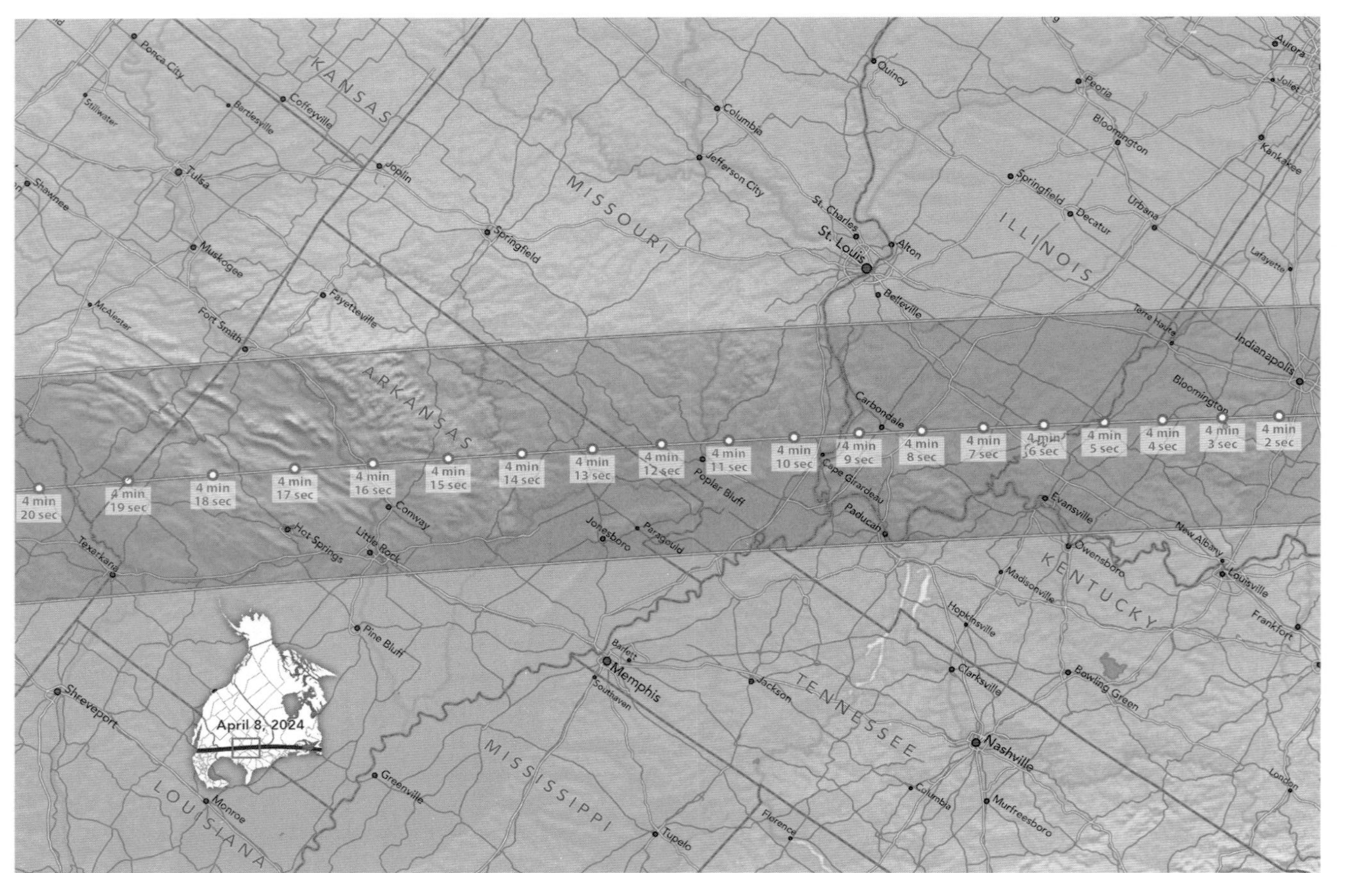

The path of totality as it crosses Arkansas, Missouri, Illinois, Kentucky, and Indiana. Points along the central line give the duration of totality in minutes and seconds. [Map ©2016 Michael Zeiler, GreatAmericanEclipse.com]

for ceremonies, some for important families. In the good soil along the river the people grew corn. The Mississippian Indians abandoned their settlement about 1450 CE, perhaps because of drought or a long cold spell or because they had used up the local wildlife and the trees upon which the settlement depended.[4]

As the central line of the 2024 eclipse slides into Illinois, it cuts across the central line of the 2017 eclipse. The crossing occurs south of Carbondale, 5 miles (8 kilometers) northwest of the town of Makanda, near the western shore of Cedar Lake in Giant City State Park.[5] The name Giant City comes from sandstone bluffs that tower like skyscrapers above a trail.

Where the two eclipse paths cross, 2017's totality lasts 2 minutes 40 seconds. Totality for 2024 lasts 4 minutes 9 seconds.[6] The central line then passes between Carbondale and Marion—party hub for celebrating two total eclipses so close together.

The eclipse of 2024 does not linger. Its central line darts across the Wabash River into Indiana, rushes into Vincennes, and dashes up 4th Street through the heart of town at 2,000 miles per hour (3,200 kilometers per hour). The central line passes a block south of the Red Skelton Museum of American Comedy. The great comedian was born in Vincennes.

Up the Ohio River and along the eclipse route about 130 miles (210 kilometers) from Paducah, Kentucky is Evansville, Indiana. Both are river cities and both are situated near thousand-year-old American Indian ceremonial and trading centers that are National Historic Landmarks. Kincaid Mounds, near Paducah, and Angel Mounds, near Evansville, were created by Native Americans of the Mississippian Culture who built cities throughout the South and Midwest. Their trade in pottery and other products reached east to the Atlantic coast and west to the Rocky Mountains. Angel Mounds was home to about 1,000 people. Like Kincaid Mounds, it rose about 1050 CE and was abandoned about 1450.

Almost on the central line of the eclipse—less than 1 second of totality to the south—is Bloomington, Indiana, and the main campus of Indiana University. Most of the buildings are made of limestone quarried a few miles south of town. This Indiana limestone is also the face of the Empire State Building, the Pentagon, and 35 of the 50 state capitol buildings in the United States.

Indianapolis, capital of Indiana, lies just north of the central line, but close enough that the entire metropolitan area of 1.8 million people lies inside the zone of totality. The Sun and Moon will look down on the Indianapolis Motor Speedway seven weeks before the Indianapolis 500 race is run. The highest speed for one lap in an Indianapolis 500 race was 236 mph (380 km/h). As the eclipse rolls over Indianapolis, its shadow is traveling 1,990 mph (3,200 km/h). No pit stop.

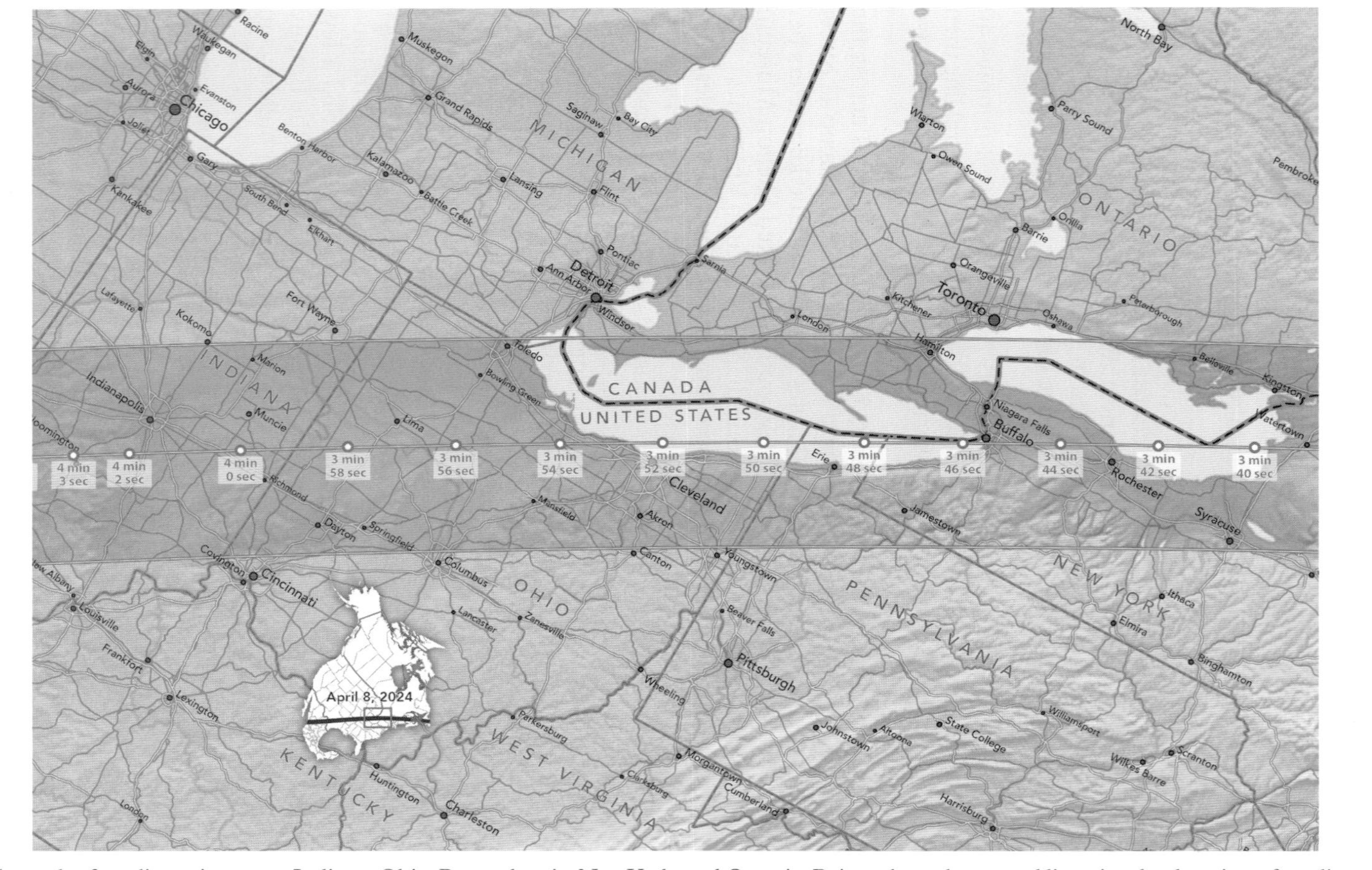

The path of totality as it crosses Indiana, Ohio, Pennsylvania, New York, and Ontario. Points along the central line give the duration of totality in minutes and seconds. [Map ©2016 Michael Zeiler, GreatAmericanEclipse.com]

Ohio-o-o-h

Just before the central line of the 2024 total eclipse leaves Indiana and enters Ohio, the duration totality falls below 4 minutes. The path of totality skirts just northwest of Cincinnati and Columbus, but it does honor Dayton, birthplace of the airplane, with 2 minutes 44 seconds of totality. Wilbur and Orville Wright would be delighted that millions of people have used their invention to fly around the world to station themselves along eclipse paths. And more. April showers are frequent along much of the 2024 eclipse route though the United States and Canada. The Wright brothers made it possible for people to take to the air when necessary to see total eclipses above the clouds.

The central line of the 2024 eclipse passes very close to downtown Cleveland—just 9 miles (15 kilometers) offshore in Lake Erie. Along the shoreline, standing side by side, are the Cleveland Browns' football stadium, the Great Lakes Science Center, the Rock and Roll Hall of Fame, and a World War II submarine. Professional sports teams in Cleveland may not have won a championship since 1964,[7] but the celestial powers of the Sun and Moon are more gracious, presenting Cleveland with 3 minutes 50 seconds of totality.

As the eclipse of 2024 approaches Cleveland, it clips the southeasternmost corner of Michigan, but misses Detroit, skips Canadian customs at the border, and enters Ontario, bringing roughly 2 minutes of totality for Canadians along the edge of Lake Erie from Leamington to Hamilton. Almost all of Lake Erie falls under the shadow of the Moon.

Along Lake Erie's southern shore, the path of totality visits northern Pennsylvania. Halfway between Cleveland and Buffalo, Erie, Pennsylvania will thrill to an eerie darkness lasting 3 minutes 42 seconds.

Over the Falls

The central line of the 2024 eclipse climbs out of Lake Erie and stampedes straight through Buffalo, New York, 2½ miles (4 kilometers) south of downtown. With 3 minutes 45 seconds of totality, downtown is less than 1 second off the central line.

Running north out of town is the very short Niagara River, only 36 miles (58 kilometers) in length. It drains Lake Erie into Lake Ontario. But halfway downstream, it flows around an island and tumbles off two cliffs. One cliff is on the American side; the other, horseshoe in shape, is mostly in Canada. Together they are Niagara Falls. The water drops 170 feet (52 meters) with the highest flow rate of any waterfall in the world. The water in Lake Superior, Lake Michigan, Lake Huron, and Lake Erie—one-fifth

Eclipse Times* for the 2024 Total Solar Eclipse in the United States[b]

State	City	Partial Begins	Total Begins	Total Ends	Partial Ends	Max. Eclipse	Sun Alt.	Eclipse Mag.	Duration Totality
Arkansas	Hot Springs	12:32 pm	01:49 pm	01:53 pm	03:10 pm	01:51 pm	62	1.055	03m 36s
	Jonesboro	12:38 pm	01:56 pm	01:58 pm	03:15 pm	01:57 pm	59	1.055	02m 21s
	Little Rock	12:34 pm	01:52 pm	01:54 pm	03:12 pm	01:53 pm	61	1.055	02m 28s
California	Los Angeles	10:06 am	–	–	12:22 pm	11:12 am	55	0.579	–
	San Francisco	10:14 am	–	–	12:16 pm	11:13 am	50	0.448	–
Colorado	Denver	11:28 am	–	–	01:54 pm	12:40 pm	58	0.714	–
D.C.	Washington	02:04 pm	–	–	04:33 pm	03:21 pm	47	0.890	–
Florida	Miami	01:48 pm	–	–	04:13 pm	03:02 pm	60	0.557	–
Georgia	Atlanta	01:46 pm	–	–	04:21 pm	03:05 pm	57	0.846	–
Illinois	Carbondale	12:43 pm	01:59 pm	02:03 pm	03:18 pm	02:01 pm	57	1.054	04m 08s
	Chicago	12:51 pm	–	–	03:22 pm	02:08 pm	52	0.942	–
Indiana	Bloomington	01:49 pm	03:05 pm	03:09 pm	04:22 pm	03:07 pm	54	1.054	04m 02s
	Evansville	01:46 pm	03:03 pm	03:06 pm	04:20 pm	03:04 pm	56	1.054	03m 04s
	Indianapolis	01:51 pm	03:06 pm	03:10 pm	04:23 pm	03:08 pm	53	1.054	03m 50s
	Marion	01:52 pm	03:08 pm	03:10 pm	04:24 pm	03:09 pm	52	1.054	02m 18s
	Muncie	01:52 pm	03:08 pm	03:11 pm	04:24 pm	03:09 pm	52	1.054	03m 46s
	Terre Haute	01:48 pm	03:04 pm	03:07 pm	04:21 pm	03:06 pm	54	1.054	02m 54s
Kentucky	Paducah	12:43 pm	02:01 pm	02:02 pm	03:19 pm	02:02 pm	57	1.055	01m 40s
Louisiana	New Orleans	12:30 pm	–	–	03:09 pm	01:50 pm	65	0.846	–
Maine	Presque Isle	02:22 pm	03:32 pm	03:35 pm	04:41 pm	03:34 pm	35	1.050	02m 49s
Massachusetts	Boston	02:16 pm	–	–	04:39 pm	03:30 pm	40	0.932	–
Michigan	Detroit	01:58 pm	–	–	04:28 pm	03:14 pm	49	0.991	–
Missouri	Cape Girardeau	12:42 pm	01:58 pm	02:02 pm	03:17 pm	02:00 pm	57	1.055	04m 07s
New York	Buffalo	02:05 pm	03:18 pm	03:22 pm	04:32 pm	03:20 pm	46	1.052	03m 45s
	Jamestown	02:04 pm	03:18 pm	03:21 pm	04:32 pm	03:19 pm	46	1.052	02m 52s
	New York	02:11 pm	–	–	04:36 pm	03:26 pm	43	0.911	–
	Plattsburgh	02:14 pm	03:26 pm	03:29 pm	04:37 pm	03:27 pm	40	1.051	03m 33s
	Rochester	02:07 pm	03:20 pm	03:24 pm	04:33 pm	03:22 pm	44	1.052	03m 39s
	Syracuse	02:09 pm	03:23 pm	03:24 pm	04:35 pm	03:24 pm	43	1.052	01m 28s
Ohio	Akron	01:59 pm	03:14 pm	03:17 pm	04:29 pm	03:16 pm	49	1.053	02m 49s
	Cleveland	01:59 pm	03:14 pm	03:18 pm	04:29 pm	03:16 pm	49	1.053	03m 49s
	Dayton	01:53 pm	03:09 pm	03:12 pm	04:26 pm	03:11 pm	52	1.054	02m 42s
	Lima	01:55 pm	03:10 pm	03:14 pm	04:26 pm	03:12 pm	51	1.053	03m 51s
	Mansfield	01:57 pm	03:12 pm	03:16 pm	04:28 pm	03:14 pm	50	1.053	03m 15s
Ohio	Toledo	01:57 pm	03:12 pm	03:14 pm	04:27 pm	03:13 pm	50	1.053	01m 48s
Pennsylvania	Erie	02:02 pm	03:16 pm	03:20 pm	04:31 pm	03:18 pm	47	1.052	03m 42s
	Philadelphia	02:08 pm	—	—	04:35 pm	03:24 pm	45	0.900	—
Texas	Austin	12:17 pm	01:36 pm	01:38 pm	02:58 pm	01:37 pm	67	1.056	01m 55s
	Dallas	12:23 pm	01:41 pm	01:44 pm	03:03 pm	01:43 pm	65	1.056	03m 48s
	Fort Worth	12:23 pm	01:40 pm	01:43 pm	03:02 pm	01:42 pm	65	1.056	02m 39s
	Houston	12:20 pm	—	—	03:01 pm	01:40 pm	67	0.943	—
	San Antonio	12:14 pm	—	—	02:56 pm	01:34 pm	68	0.998	—
	Texarkana	12:29 pm	01:47 pm	01:49 pm	03:08 pm	01:48 pm	63	1.056	02m 28s
	Waco	12:20 pm	01:38 pm	01:42 pm	03:01 pm	01:40 pm	66	1.056	04m 12s
Utah	Salt Lake City	11:26 am	–	–	01:41 pm	12:32 pm	55	0.577	–
Vermont	Burlington	02:14 pm	03:26 pm	03:29 pm	04:37 pm	03:28 pm	40	1.051	03m 15s
	Montpelier	02:15 pm	03:28 pm	03:29 pm	04:38 pm	03:28 pm	40	1.051	01m 41s
Washington	Seattle	10:39 am	–	–	12:21 pm	11:29 am	45	0.310	–

*All times are in local time (including daylight saving time)

[b]Eclipse times calculations by Fred Espenak, www.eclipsewise.com.

of all the fresh water in the world—takes the plunge over Niagara Falls. Weather permitting, the Sun in eclipse over Niagara Falls for 3 minutes 31 seconds might be a prize photo and an indelible memory.

Like it did for Lake Erie, the eclipse gives almost all the Lake Ontario a bath of totality, but gives Toronto, largest city in Canada, not even a splash.

Between Buffalo and Rochester, New York, the central line of the eclipse closely traces the westernmost 80 miles (130 kilometers) of the Erie Canal. The canal, started in 1817 and completed in 1825, connected Albany to Buffalo (363 miles [584 kilometers]), allowing west-bound passengers and freight to travel by water from New York City up the Hudson River, along the canal, and into the Great Lakes, an engineering triumph of its age that did much to settle and unify the western territories of the young United States.

Rushing on beyond Buffalo, the central line of the eclipse passes just north of Rochester, New York, bringing 3 minutes 39 seconds of totality, and promptly dives into Lake Ontario. To the north of the central line, at the northeastern end of Lake Ontario, is Kingston, Ontario. Meanwhile, to the south and away from the Lake Ontario, the eclipse clips the nails of the Finger Lakes in upstate New York and takes in Syracuse, near the southern limit of totality. Syracuse has the dubious distinction of being the snowiest city in the United States, far snowier than Alaskan and Rocky Mountain cities. Syracuse even edges out Buffalo and Rochester. The reason for so much snow is the lake effect. The Great Lakes, especially Lake Ontario because of its depth, store their summer warmth. Cold Canadian air moving southeast over the slightly warmer waters of Lakes Erie and Ontario gathers moisture from the lakes and dumps it on the cooler land east and southeast of the lakes. This lake-effect snow usually ends by March. Besides, would the elements really be so cruel as to snow out a total solar eclipse in April?

The central line of the eclipse emerges from the east end of Lake Ontario and passes just southeast of Watertown, New York, heading toward Lake Champlain. To the south of the central line are the Adirondack Mountains of northern New York State, headwaters of the Hudson River, which flows south to New York City and the sea.

To the north of the central line, the path of totality moves into the province of Quebec and across the southern half of Montreal, including downtown. In the midst of Montreal is a hilltop with three peaks—Mount Royal—from which the city takes its name. There, totality lasts 1 minute 17 seconds. The metropolitan areas of Toronto and Montreal together account for almost 30% of Canada's population.

While the northern edge of totality is enveloping Montreal, the central line, still offering 3 minutes 34 seconds of totality, is passing just northwest of downtown Plattsburgh, New York, on the western shore of Lake Champlain. The lake forms the border between New York and Vermont, with the northernmost tip of the lake resting in Canada.

Saros 139—Portrait of a Young Superstar

Getting strong now. If the 2024 total eclipse of the Sun could sing, it and its eclipse family—its saros series—might be chanting the song from *Rocky*. The 2024 eclipse belongs to up-and-coming saros series 139.

Saros 139 gave birth to its first eclipse—a weak partial (as saros series always do)—on May 17, 1501, near the North Pole. The odd number of this saros series indicates that the Moon is at its ascending node as it blocks the Sun to cause each of these eclipses. Saros 139 began creating eclipses in the Arctic, with subsequent eclipses migrating southward. Six partial eclipses followed the first in 1501, each covering more of the Sun's face.

In 1627, a strange eclipse occurred. It started as an annular, became total for just an instant, and then returned to being annular. The spherical shape of Earth had brought its surface just close enough to the Moon so that the Moon appeared large enough to hide the entire disk of the Sun from view for just a moment. An eclipse that is annular in some places and total in others is called a hybrid—the rarest type of eclipse. Only 4.5% of eclipses are hybrids. Saros 139 produced 11 more hybrids. With each eclipse, there was less annularity and more totality.

Then, in 1843, saros 139 generated its first eclipse that was total throughout its path. At its peak, totality lasted 1 minute 43 seconds. Thereafter, each eclipse created by saros 139 was total, and the length of totality was increasing.

In 1970, a saros 139 total eclipse lasting 3 minutes 28 seconds visited the east coast of the United States, an event vividly remembered by some eclipse veterans who tell their stories in this book. Saros 139 presented eclipse chasers with totals in 1988

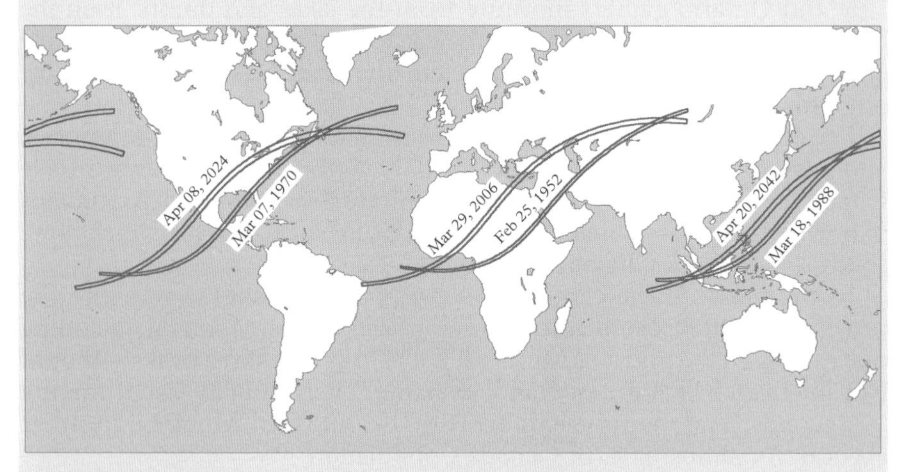

The paths of six central eclipses of saros 139. The tracks shift west with each succeeding eclipse and include total eclipses in 1952, 1970, 1988, 2006, 2024, and 2042. Note how the path of the 2024 eclipse through the United States resembles the one in 1970. [Map and eclipse calculations by Fred Espenak]

in Indonesia and 2006 in Africa. With the dawn of the 21st century, the duration of totality passed the 4-minute mark.

And now comes the total eclipse of April 8, 2024, with totality lasting 4 minutes 28 seconds. It is the 30th eclipse of the 71 that saros 139 will produce. And with each return—2042, 2060, 2078 (across Mexico and the southeastern United States)—the length of totality keeps on growing. Saros 139 is in the prime of its life.

On the 39th eclipse in the series, on July 16, 2186, a total eclipse of saros 139 will last 7 minutes 29 seconds, the longest totality on record in the 8,000 years from 3000 BCE to 5000 CE. That duration is only 3 seconds off the theoretical maximum of 7 minutes 32 seconds. The eclipse peaks over the Atlantic Ocean, but Colombia and Venezuela will experience more than 7 minutes of totality. Total eclipses last their longest in the tropics.

After 2186, as eclipses of saros 139 drift farther south, the duration of totality will dwindle. In 2294, a saros 139 total eclipse will slip below the 5-minute mark. Yet that's still a very long eclipse. Most saroses never generate an eclipse that long.

In 2601, deep in the southern hemisphere, there will be a total eclipse lasting 35 seconds. It will be the last total for saros 139. Saros 139, in old age, will produce 9 more partials near the South Pole. Saros 139 ends it career in partial eclipse on July 3, 2763.

In a span of 1,262 years, saros 139 will have created 71 eclipses, starting with 7 partials, then 12 hybrids, then 43 totals (including one of epic duration), and concluding with 9 partials. It is a career more than worthy of the Eclipse Hall of Fame.[a]

[a] For additional information about saros 139, see http://eclipsewise.com/solar/SEsaros/SEsaros139.html

To the south of the central line, on the eastern shore of Lake Champlain, lies Burlington, Vermont's largest city (metropolitan area: 215,000). Forty miles (64 kilometers) to the southeast but still within the path of totality is Montpelier, capital of Vermont. Its population, 8,000, makes it the smallest of all the American state capitals. Burlington gets 3 minutes 16 seconds of totality; Montpelier gets 1 minute 40 seconds.

The Line Crosses

For nearly 2,000 miles, the central line of the 2024 eclipse has been traveling through the United States. Since it approached Cleveland, the path of totality has been skimming along the border with Canada, dispensing totality to some Canadian cities and towns north of the central line along Lakes Erie and Ontario.

Eclipse Times* for the 2024 Total Solar Eclipse in Canada and Mexico[a]

Country/ Provence	City	Partial Begins	Total Begins	Total Ends	Partial Ends	Max. Eclipse	Sun Alt.	Eclipse Mag.	Duration Totality
CANADA									
Ontario	Hamilton	02:04 pm	03:18 pm	03:20 pm	04:31 pm	03:19 pm	46	1.052	01m 52s
	Belleville	02:08 pm	03:22 pm	03:24 pm	04:34 pm	03:23 pm	44	1.052	02m 00s
	Kingston	02:09 pm	03:22 pm	03:25 pm	04:34 pm	03:24 pm	43	1.052	03m 02s
	Toronto	02:05 pm	–	–	04:32 pm	03:20 pm	45	0.997	–
	Brockville	02:11 pm	03:23 pm	03:26 pm	04:35 pm	03:25 pm	42	1.051	02m 47s
	Cornwall	02:13 pm	03:25 pm	03:27 pm	04:36 pm	03:26 pm	41	1.051	02m 10s
Quebec	Drummondville	02:16 pm	03:29 pm	03:29 pm	04:38 pm	03:29 pm	39	1.051	00m 32s
	Montreal	02:14 pm	03:27 pm	03:28 pm	04:37 pm	03:28 pm	40	1.051	01m 16s
	Sherbrooke	02:17 pm	03:28 pm	03:31 pm	04:38 pm	03:29 pm	39	1.050	03m 25s
New Brunswick	Fredericton	03:24 pm	04:34 pm	04:36 pm	05:42 pm	04:35 pm	35	1.049	02m 18s
Newfoundland	Channel Port Aux Basques	04:03 pm	05:10 pm	05:13 pm	06:15 pm	05:11 pm	28	1.048	02m 45s
	Gander	04:07 pm	05:13 pm	05:15 pm	06:16 pm	05:14 pm	25	1.047	02m 14s
MEXICO									
	Durango	11:55 am	01:12 pm	01:16 pm	02:37 pm	01:14 pm	70	1.057	03m 47s
	Mazatlan	09:51 am	11:07 am	11:12 am	12:32 pm	11:10 am	69	1.057	04m 18s
	Torreon	12:00 am	01:17 pm	01:21 pm	02:41 pm	01:19 pm	70	1.057	04m 12s

*All times are in local time (including daylight saving time)

[a]Eclipse times calculations by Fred Espenak, www.eclipsewise.com.

Now, for the first time, the central line of the eclipse enters Canada, just east of the tiny village of East Richford, Vermont. There, a two-lane bridge crosses the Missisquoi River. On opposite sides of the border, Canadian and American officials man inspection stations. Meanwhile, East Richford Slide Road, the other street in town, bends eastward to the East Richford Cemetery. The cemetery itself is half on the American side, half on the Canadian, with graves straddling the border. The central line goes right through the middle of the tiny cemetery, while, a few yards away, East Richford Slide Road strays across the Canadian border, stays in Canada for 200 yards (200 meters), and then nonchalantly slips back into the United States with no inspections.

The central line of the eclipse, however, remains in Canada for 120 miles (194 kilometers)—passing south of Granby and Sherbrooke, Quebec. Sherbrooke receives 3 minutes 25 seconds of totality, just 5 seconds less than the central line is now providing.

South of the central line, the northernmost 60 miles (100 kilometers) of New Hampshire enjoys some totality. Pittsburg, population 869, the northernmost town in New Hampshire, proudly proclaims that it has more moose than people. Both get 3 minutes 15 seconds of total eclipse.

Then the central line's brief foray into Canada ends as it reenters the United States for a farewell visit that crosses northcentral Maine. After trekking across the United States from Texas to Maine, the central line of the 2024 eclipse passes 7 miles (12 kilometers) north of Mt. Katahdin, tallest mountain in Maine. Mt. Katahdin rises nearly a mile above sea level. It is the northern end of the 2,200-mile (3,500-kilometer) Appalachian Trail, the longest hiking-only trail in the world. The Appalachian Trail traverses 14 states. The trail of 2024 totality traverses 14 states. Both are beloved by their followers. On its way out of Maine, the eclipse enshrouds Presque Isle with 2 minutes 41 seconds of totality. The town is located on a peninsula where Presque Isle Stream angles into the Aroostock River, making the city "almost an island"—Presque Isle.

Going from America

Meanwhile, the eclipse's central line passes 9 miles (15 kilometers) north of Houlton, Maine. Houlton marks the northern end of Interstate Highway 95, America's east coast expressway. It runs 1,920 miles (3,090 kilometers), the longest north-south *Interstate* highway in the country. At Houlton, I-95 crosses US Route 1. Route 1 was the principal east coast highway in the United States until I-95 replaced it. But US 1 starts farther south—in Key West—and continues north past Houlton. It is 23% longer than I-95 and holds the distinction of being the longest north-south *road* in the United States.

Almost side by side, I-95 and the central line of the 2024 eclipse—carrying with it 3 minutes 22 seconds of totality—cross from Maine into Canada, ending the eclipse's journey in the United States. The central line of the 2024 eclipse travels a total of 2,040 miles (3,280 kilometers) within the United States, more than Mexico (535 miles [860 kilometers]) and Canada (810 miles [1,300 kilometers]) combined.

On its way across the Canadian province of New Brunswick, the eclipse visits its capital city, Fredericton, with so many church spires that its 106,000 citizens have nicknamed their town the Celestial City.[8] Two celestial bodies wink back for 2 minutes 17 seconds.

The southern limit of the eclipse passes through the northern outskirts of Moncton, largest city in New Brunswick. Moncton is just up the Petitcodiac River from the eastern end of the Bay of Fundy, famous for its extreme tides—yet another collaboration of the Moon and Sun.

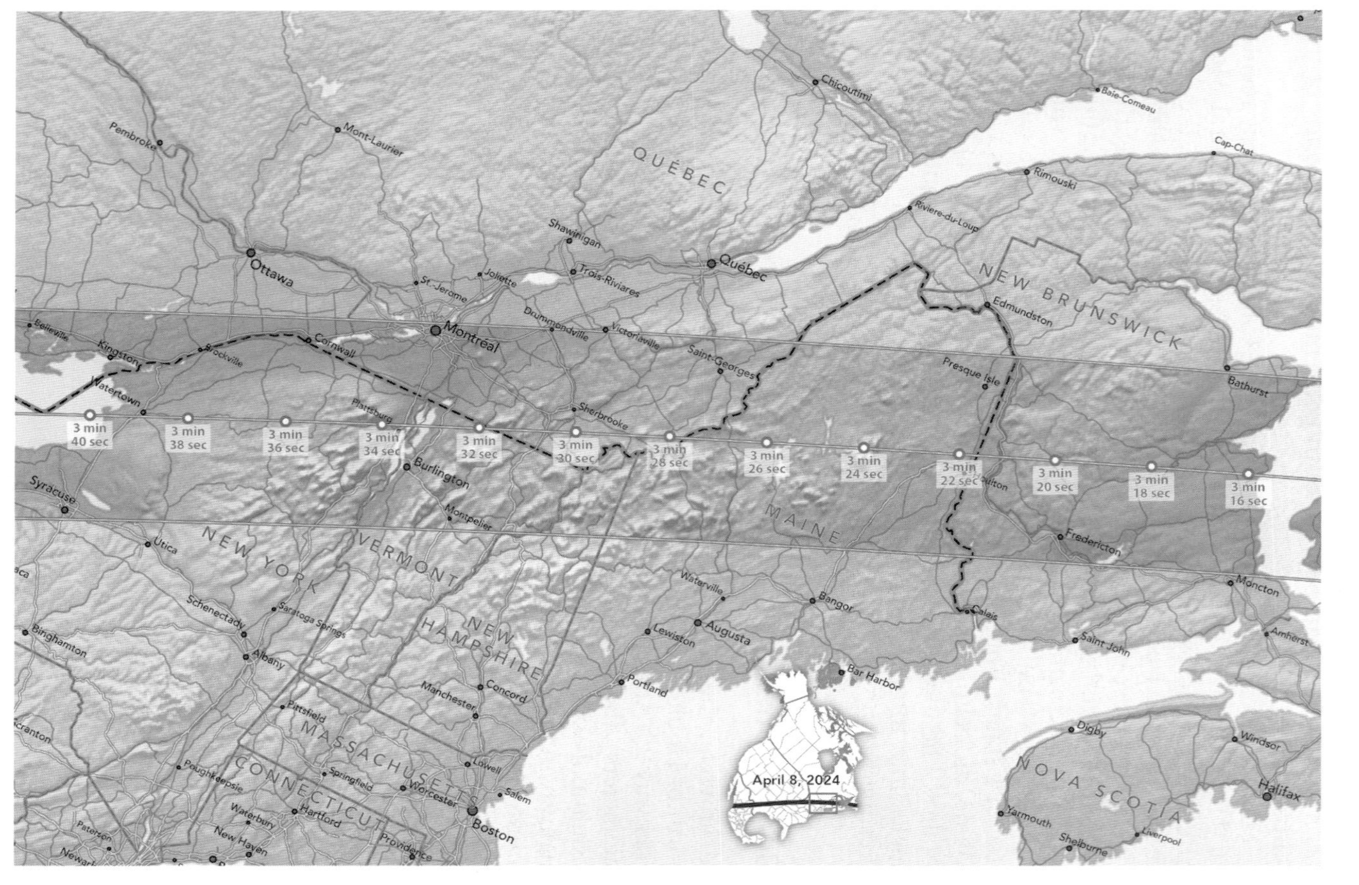

The path of totality as it crosses New York, Vermont, New Hampshire, Maine, Quebec, and New Brunswick. Points along the central line give the duration of totality in minutes and seconds. [Map ©2016 Michael Zeiler, GreatAmericanEclipse.com]

Total Eclipses for America in the 21st Century

If the Moon has been stingy about using the Sun to cast its shadow on the United States in recent decades, the Sun and Moon will be more generous through the remainder of the 21st century—especially between 2017 and 2052.

In 2017 and 2024, less than 7 years apart, total eclipses cross the country diagonally to one another. Twenty years will then elapse until the United States has totality again, but it will return in a burst of three total eclipses over a period of 7½ years, with two eclipses less than a year apart:

August 23, 2044	Montana and North Dakota	2 minutes 4 seconds
August 12, 2045	California through Florida	6 minutes 6 seconds
March 30, 2052	Louisiana through South Carolina	4 minutes 8 seconds

Three more eclipses will visit the United States before the century is out:

May 11, 2078	Texas through North Carolina	5 minutes 40 seconds
May 1, 2079	Maryland through Maine	2 minutes 55 seconds
Sept. 14, 2099	Montana through North Carolina	5 minutes 18 seconds

At full moon and new moon, when the Moon and Sun combine their gravitational effects on ocean waters, Bay of Fundy high tides rise 50 feet (15 meters) above low tides.

The central line of the eclipse then leaps into the Gulf of St. Lawrence, part of the Atlantic Ocean, and barely misses the northern tip of Prince Edward Island, although the southern portion of totality laps over 40% of the island. Prince Edward Island, smallest of the 13 Canadian provinces and territories, is about the size of Delaware, the second smallest American state. Near the northern tip of the island, where totality lasts longest—3 minutes 14 seconds—is a tiny settlement called Seacow Pond. This name does not refer to the manatee (or sea cow) of Florida, the threatened marine mammal that prefers warm waters. Instead, Seacow Pond honors the Atlantic walrus, sometimes called a sea cow because it makes mooing sounds.[9] Walruses, also threatened with extinction, have tusks 3 feet (1 meter) long that they use to pull themselves up onto ice floes. That talent is noted in their scientific name: *Odobenus rosmarus*—tooth-walking sea-horse.

Midway between Prince Edward Island and Newfoundland, the eclipse of 2014 covers all of the Magdalen Islands, part of Quebec, a prime site where eco-tourists go in late February to see newborn white-coated harp seals.

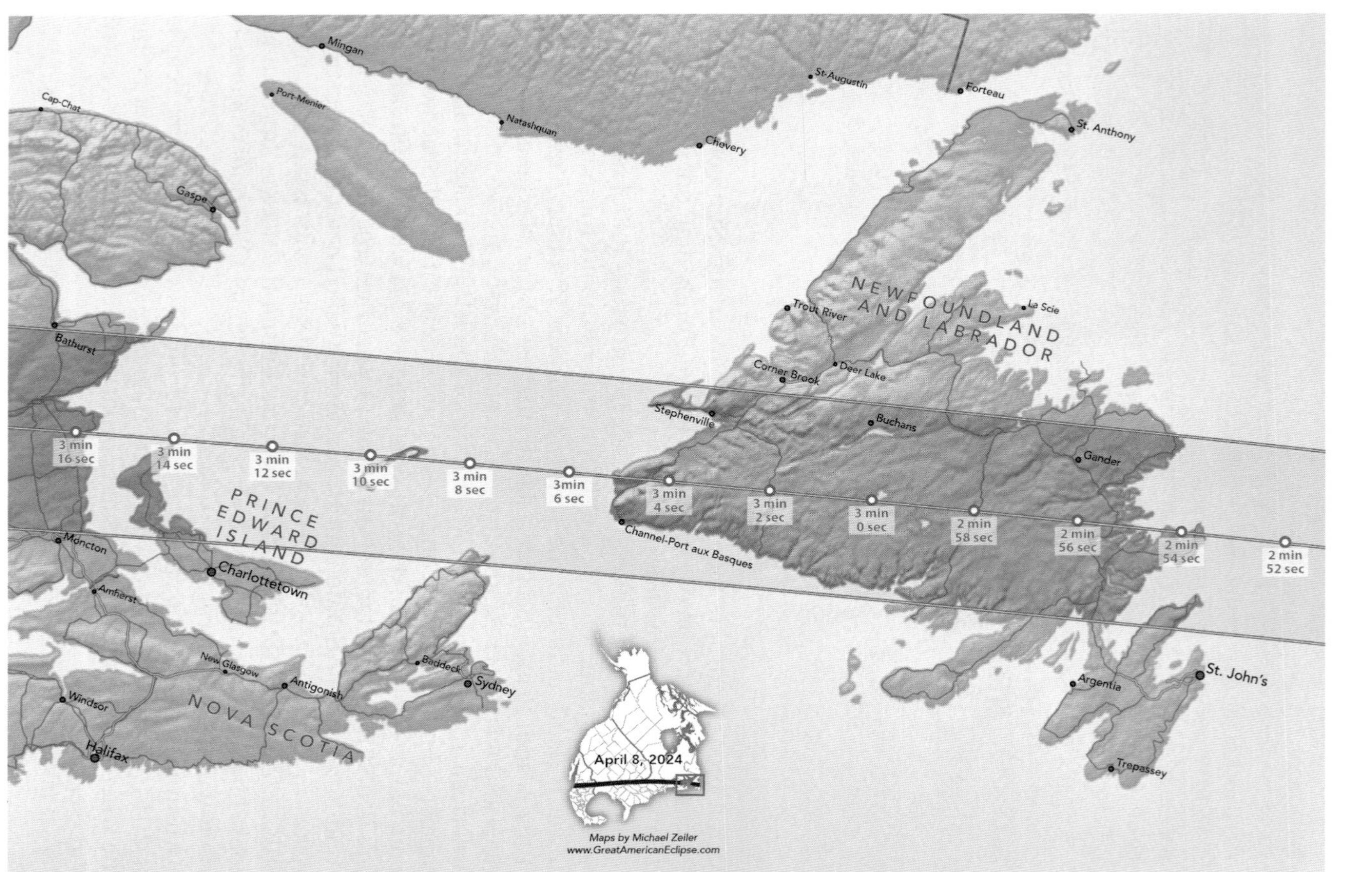

The path of totality as it crosses Prince Edward Island and Newfoundland. Points along the central line give the duration of totality in

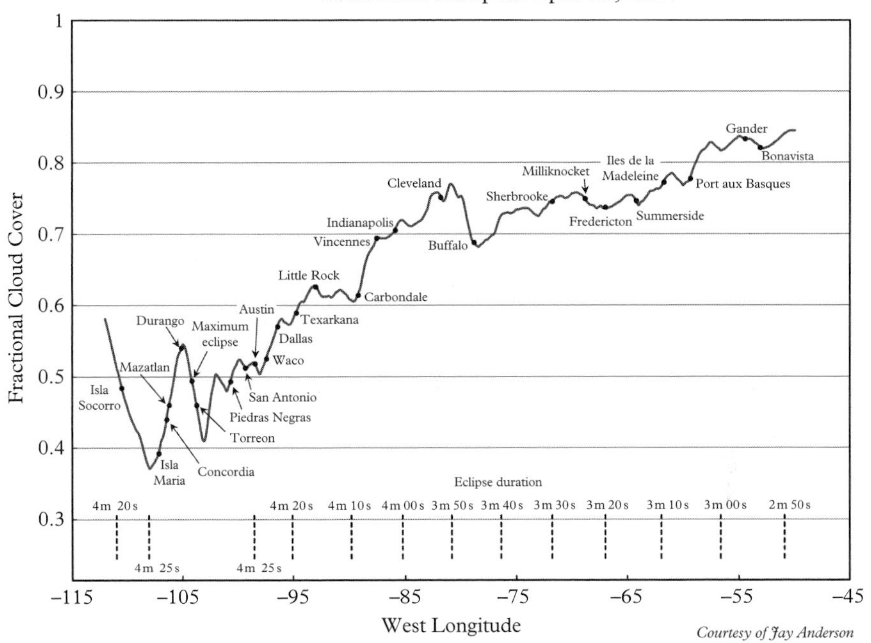

The average cloud amount along the central line of the 2024 eclipse path has been derived from 20 years of satellite imagery. The locations of cities and towns are indicated along the path. [Source: Jay Anderson and Patmos-X: CIMMS/SSEC]

The southern limit of totality then catches the northern tip of Cape Breton Island, part of Nova Scotia.

The final landfall of the 2024 eclipse is Newfoundland. As it moves across the island, the duration of totality subsides from 3 minutes 5 seconds to 2 minutes 54 seconds. Most of Newfoundland's cities and towns lie on or near the coast—primarily the east coast—and that's where the roads are too.

Gander, Newfoundland played a major role in the development of aviation. Lying at the northeast corner of North America, closest to Europe, Gander was a crucial refueling stop for the first flights from the United States and Canada to England and France and back. During World War II, by refueling in Gander, fighters and bombers could ferry themselves from North America to England, ready for missions against Nazi forces. Thus Gander became known as the Crossroads of the World. On April 8, 2024, the Moon and Sun cross paths there, providing 2 minutes 15 seconds of totality (at the airport). Transatlantic flights no longer need to stop for fuel in Gander, but the city still provides runways and hospitality for intercontinental flight emergencies. So taken with aviation are Ganderites that they have named their streets Charles Lindbergh, Amelia Earhart, Chuck Yeager … pilots who paved the way for man to roam the Moon—and for eclipse chasers to roam the Earth.

The Sun and Moon in total eclipse leave Newfoundland with a nod to two villages on the southeast coast—Sunnyside and Come By Chance. Then the central line of the eclipse leaves Newfoundland—and land altogether—at Maberly, an east coast village so small it doesn't make the list of provincial towns. The eclipse will spend its shrinking path and time of totality racing ever faster eastward across the Atlantic, to fall off the edge of the Earth at sundown before it can reach Europe.

The party is over. For the next 20 years, people in the United States, Canada, and Mexico will have to travel the world to stand in the shadow of the Moon. After April 8, 2024, there will be no total eclipses of the Sun in North America until August 23, 2044.

NOTES AND REFERENCES

1. Edwin Dunkin: *Autobiography*, unpublished (compiled by Peter Hingley, Royal Astronomical Society). Dunkin is referring to the total solar eclipse of July 28, 1851, as seen from within the northern limit of the path of totality in Scandinavia.
2. The *Metropolis News* changed its name in 1972 to the *Metropolis Planet*.
3. The mounds are only 6 miles from Paducah, but the road takes one north and east around the mounds, increasing the distance to 20 miles (32 kilometers).
4. John E. Schwegman: "A Prehistoric Cultural and Religious Center in Southern Illinois." http://www.southernmostillinoishistory.net/kincaid-mounds.html
5. Latitude 37° 38′ 42″ north; longitude 89° 16′ 37″ west.
6. The 2017 eclipse reaches its longest duration (2 minutes 40.1 seconds) at a point 6 miles (10 kilometers) southeast of the location in Giant City State Park where the central lines of the 2017 and 2024 eclipses cross. The 2024 reaches longest duration in Mexico.
7. Emergency edit in press: In honor of the upcoming eclipse, in 2016 the Cleveland Cavaliers won the National Basketball Association championship to break Cleveland's winless curse.
8. http://www.fredericton.ca/en/communityculture/churches_placesofworship.asp.
9. Todd Dupuis, Assistant Deputy Minister of Environment, Prince Edward Island, personal correspondence, December 2, 2015. Jeremy Berlin: "The Unexpected Walrus," *National Geographic,* December 2013 (online).

Easter Island's Anakena Beach offered a surreal landscape of Maoi statues and palm trees for this time-lapse sequence of the total solar eclipse of July 11, 2010. [Canon 5D Mark II, Canon 16–35 zoom at 24 mm, solar filter for partial, no filter for total. ©2010 Blanchard Guillaume]

A stunning composite of the total solar eclipse of August 1, 2008 was produced using 28 exposures. Features on the dark side of the Moon are also visible. The images were processed and combined using Photomatix and Photoshop. [Nikon D300, Borg 77 mm ED refractor, fl = 500 mm, f/6.5, exposures: 1/4000 to two seconds. ©2008 Alson Wong]

Thin clouds provide a screen for the projection of shadow bands during the third contact diamond ring of the Australian eclipse of November 14, 2012. [Canon 50D, 70–200 mm zoom, ISO 400, exposure 1/1000 second. ©2012 Stephen Mudge]

The Great Wall of China at Jiayuguan Fort frames the sky during the total eclipse of August 1, 2008. A composite of eight exposures reveals the Sun's corona as well as the planets Mercury, Venus, Mars, and Saturn. [Canon 450D, Canon 10–22 mm, exposures: 1/125 at f8 to one second at f5.6. ©2008 Terry Cuttle]

Easter Island statures (maoi) bare mute testimony to the spectacle of totality from Anakena Beach during the total solar eclipse of July 11, 2010. [Canon 5D Mark II, 14 mm lens. ©2010 Stephane Guisard]

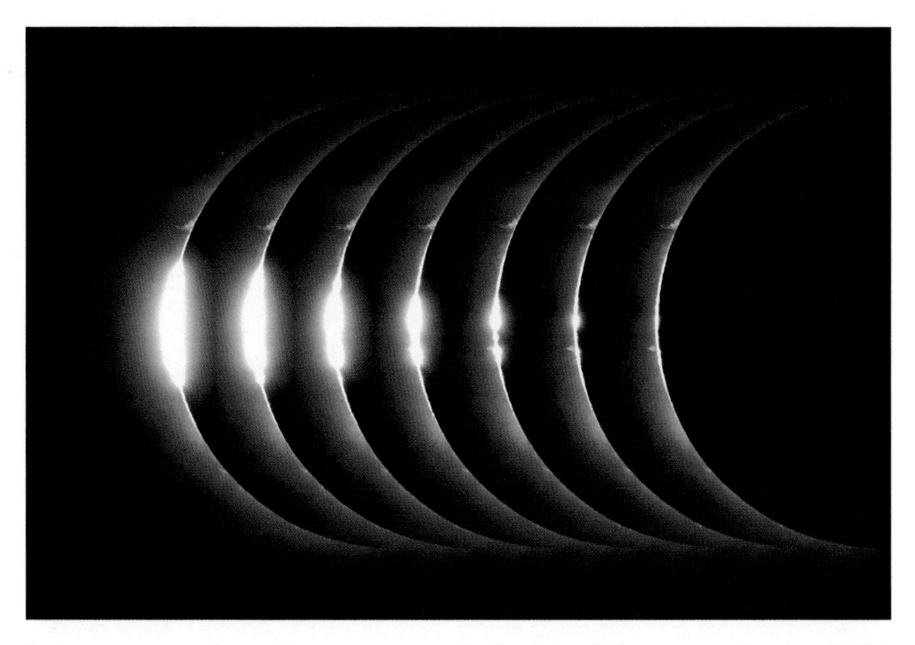

A time sequence of seven images captures the breakup of the crescent Sun into Baily's beads immediately preceding second contact. The total eclipse of March 29, 2006 was photographed from Jalu, Libya. [Nikon D200, Vixen 90 mm fluorite refractor, fl = 810 mm, f/9 1/1000s, ISO 200. ©2006 Fred Espenak]

The Moon barely covered the Sun during the total eclipse of November 3, 2013 from Pokero, Uganda. This offered photographers the rare chance to record prominences and the chromosphere surrounding the entire limb of the Moon. [Nikon D800, Televue 102 refractor, fl = 1500 mm, f/15, 1/800 second. ©2013 Jaime M. P. Vilinga]

A low-altitude total eclipse offers the perfect opportunity for a "selfie" with totality. The total eclipse of August 1, 2008 was photographed from the deserts of northern China. [Canon G9, ISO 400, f/5.6, autoexposure. ©2008 Fred Espenak]

The Altiplano of Chile offered a unique landscape complete with volcanoes for the total solar eclipse of November 3, 1994. [35 mm SLR, 28 mm lens, f/2.8, 2s, Ektachrome 100 slide film. ©1994 Alan Dyer]

Amazing detail in the prominences and chromoshpere are visible in this high-magnification image of the July 11, 2010 total eclipse from Easter Island. [Nikon D700, Borg 100ED and 2x teleconverter, fl = 1280 mm, ISO 800, f/12.8 at 1/5000 second. ©2010 Dave Kodama]

The diamond ring effect is photographed with the Sun just clearing the high Andes Mountains. Excited eclipse watchers are seen in the foreground of this dramatic image of the July 11, 2010 total eclipse from near El Calafate, Argentina. [Nikon D300, Nikon 24–120 zoom, auto-exposure. ©2010 Charles Fulco]

Details in the corona are revealed in this Photoshop composite that was processed to bring out fine structure. The images were obtained during the March 29, 2006 eclipse from the Sahara Desert in northern Libya. [Nikon D200 DSLR, TeleVue Ranger 70 mm refractor, fl = 476 mm, f/6.8, composite of 22 exposures: 1/1000 to two seconds, ISO 200. ©2006 Fred Espenak]

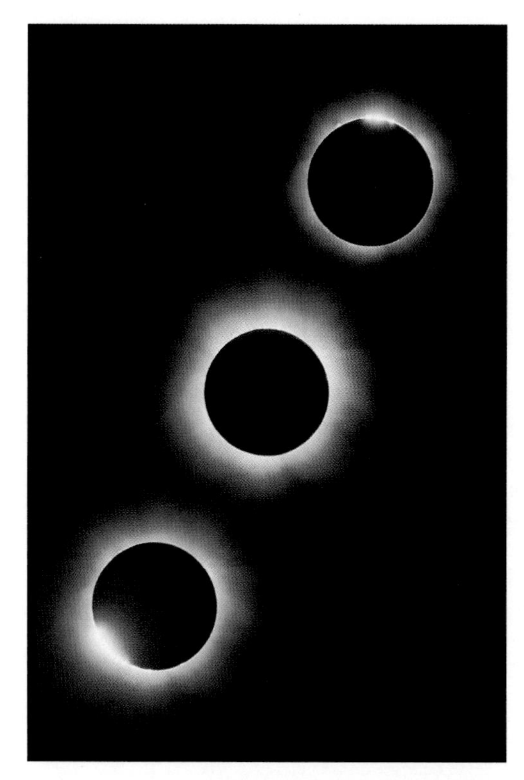

Totality is framed by two diamond rings in this composite of three images shot from Lake Hazar, Turkey during the August 11, 1999 total eclipse. [Nikon 6006 SLR, Celestron C90 Maksutov, fl = 1000 mm, f/11, auto-exposure, Kodak Royal Gold 400 negatives. ©1999 Patricia Totten Espenak]

Sometimes a hole in the clouds appears at just the right moment as in this photo shot from a sailing ship in the Atlantic during the total solar eclipse of November 3, 2013. [Canon 5D Mark II, 19 mm, f/2.8, 1/60 second. ©2013 Alan Dyer]

Baily's beads are captured as totality ends during the November 3, 2013 total solar eclipse from Uganda. This complex photo was produced from images shot with two cameras and processed with custom software. [Canon 6D and 350D, lenses: Rubinar 10/1000 mm, 3M-5CA 8/500 mm, ISO 250 and 100, 23 exposures: 1/500 to one second. © 2014 Miloslav Druckmüller]

The frigid landscape of Svalbard served as the backdrop for this wide-angle image of the March 20, 2015 total eclipse of the Sun. [Sony A550, Sigma 10–20 zoom, f/4.5, 1/10 second. ©2015 Sarah Marwick]

A MOMENT OF TOTALITY

The Extinction of Total Solar Eclipses

In 1695, Edmond Halley discovered that eclipses recorded in ancient history did not match calculations for the times or places of those eclipses. Starting with records of eclipses in his day and the observed motion of the Moon and Sun, he used Isaac Newton's new theory of universal gravitation (1687) to calculate when and where ancient eclipses should have occurred and then compared them with eclipses actually observed more than 2,000 years earlier. They did not match. Halley had great confidence in the theory of gravitation and resisted the temptation to conclude that the force of gravity was changing as time passed. Instead, he proposed that the length of a day on Earth must be slowly increasing. The Earth's rotation must be slowing down.

If the Earth's rotation had slowed down slightly, the Moon must have gained angular momentum to conserve the total angular momentum of the Earth-Moon system. This boost in angular momentum for the Moon would have caused it to spiral slowly outward from the Earth to a more distant orbit where it travels more slowly. If, 2,000 years earlier, the Earth had been spinning a little faster and the Moon had been a little closer and orbiting a little faster, then eclipse theory and observation would match. Scientists soon realized that Halley was right.

But what would cause the Earth's spin to slow? Tides. The gravitational attraction of the Moon is the principal cause of the ocean tides on Earth. As the shallow continental shelves (primarily in the Bering Sea) collide with high tides, the Earth's rotation is retarded. The slower spin of the Earth causes the Moon to edge farther from our planet.

From 1969 to 1972, the Apollo astronauts left a series of laser reflectors on the Moon's surface. Since then, scientists on Earth have been bouncing powerful lasers off these reflectors. By timing the round trip of each laser pulse, the Moon's distance can be measured to an accuracy of several inches. The Moon is receding from the Earth at the rate of about 1.5 inches (3.8 centimeters) a year. As the Moon recedes from Earth, its apparent disk becomes smaller. Total eclipses become rarer; annular eclipses more frequent. Total eclipses are moving toward extinction. When the Moon's mean distance from the Earth has increased by 14,550 miles (23,410 kilometers), the Moon's apparent disk will be too small to cover the entire Sun, even when the Moon's elliptical orbit carries it closest to Earth. Total eclipses will no longer be possible.

How long will that take? With the Moon receding at 1.5 inches a year, the last total solar eclipse visible from the surface of the Earth will take place 620 *million* years from now. There is still time to catch one of these majestic events.

17

Epilogue: Eclipses—Cosmic Perspective, Human Perspective

No one should pass through life without seeing a total solar eclipse.

Leif Robinson (1999)[1]

The Moon's shadow has darkened space for 4½ billion years, since the Sun formed and began to shine and the planets and their moons formed and could not shine. From the newly formed Sun, light streamed outward in all directions. Here and there it illuminated a body of rock or gas or ice. The Sun's dominant light identified that world as one of its own children.

A little over 8 minutes of light-travel outbound from the Sun, a portion of the light encountered two small dark bodies and bathed the sunward half of each in brightness. The surrounding flood of unimpeded light sped on, leaving a long cone of darkness—a shadow—behind the planet and its one large moon.

The two worlds lay the same average distance from the Sun. Not long before—measured on a cosmic timescale—Earth and Moon had been a single body, but they were separate and utterly different now.

A Cosmic Birth

The Sun, planets, moons, asteroids, and comets had all begun within a cloud of gas and dust, a cloud so large that there was material enough to make dozens or hundreds of solar systems. Where the density was great enough, fragments of the cloud began to condense by gravity. At the heart of each fragment, a star was coalescing. Near those stars-to-be, other bodies, too low in mass ever to reach starhood, also began to form. At first they grew by gentle collisions and adhesions, gathering up a grain of ice, a fleck of dust along their paths around the Sun until icy

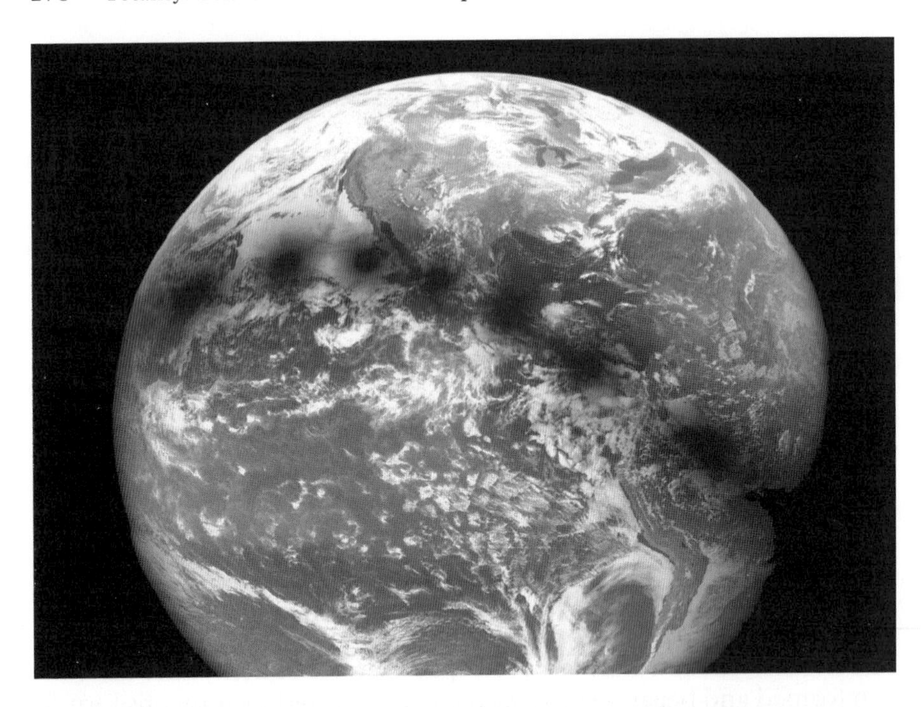

A composite of eight NASA GOES-7 weather satellite images captures the path of the Moon's shadow during the total solar eclipse of July 11, 1991. [Photo courtesy NASA & Dr. William Emery and Timothy D. Kelley, University of Colorado]

or rocky planetesimals had taken shape—the beginnings of planets. And still they gathered dust and small debris until they were so massive that gravity became their prime means of growth, gathering to them still more materials—and other planetesimals. The number of small bodies declined. The size of a few large bodies increased. The planets had been formed by convergence.

Convergence brought near disaster as well. Another planet-size body (the size of Mars today) wandered across the path of a body that would become the Earth. No living thing witnessed the collision. No living thing could have survived the collision that obliterated the smaller world and nearly shattered the Earth. The Earth recoiled from the impact by spewing molten fragments of the intruder and its own crust and mantle outward. Some escaped from Earth; others rained down from the skies, pelting the surface in a rock storm of unimaginable proportions. But many of the fragments, caught in the Earth's gravity, stayed aloft, orbiting the Earth as the Earth orbited the Sun. Quickly, in a century or less, the fragments joined together by collisions and accretion to form a new world circling the first. That new world was the Moon.[2]

From convergence had come divergence. The two worlds, sprung from one, continued to diverge. Both were the same distance from the Sun, but the Earth was 81 times more massive than the Moon. That mass allowed the Earth to hold an atmosphere by gravity, while the Moon could not.

The eons passed. Life arose on Earth and covered the planet. Plants and animals responded to the tides raised by the Moon. The lunar tides slowed the Earth and caused the Moon to spiral slowly outward, diverging ever farther from the Earth in distance and ever further from the Earth in environment as well.

The lifeless Moon withdrew until today its shadow can just barely reach the Earth. As shadows in the universe go, this one is of no great size: a cone of darkness extending at most only 236,000 miles (379,900 kilometers) in length before dwindling to a point. It is long enough to touch the Earth only occasionally and very briefly and with a single narrow stroke.

The black insubstantial cone reaches out, but for most of the time there is nothing to touch. The shadow sweeps on through space unseen, unnoticed.

Awesome totality—June 21, 2001, from Zambia. [Nikon 8008 SLR, MF-21 program back, Sigma 18–35 mm at 18 mm, F/5.6, auto-exposure, Kodak Royal Gold 200 negatives. [©2001 Fred Espenak]

Contact

Yet now ahead lies the Earth. It is a special day. Suddenly the Moon's shadow becomes visible as it collides with a world of rock and water and air. The shadow swoops in from the heavens, silently darkening the sky where it alights upon the Earth and begins its ceaseless rush across the planet's surface.

On these occasions, the people of Earth have gathered and still gather scientific knowledge from the Sun. And more. For long before we drew information from total eclipses, we stared in wonder at them. And long after all knowledge from eclipses has been gleaned, people will still travel to the ends of the Earth to treasure their majesty and beauty.

So compelling is a solar eclipse that when at last the Moon has drifted too far away to touch the Earth with darkness, the beings of that era may use their technology to gently, gradually halt and then reverse the Moon's retreat—bring it closer once again—so that they too will see the sight that their books and visual records can only hint at: a total eclipse of the Sun.

NOTES AND REFERENCES

1. Epigraph: Leif Robinson: Foreword to *Totality: Eclipses of the Sun*, 2nd edition, by Mark Littmann, Fred Espenak, and Ken Willcox (New York: Oxford University Press, 1999).
2. The leading theory of the Moon's formation involves a collision between a large planetesimal and the proto Earth.

Appendix A

Maps for Every Solar Eclipse 2017–2045

I doubt if the effect of witnessing a total eclipse ever quite passes away. The impression is singularly vivid and quieting for days, and can never be wholly lost.

Mabel Loomis Todd (1900)[1]

Between the years 2017 and 2045, a 28-year period, the Moon will eclipse the Sun 64 times. This interval samples at least one eclipse from every saros series currently producing eclipse. The eclipses during this period fall into the following categories:

Eclipse Type	2017–2045	Compared to 2000 BCE–3000 CE
Total	20 = 31.3%	26.7%
Annular	22 = 34.4%	33.2%
Hybrid	2 = 3.1%	4.8%
Partial	20 = 31.3%	35.3%

The following pages offer 64 global maps, one for each eclipse.[2] The odd saddle-shaped zone in each map shows the region where the partial eclipse is visible. The magnitude of each eclipse (maximum fraction of the Sun's diameter covered) is shown in increments of 25%, 50%, and 75%. This allows you to quickly estimate the magnitude for any location within the eclipse path. For central eclipses, the path of either totality or annularity is plotted.

The Moon's penumbral shadow typically produces a zone of partial eclipse (during both partial and central eclipses) that covers 25% to 50% of the daylight hemisphere of the Earth. In comparison, the Moon's

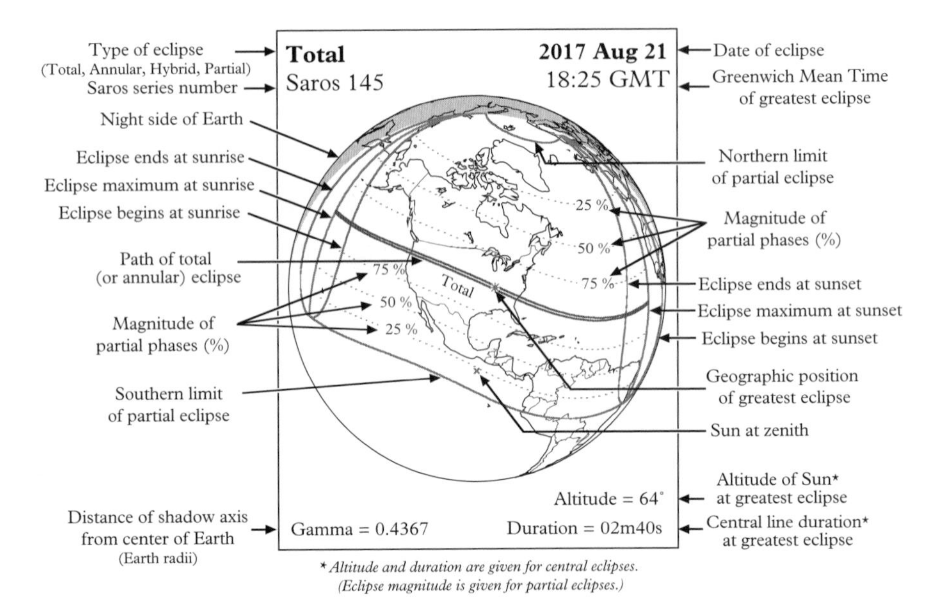

** Altitude and duration are given for central eclipses.*
(Eclipse magnitude is given for partial eclipses.)

umbral shadow (total eclipse) or antumbral shadow (annular eclipse) is much smaller: its path covers less than 1% of the Earth's surface.

Additional information on each map can be identified using the key above.

The maps show North and South America, Europe and Africa, and Asia and Australia. These detailed maps show the paths of every central solar eclipse (total, annular, and hybrid) from 2017 through 2045. The maximum duration of totality or annularity as well as a list of all countries within each central eclipse path can be found in Appendix B, which covers eclipses from 2017 through 2070..

Use *Totality: The Great American Eclipses of 2017 and 2024* to plan your own voyage into the Moon's shadow. Remember, words and pictures can never fully convey the wonder of a total eclipse of the Sun: to stand in the path of totality, in the light of the corona. You must see one for yourself—or two, or . . .

Hope to meet you there.

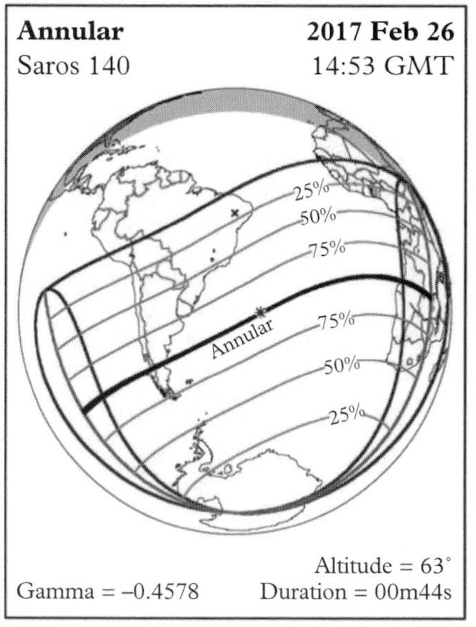
Annular
2017 Feb 26
Saros 140
14:53 GMT
25%
50%
75%
Annular
75%
50%
25%
Altitude = 63°
Gamma = –0.4578
Duration = 00m44s

Total
2017 Aug 21
Saros 145
18:25 GMT
25%
50%
75%
Total
75%
50%
25%
Altitude = 64°
Gamma = 0.4367
Duration = 02m40s

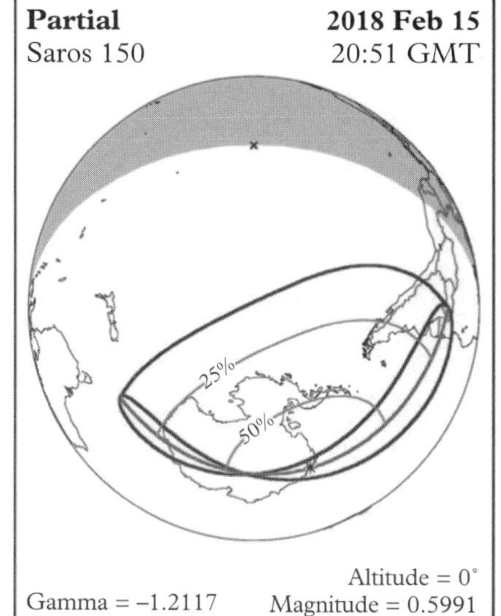
Partial
2018 Feb 15
Saros 150
20:51 GMT
25%
50%
Altitude = 0°
Gamma = –1.2117
Magnitude = 0.5991

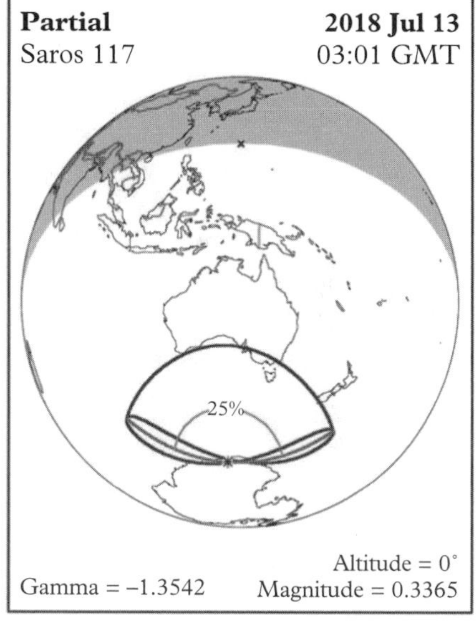
Partial
2018 Jul 13
Saros 117
03:01 GMT
25%
Altitude = 0°
Gamma = –1.3542
Magnitude = 0.3365

Partial **2018 Aug 11**
Saros 155 09:46 GMT

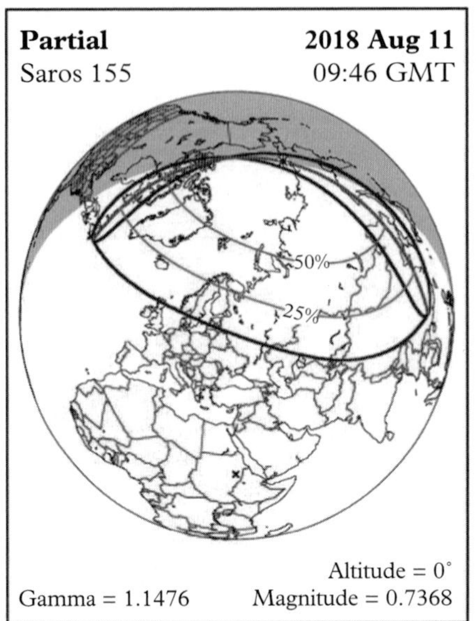

Altitude = 0°
Gamma = 1.1476 Magnitude = 0.7368

Partial **2019 Jan 06**
Saros 122 01:41 GMT

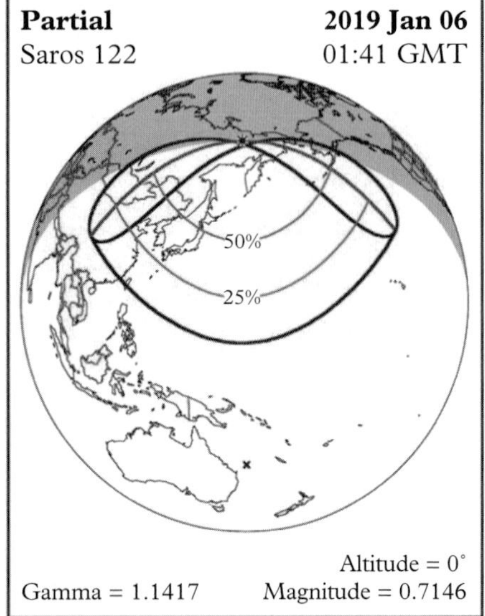

Altitude = 0°
Gamma = 1.1417 Magnitude = 0.7146

Total **2019 Jul 02**
Saros 127 19:23 GMT

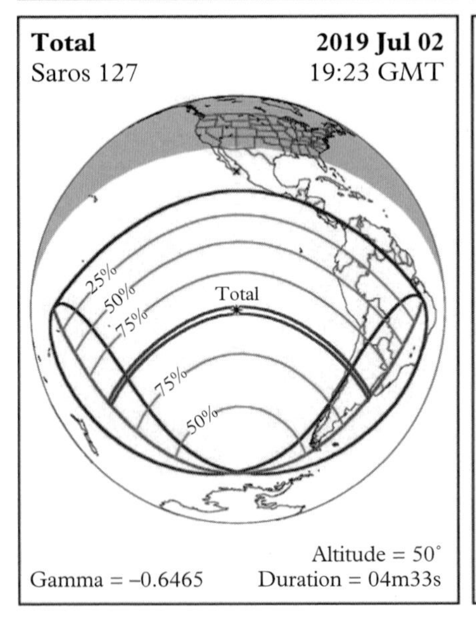

Altitude = 50°
Gamma = –0.6465 Duration = 04m33s

Annular **2019 Dec 26**
Saros 132 05:18 GMT

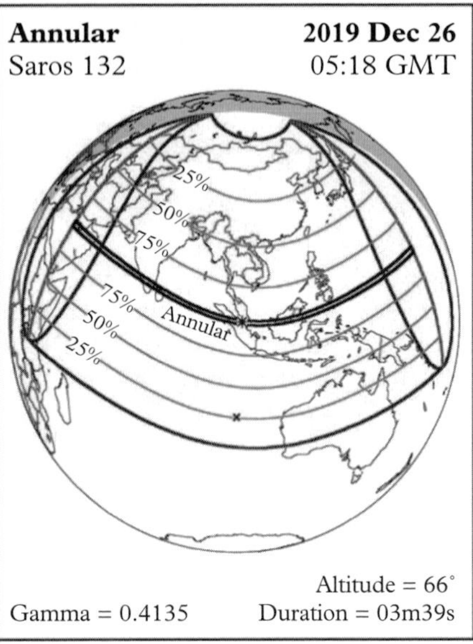

Altitude = 66°
Gamma = 0.4135 Duration = 03m39s

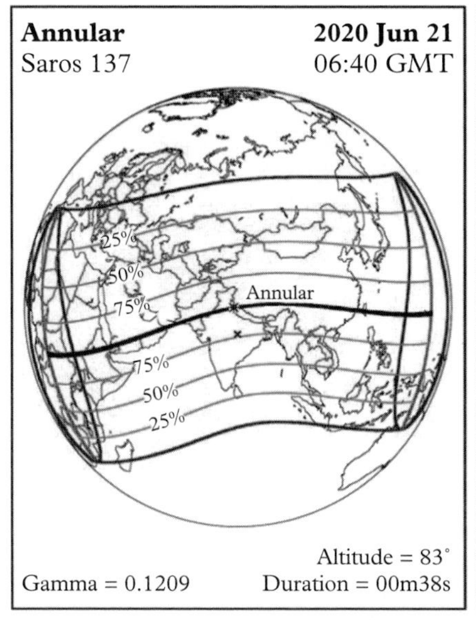
Annular
2020 Jun 21
Saros 137
06:40 GMT
25%
50%
75%
Annular
75%
50%
25%
Altitude = 83°
Gamma = 0.1209
Duration = 00m38s

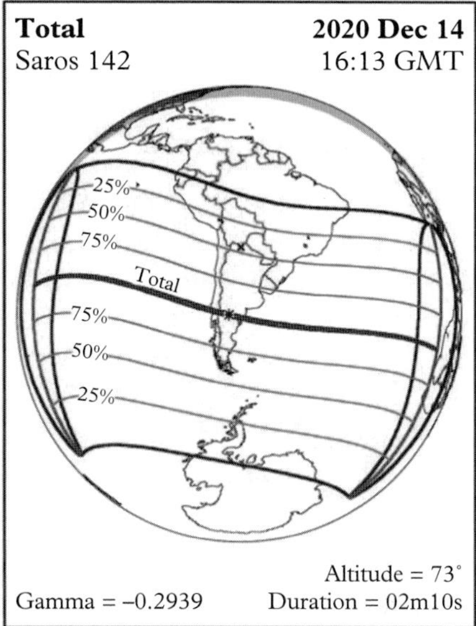
Total
2020 Dec 14
Saros 142
16:13 GMT
25%
50%
75%
Total
75%
50%
25%
Altitude = 73°
Gamma = −0.2939
Duration = 02m10s

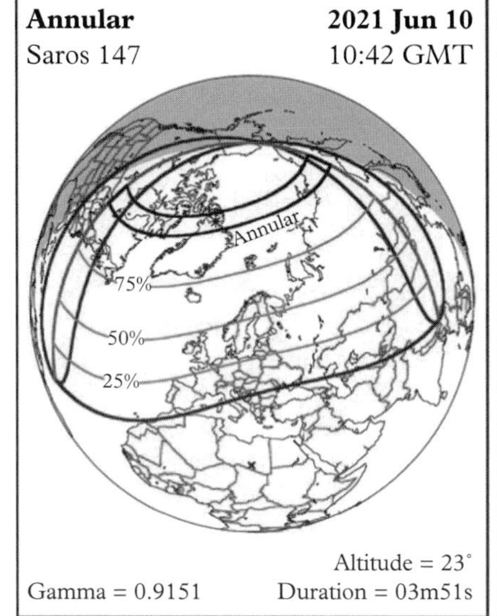
Annular
2021 Jun 10
Saros 147
10:42 GMT
Annular
75%
50%
25%
Altitude = 23°
Gamma = 0.9151
Duration = 03m51s

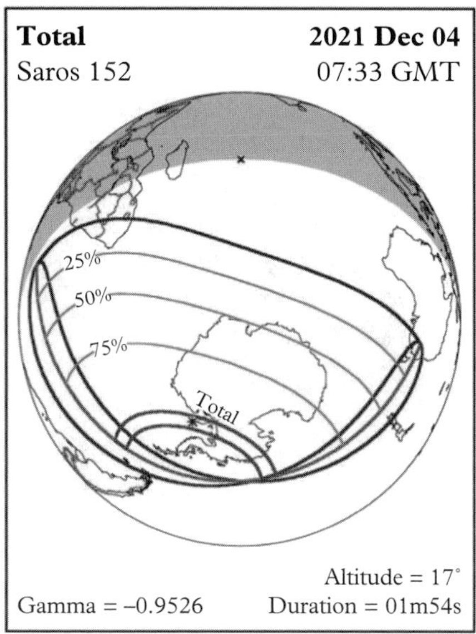
Total
2021 Dec 04
Saros 152
07:33 GMT
25%
50%
75%
Total
Altitude = 17°
Gamma = −0.9526
Duration = 01m54s

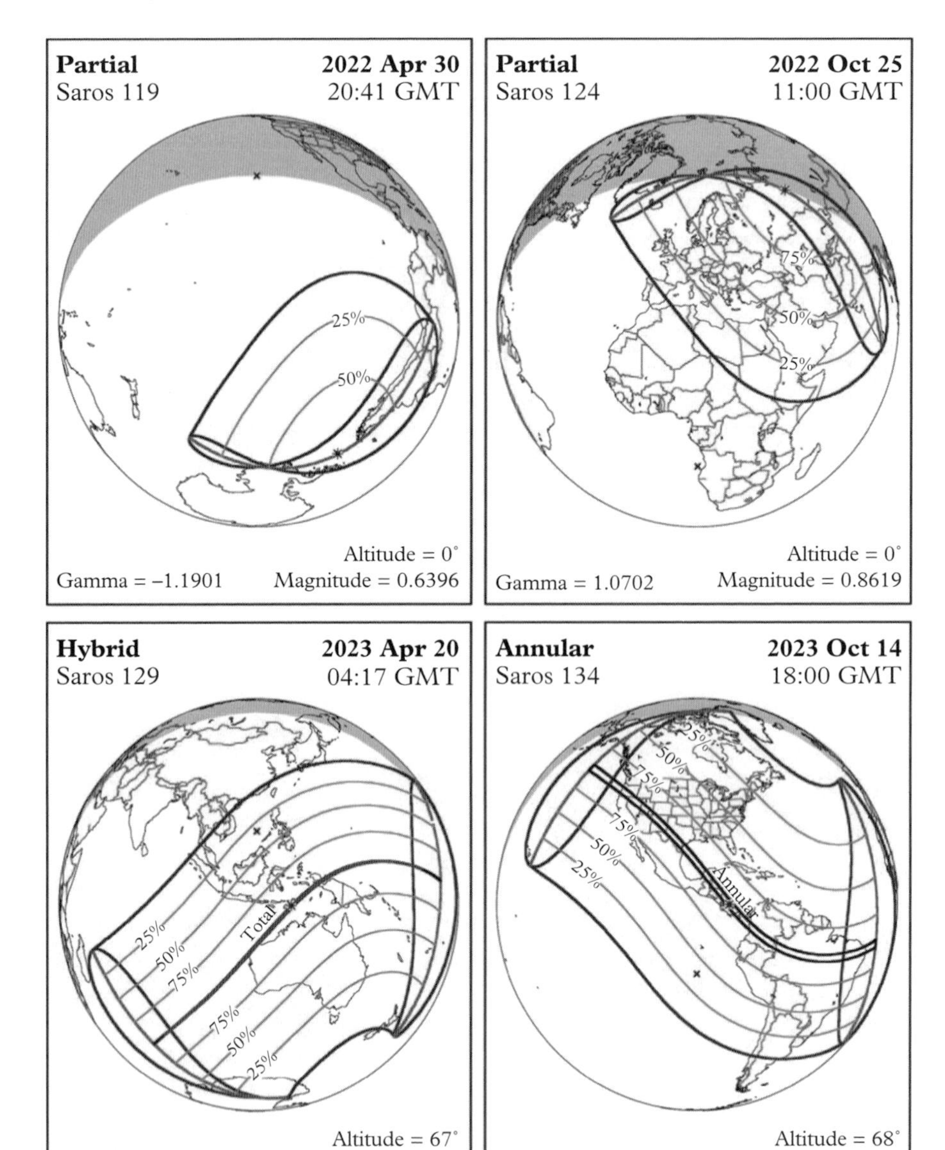
Partial
Saros 119
2022 Apr 30
20:41 GMT
25%
50%
Altitude = 0°
Gamma = –1.1901
Magnitude = 0.6396
Partial
Saros 124
2022 Oct 25
11:00 GMT
75%
50%
25%
Altitude = 0°
Gamma = 1.0702
Magnitude = 0.8619
Hybrid
Saros 129
2023 Apr 20
04:17 GMT
25%
50%
75%
Total
75%
50%
25%
Altitude = 67°
Gamma = –0.3952
Duration = 01m16s
Annular
Saros 134
2023 Oct 14
18:00 GMT
25%
50%
75%
75%
50%
25%
Annular
Altitude = 68°
Gamma = 0.3753
Duration = 05m17s

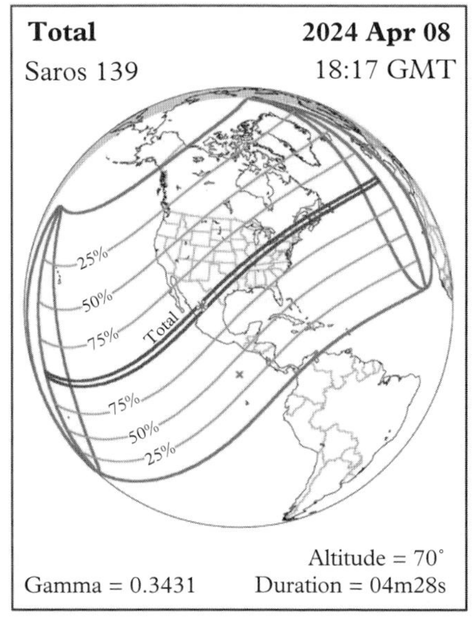
Total
2024 Apr 08
Saros 139
18:17 GMT
25%
50%
75%
Total
75%
50%
25%
Altitude = 70°
Gamma = 0.3431
Duration = 04m28s

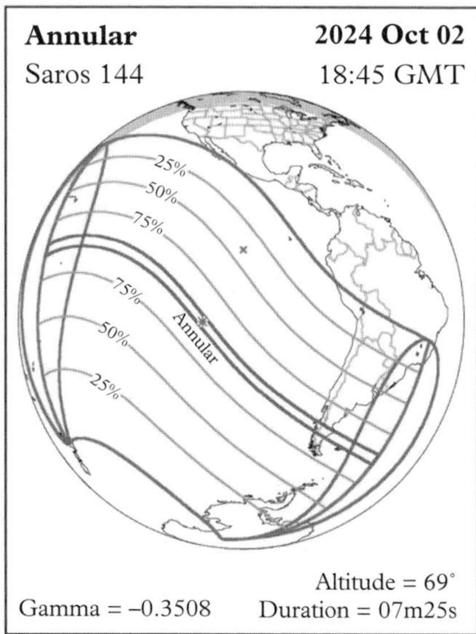
Annular
2024 Oct 02
Saros 144
18:45 GMT
25%
50%
75%
75%
50%
25%
Annular
Altitude = 69°
Gamma = –0.3508
Duration = 07m25s

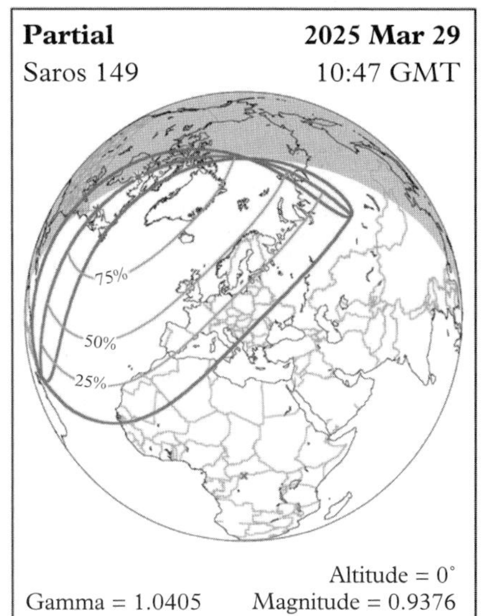
Partial
2025 Mar 29
Saros 149
10:47 GMT
75%
50%
25%
Altitude = 0°
Gamma = 1.0405
Magnitude = 0.9376

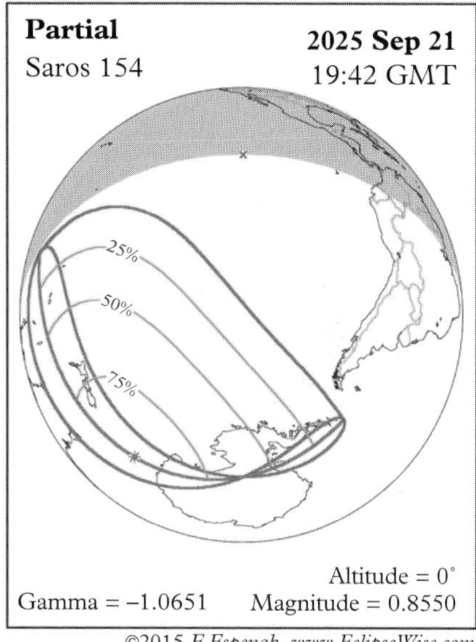
Partial
2025 Sep 21
Saros 154
19:42 GMT
25%
50%
75%
Altitude = 0°
Gamma = –1.0651
Magnitude = 0.8550

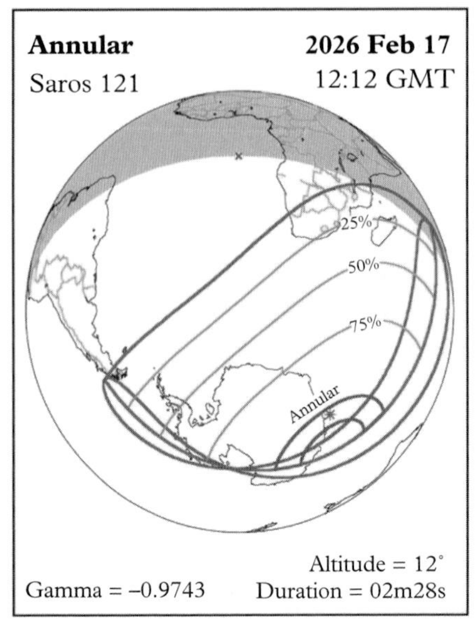
Annular
2026 Feb 17
Saros 121
12:12 GMT
25%
50%
75%
Annular
Altitude = 12°
Gamma = −0.9743
Duration = 02m28s

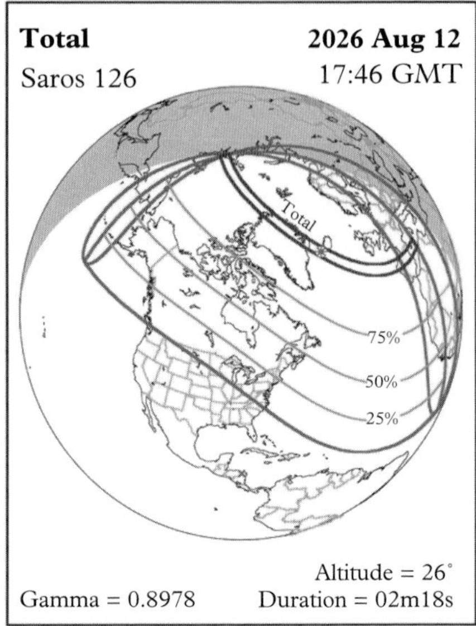
Total
2026 Aug 12
Saros 126
17:46 GMT
Total
75%
50%
25%
Altitude = 26°
Gamma = 0.8978
Duration = 02m18s

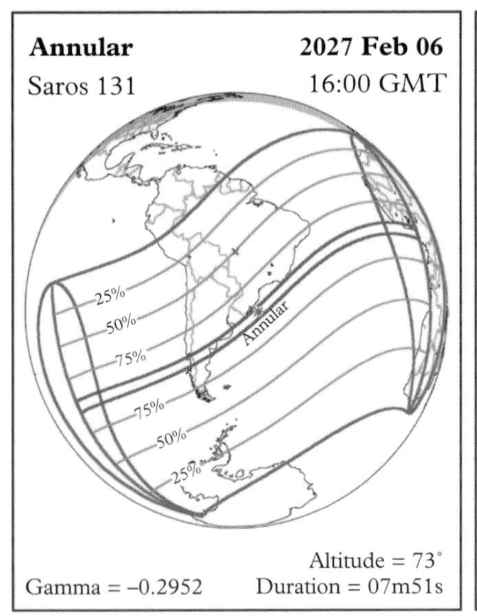
Annular
2027 Feb 06
Saros 131
16:00 GMT
25%
50%
75%
Annular
75%
50%
25%
Altitude = 73°
Gamma = −0.2952
Duration = 07m51s

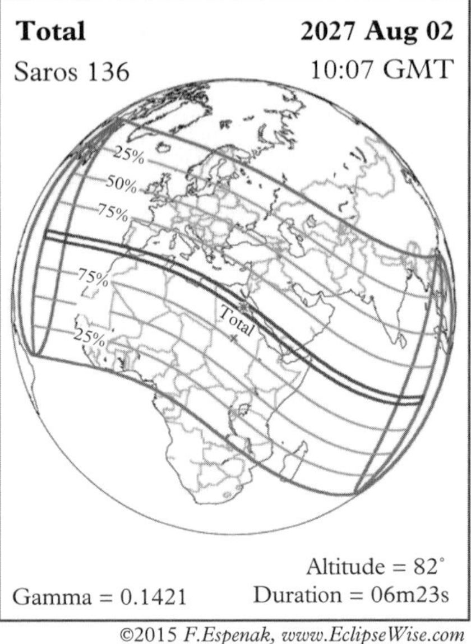
Total
2027 Aug 02
Saros 136
10:07 GMT
25%
50%
75%
75%
Total
25%
Altitude = 82°
Gamma = 0.1421
Duration = 06m23s

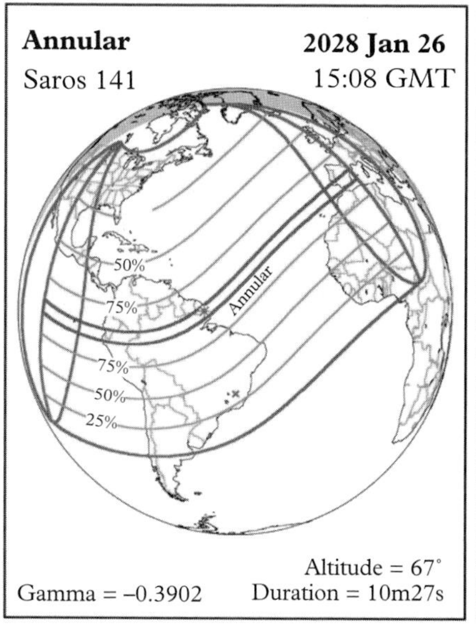
Annular
2028 Jan 26
Saros 141
15:08 GMT
50%
75%
Annular
75%
50%
25%
Altitude = 67°
Gamma = –0.3902
Duration = 10m27s

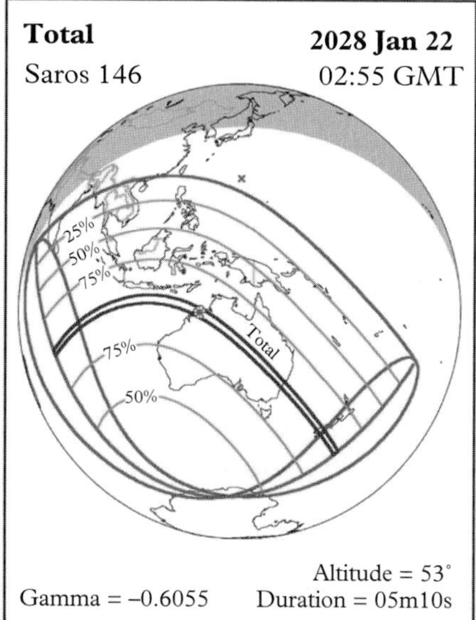
Total
2028 Jan 22
Saros 146
02:55 GMT
25%
50%
75%
Total
75%
50%
Altitude = 53°
Gamma = –0.6055
Duration = 05m10s

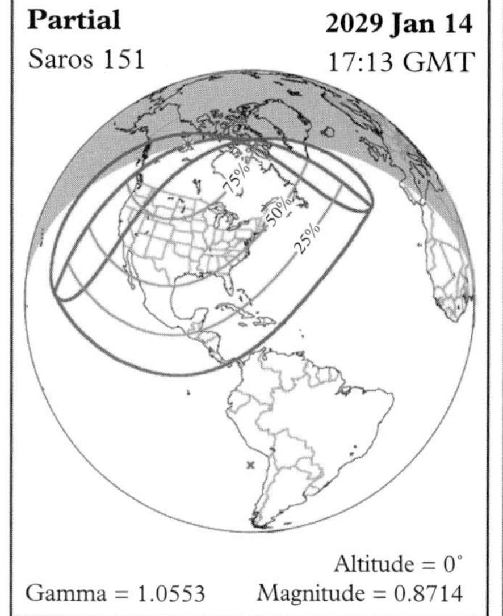
Partial
2029 Jan 14
Saros 151
17:13 GMT
75%
50%
25%
Altitude = 0°
Gamma = 1.0553
Magnitude = 0.8714

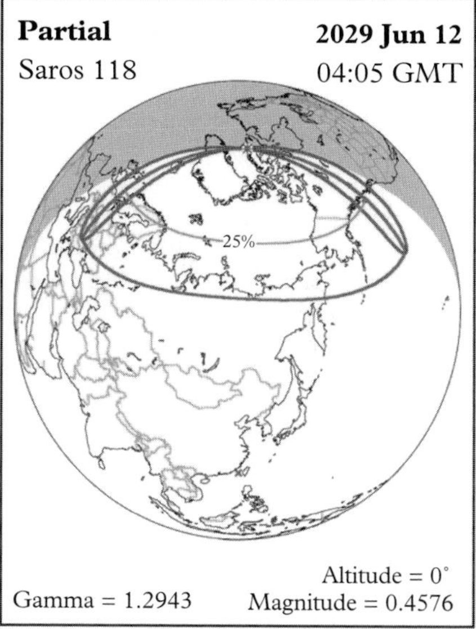
Partial
2029 Jun 12
Saros 118
04:05 GMT
25%
Altitude = 0°
Gamma = 1.2943
Magnitude = 0.4576

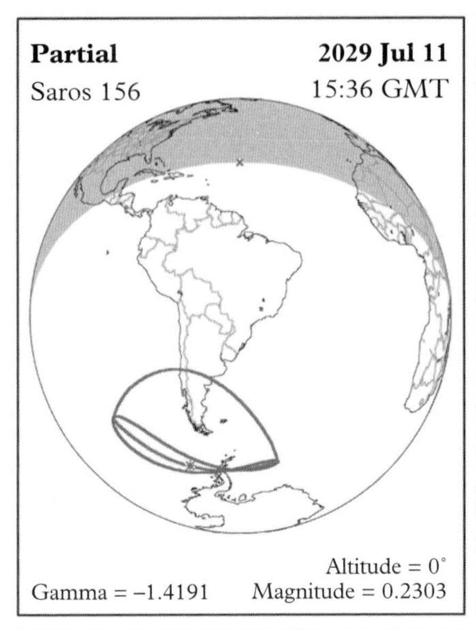
Partial
2029 Jul 11
Saros 156
15:36 GMT
Altitude = 0°
Gamma = –1.4191
Magnitude = 0.2303

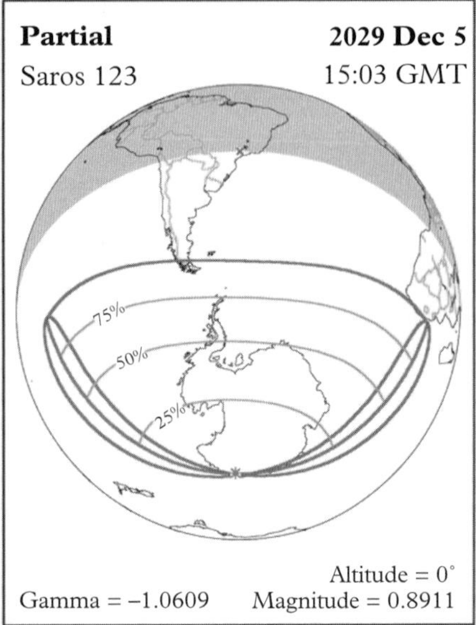
Partial
2029 Dec 5
Saros 123
15:03 GMT
75%
50%
25%
Altitude = 0°
Gamma = –1.0609
Magnitude = 0.8911

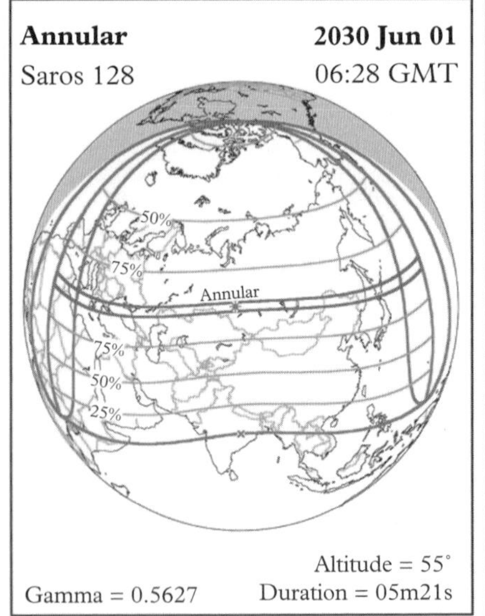
Annular
2030 Jun 01
Saros 128
06:28 GMT
50%
75%
Annular
75%
50%
25%
Altitude = 55°
Gamma = 0.5627
Duration = 05m21s

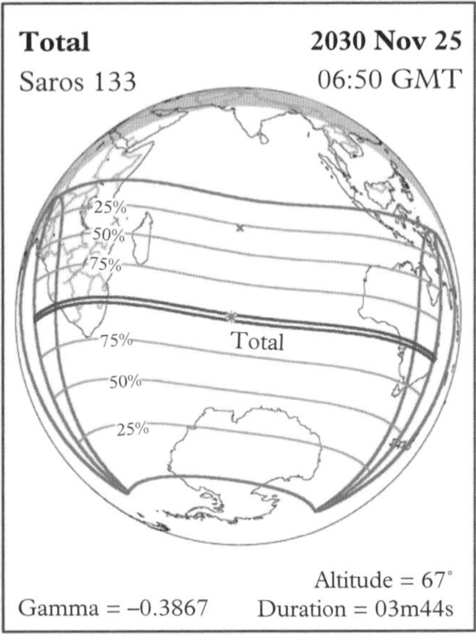
Total
2030 Nov 25
Saros 133
06:50 GMT
25%
50%
75%
75%
Total
50%
25%
Altitude = 67°
Gamma = –0.3867
Duration = 03m44s

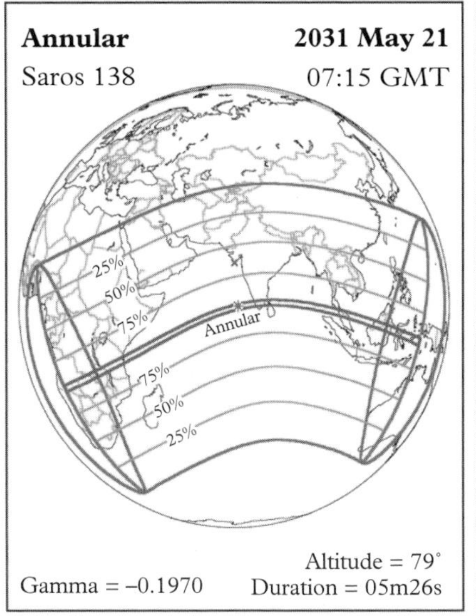
Annular
2031 May 21
Saros 138
07:15 GMT
25%
50%
75%
Annular
75%
50%
25%
Altitude = 79°
Gamma = –0.1970
Duration = 05m26s

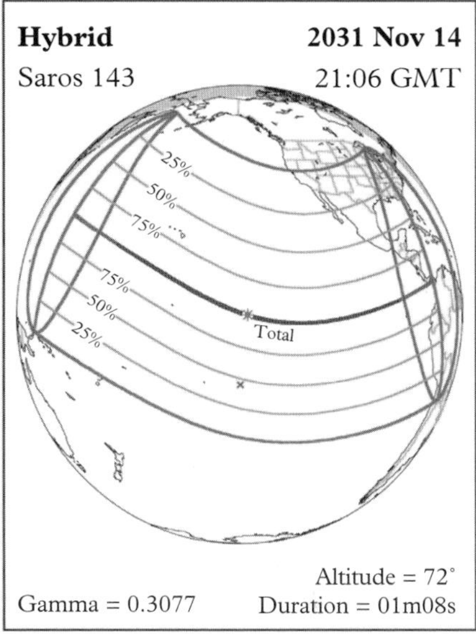
Hybrid
2031 Nov 14
Saros 143
21:06 GMT
25%
50%
75%
75%
50%
25%
Total
Altitude = 72°
Gamma = 0.3077
Duration = 01m08s

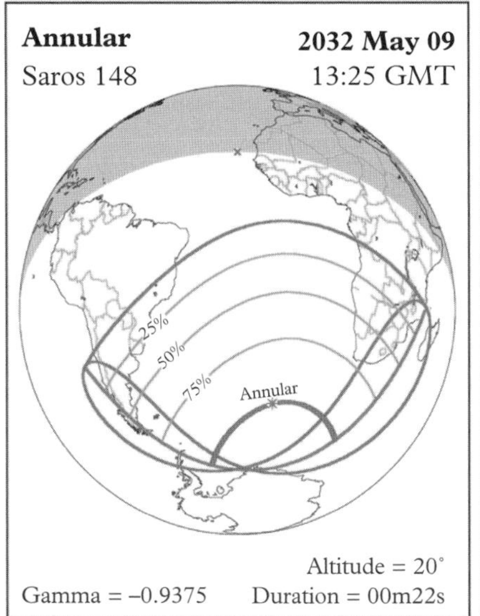
Annular
2032 May 09
Saros 148
13:25 GMT
25%
50%
75%
Annular
Altitude = 20°
Gamma = –0.9375
Duration = 00m22s

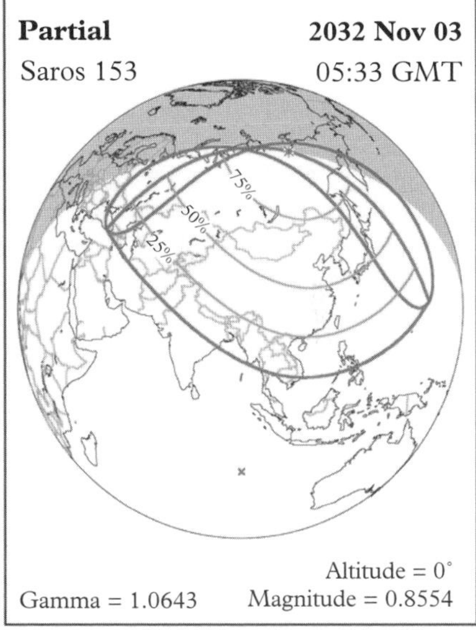
Partial
2032 Nov 03
Saros 153
05:33 GMT
75%
50%
25%
Altitude = 0°
Gamma = 1.0643
Magnitude = 0.8554

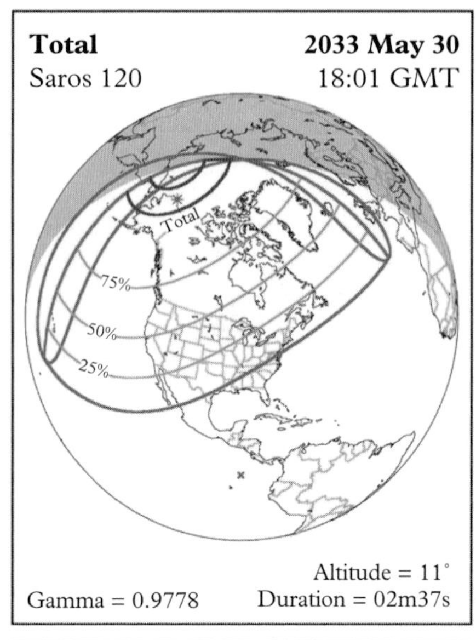
Total
2033 May 30
Saros 120
18:01 GMT
Total
75%
50%
25%
Altitude = 11°
Gamma = 0.9778
Duration = 02m37s

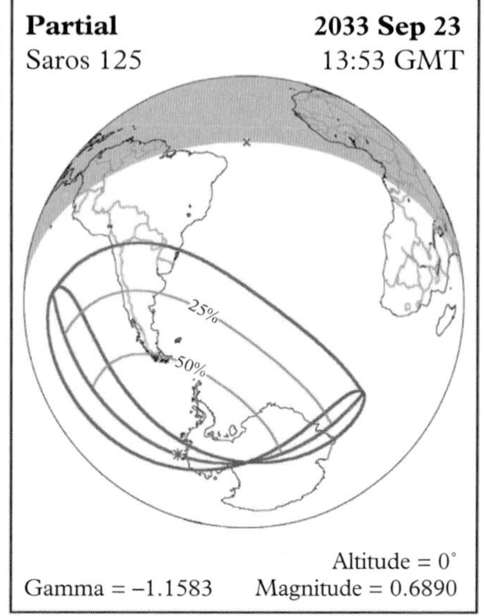
Partial
2033 Sep 23
Saros 125
13:53 GMT
25%
50%
Altitude = 0°
Gamma = –1.1583
Magnitude = 0.6890

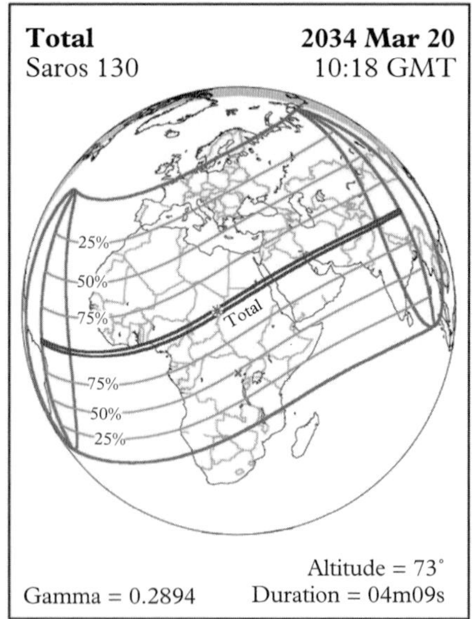
Total
2034 Mar 20
Saros 130
10:18 GMT
25%
50%
75%
Total
75%
50%
25%
Altitude = 73°
Gamma = 0.2894
Duration = 04m09s

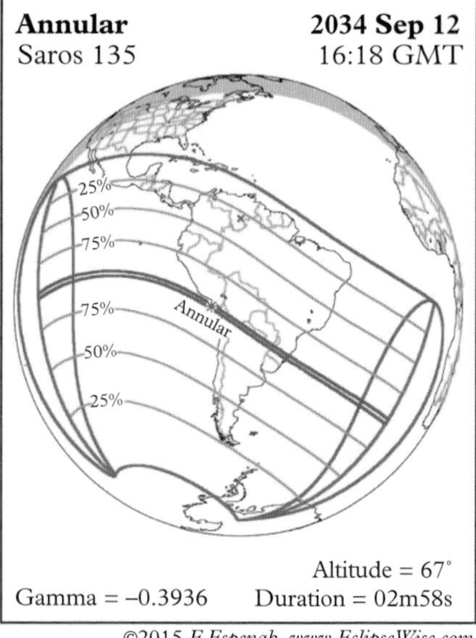
Annular
2034 Sep 12
Saros 135
16:18 GMT
25%
50%
75%
75%
Annular
50%
25%
Altitude = 67°
Gamma = –0.3936
Duration = 02m58s

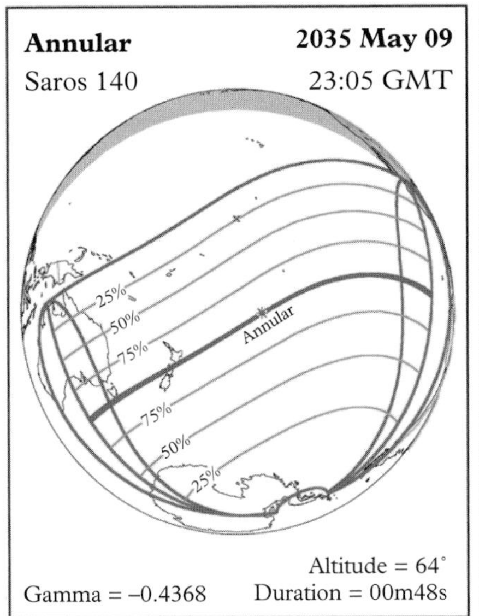
Annular
Saros 140
2035 May 09
23:05 GMT
25%
50%
75%
Annular
75%
50%
25%
Altitude = 64°
Gamma = –0.4368
Duration = 00m48s

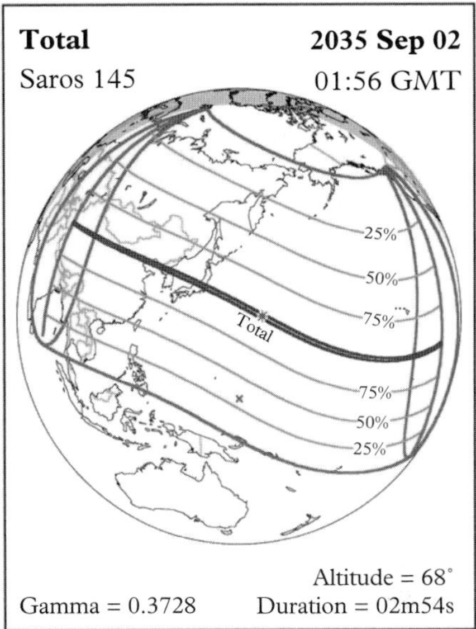
Total
Saros 145
2035 Sep 02
01:56 GMT
25%
50%
75%
Total
75%
50%
25%
Altitude = 68°
Gamma = 0.3728
Duration = 02m54s

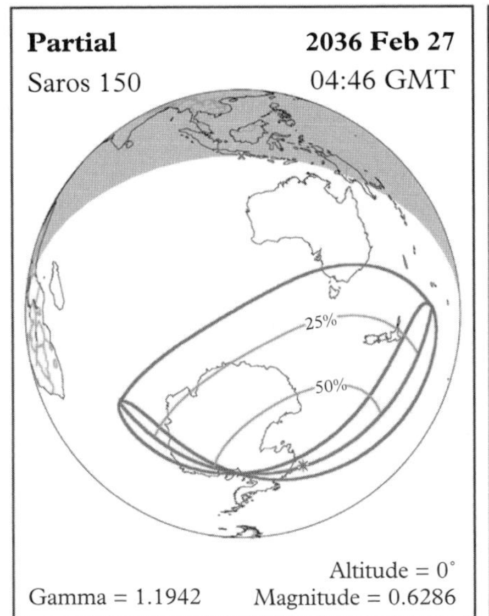
Partial
Saros 150
2036 Feb 27
04:46 GMT
25%
50%
Altitude = 0°
Gamma = 1.1942
Magnitude = 0.6286

Partial
Saros 117
2036 Jul 23
10:31 GMT
Altitude = 0°
Gamma = –1.4250
Magnitude = 0.1992

Partial **2036 Aug 21**
Saros 155 17:24 GMT

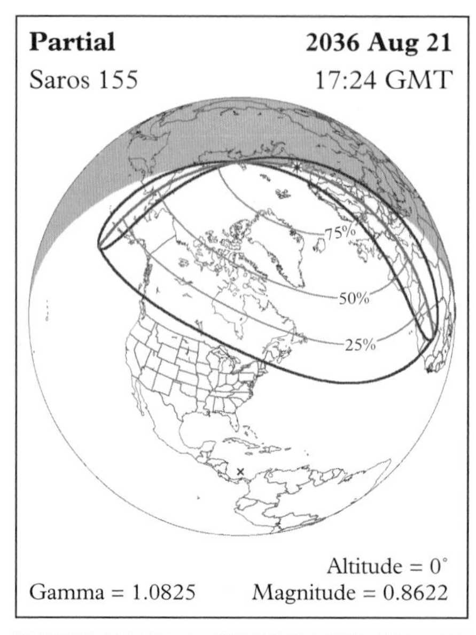

Altitude = 0°
Gamma = 1.0825 Magnitude = 0.8622

Partial **2037 Jan 16**
Saros 122 09:48 GMT

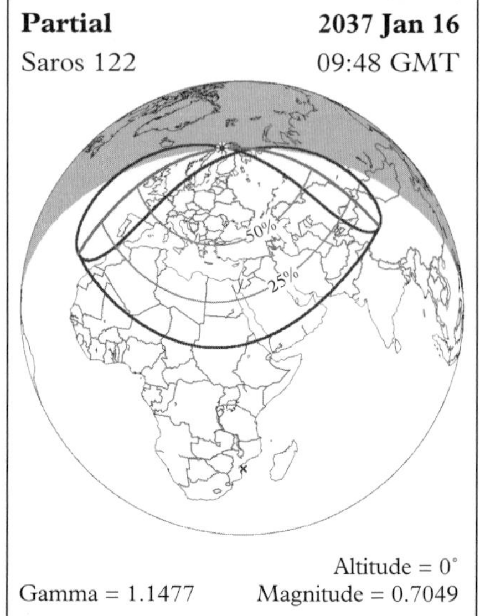

Altitude = 0°
Gamma = 1.1477 Magnitude = 0.7049

Total **2037 Jul 13**
Saros 127 02:39 GMT

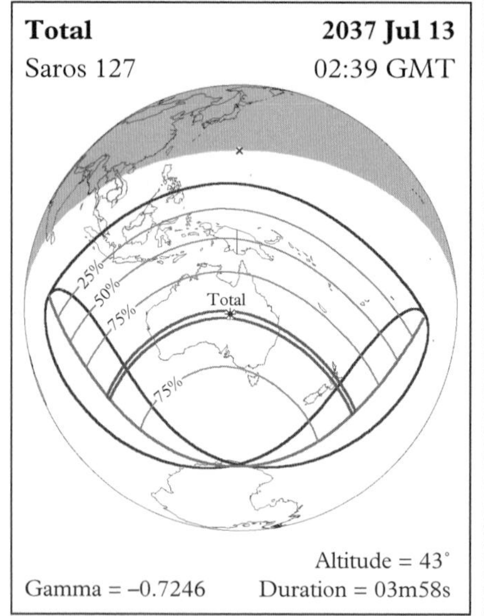

Altitude = 43°
Gamma = –0.7246 Duration = 03m58s

Annular **2038 Jan 05**
Saros 132 13:46 GMT

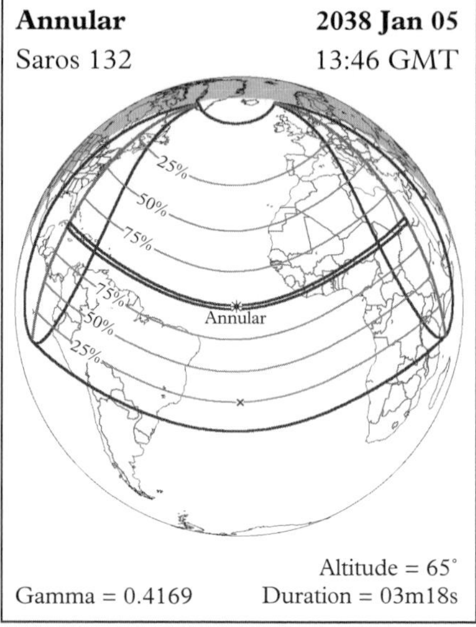

Altitude = 65°
Gamma = 0.4169 Duration = 03m18s

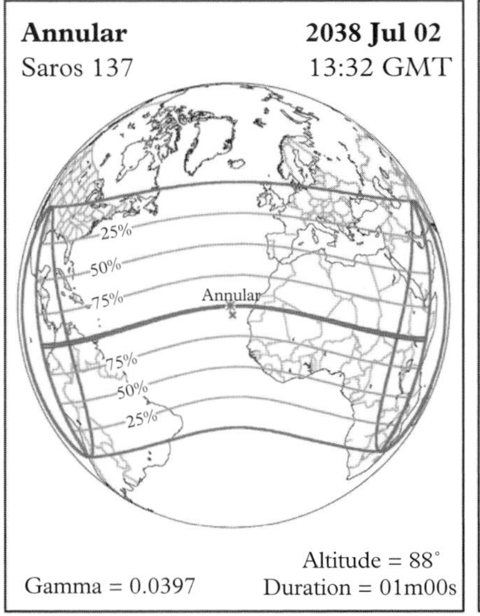
Annular
2038 Jul 02
Saros 137
13:32 GMT
25%
50%
75%
Annular
75%
50%
25%
Altitude = 88°
Gamma = 0.0397
Duration = 01m00s

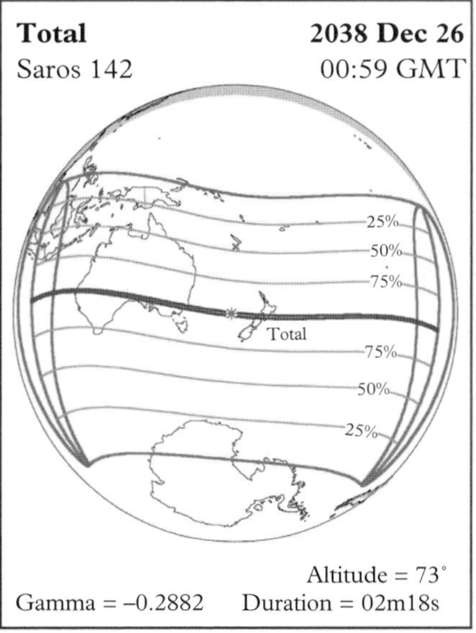
Total
2038 Dec 26
Saros 142
00:59 GMT
25%
50%
75%
Total
75%
50%
25%
Altitude = 73°
Gamma = –0.2882
Duration = 02m18s

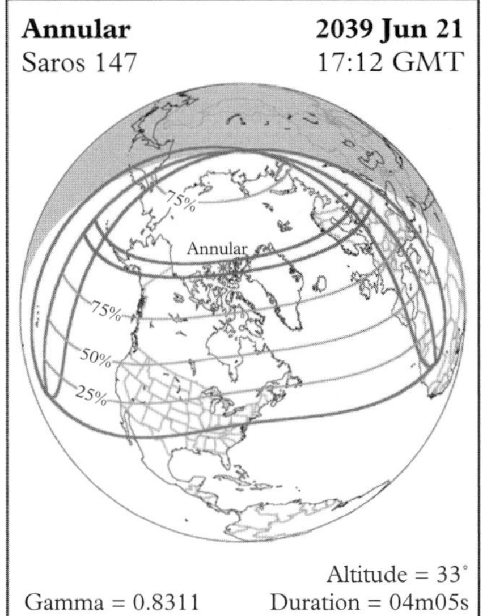
Annular
2039 Jun 21
Saros 147
17:12 GMT
75%
Annular
75%
50%
25%
Altitude = 33°
Gamma = 0.8311
Duration = 04m05s

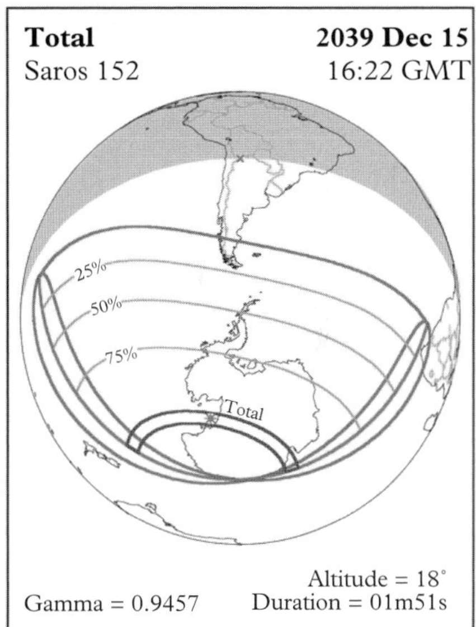
Total
2039 Dec 15
Saros 152
16:22 GMT
25%
50%
75%
Total
Altitude = 18°
Gamma = 0.9457
Duration = 01m51s

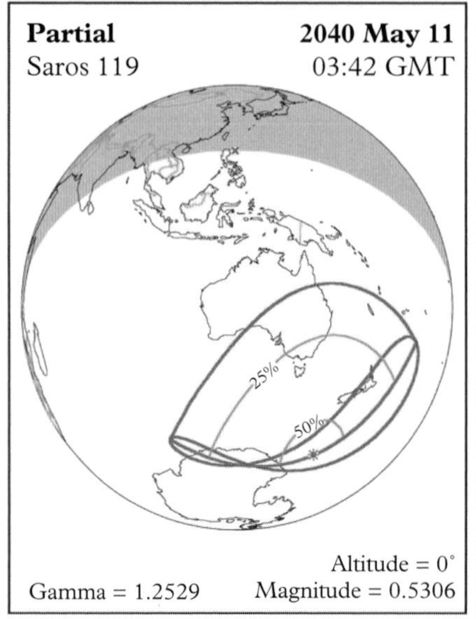
Partial
Saros 119
2040 May 11
03:42 GMT
25%
50%
Altitude = 0°
Gamma = 1.2529
Magnitude = 0.5306

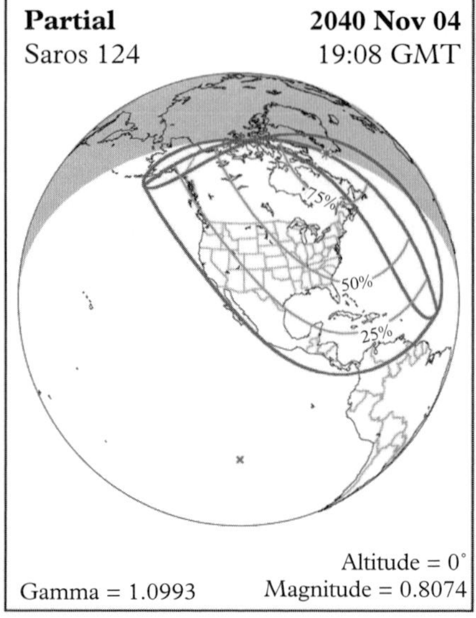
Partial
Saros 124
2040 Nov 04
19:08 GMT
75%
50%
25%
Altitude = 0°
Gamma = 1.0993
Magnitude = 0.8074

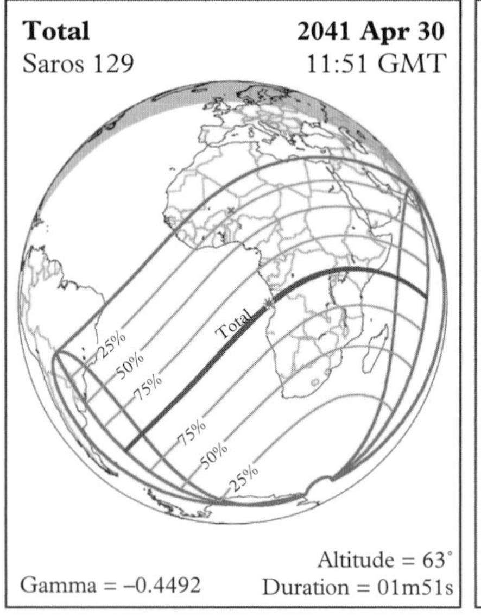
Total
Saros 129
2041 Apr 30
11:51 GMT
Total
25%
50%
75%
75%
50%
25%
Altitude = 63°
Gamma = –0.4492
Duration = 01m51s

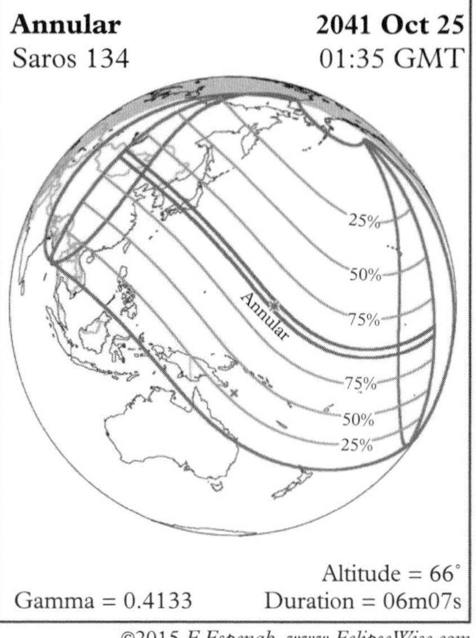
Annular
Saros 134
2041 Oct 25
01:35 GMT
25%
50%
75%
Annular
75%
50%
25%
Altitude = 66°
Gamma = 0.4133
Duration = 06m07s

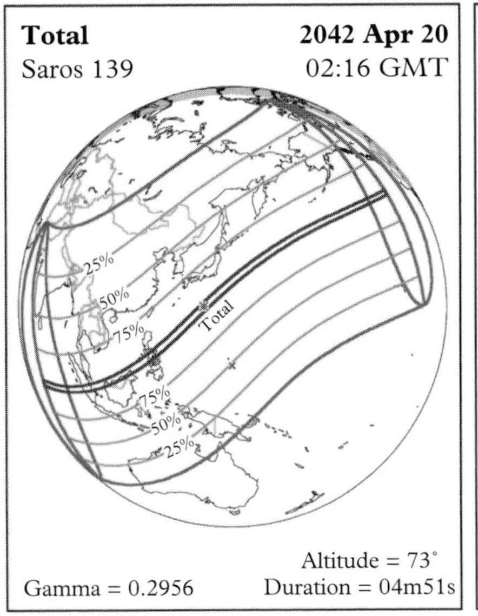
Total
2042 Apr 20
Saros 139
02:16 GMT
25%
50%
75%
Total
75%
50%
25%
Altitude = 73°
Gamma = 0.2956
Duration = 04m51s

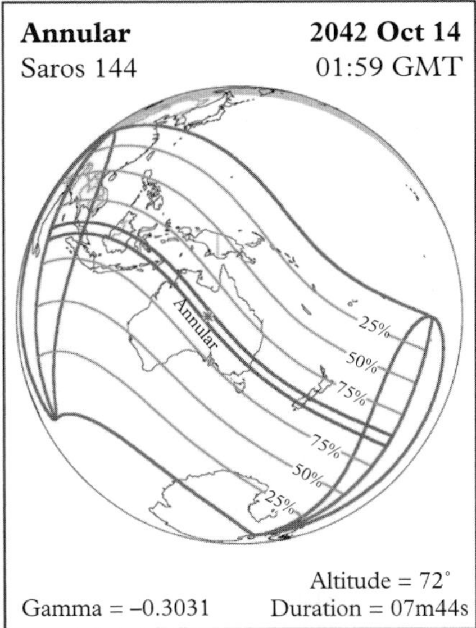
Annular
2042 Oct 14
Saros 144
01:59 GMT
Annular
25%
50%
75%
75%
50%
25%
Altitude = 72°
Gamma = –0.3031
Duration = 07m44s

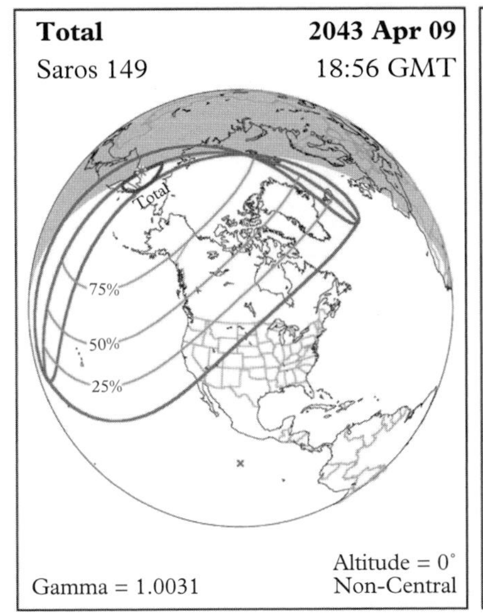
Total
2043 Apr 09
Saros 149
18:56 GMT
Total
75%
50%
25%
Altitude = 0°
Gamma = 1.0031
Non-Central

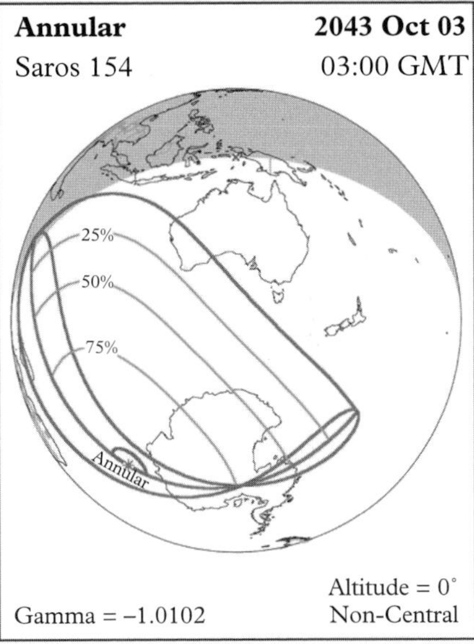
Annular
2043 Oct 03
Saros 154
03:00 GMT
25%
50%
75%
Annular
Altitude = 0°
Gamma = –1.0102
Non-Central

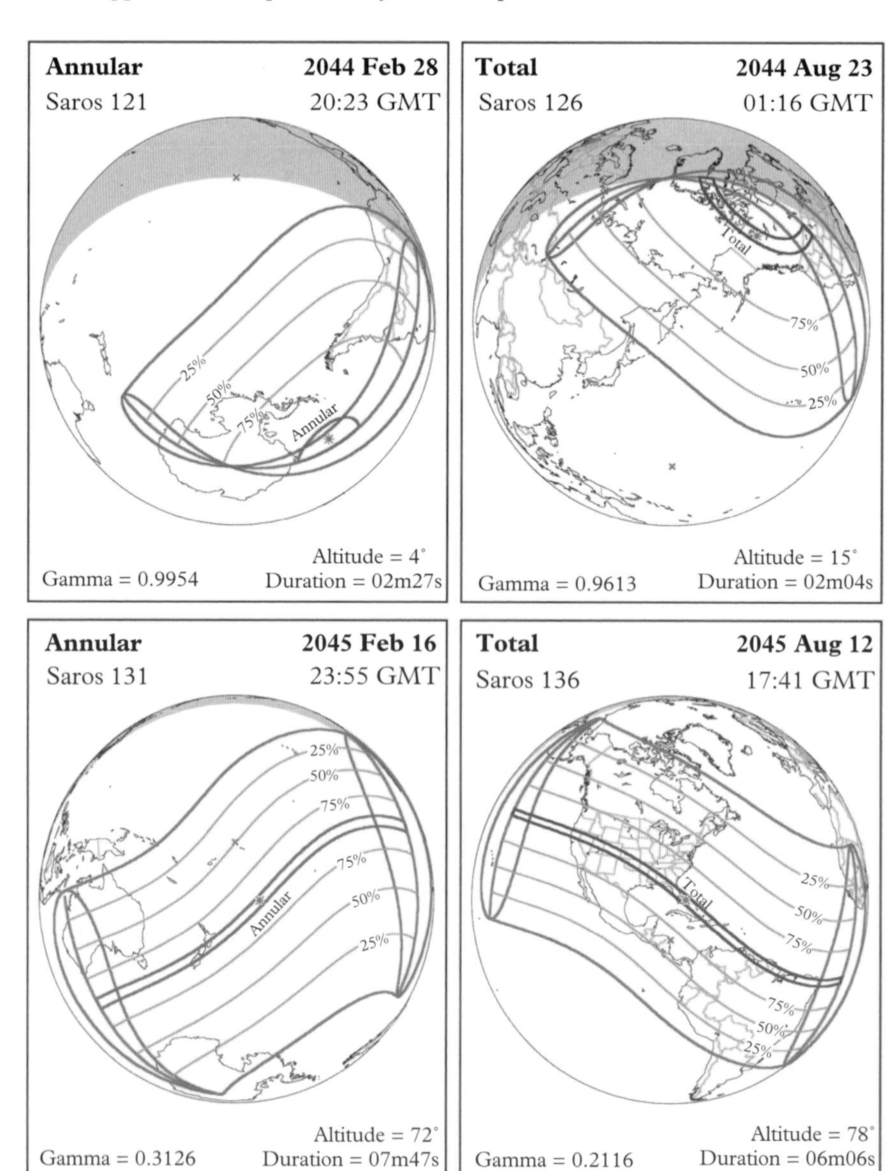
Annular
2044 Feb 28
Saros 121
20:23 GMT
25%
50%
75%
Annular
Altitude = 4°
Gamma = 0.9954
Duration = 02m27s
Total
2044 Aug 23
Saros 126
01:16 GMT
Total
75%
50%
25%
Altitude = 15°
Gamma = 0.9613
Duration = 02m04s
Annular
2045 Feb 16
Saros 131
23:55 GMT
25%
50%
75%
75%
50%
25%
Annular
Altitude = 72°
Gamma = 0.3126
Duration = 07m47s
Total
2045 Aug 12
Saros 136
17:41 GMT
25%
50%
75%
Total
75%
50%
25%
Altitude = 78°
Gamma = 0.2116
Duration = 06m06s

Central Solar Eclipses for North & South America: 2017–2045

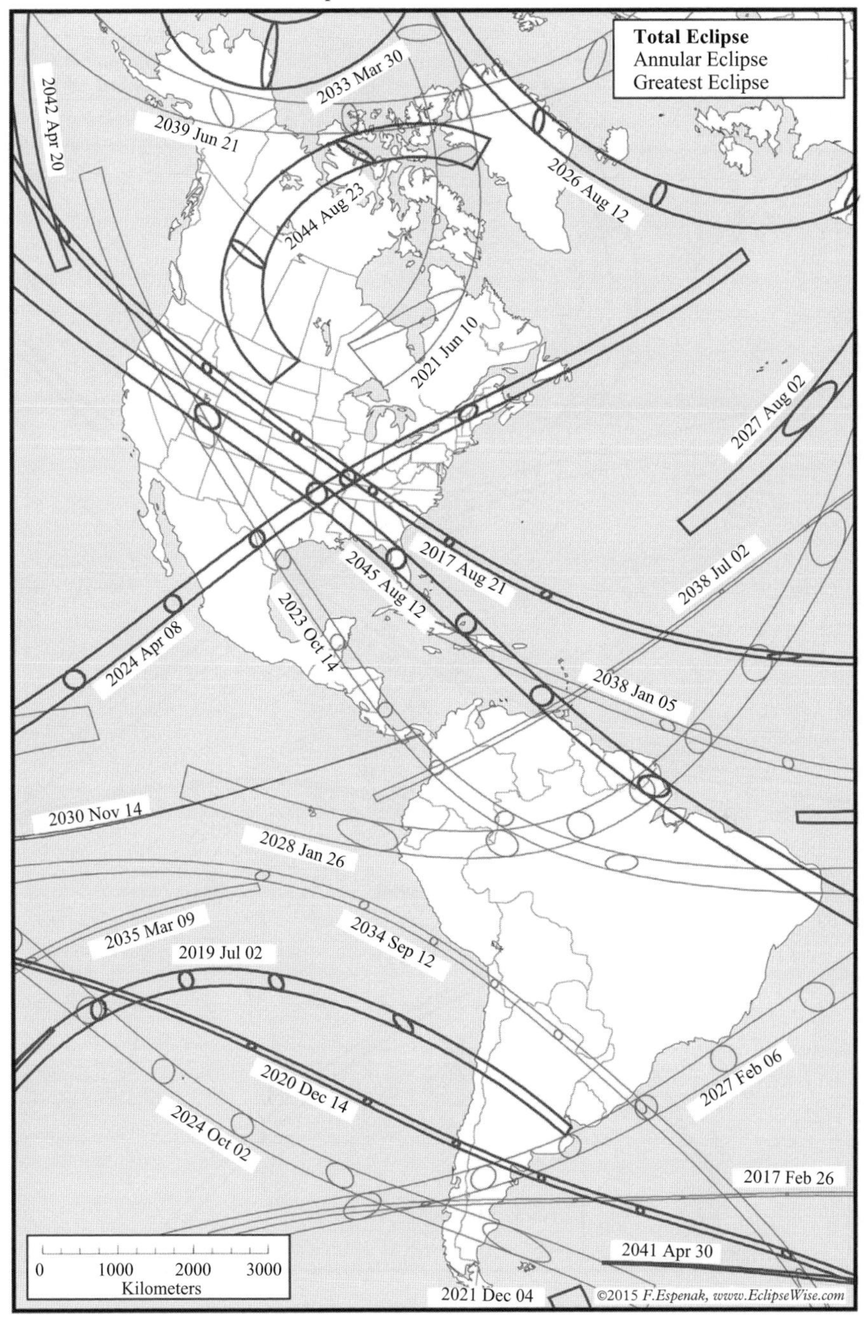

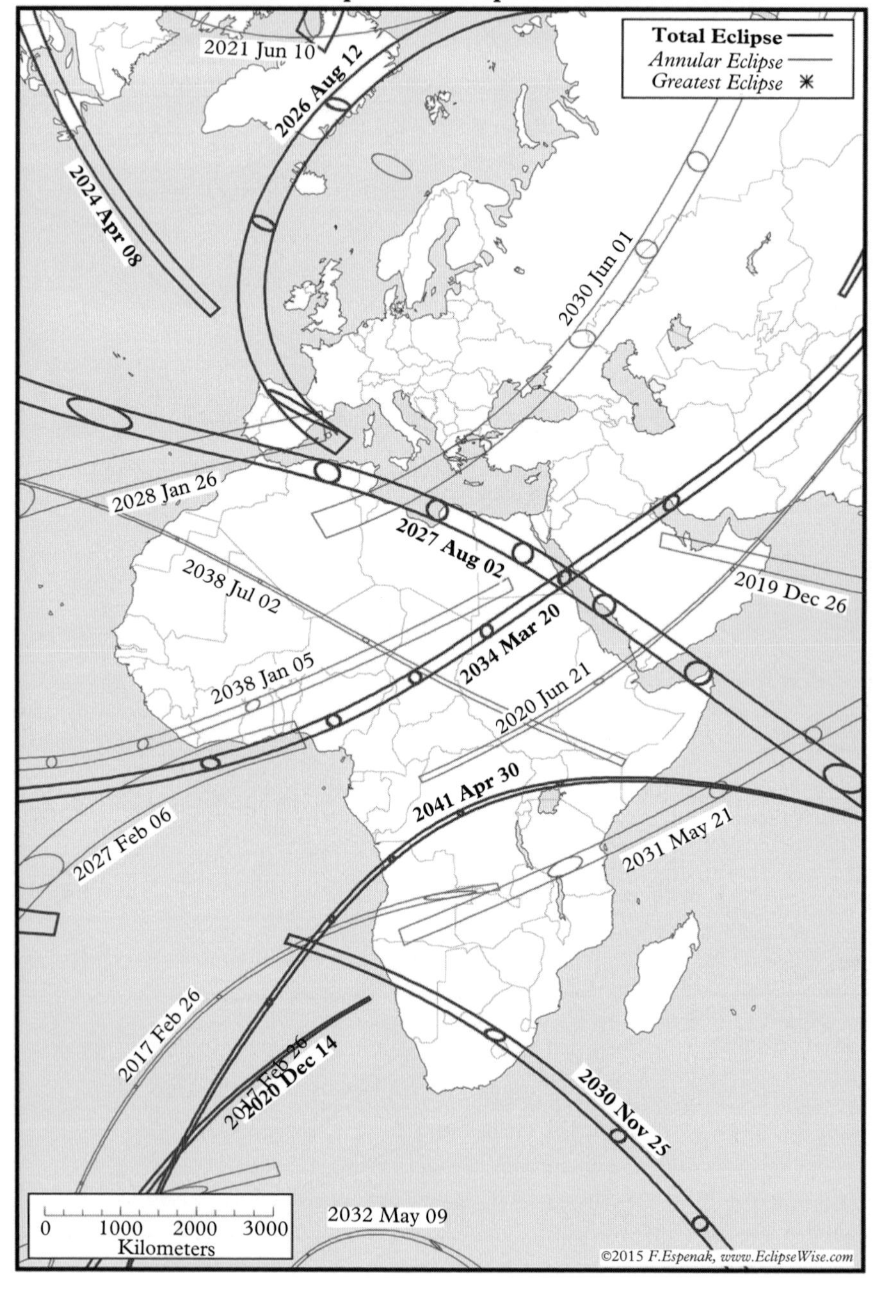
Central Solar Eclipses for Europe and Africa: 2017–2045
Total Eclipse
Annular Eclipse
Greatest Eclipse
2021 Jun 10
2026 Aug 12
2024 Apr 08
2030 Jun 01
2028 Jan 26
2027 Aug 02
2038 Jul 02
2019 Dec 26
2034 Mar 20
2038 Jan 05
2020 Jun 21
2041 Apr 30
2027 Feb 06
2031 May 21
2017 Feb 26
2020 Dec 14
2030 Nov 25
0 1000 2000 3000
Kilometers
2032 May 09
©2015 F.Espenak, www.EclipseWise.com

Central Solar Eclipses for Asia and Australia: 2017–2045

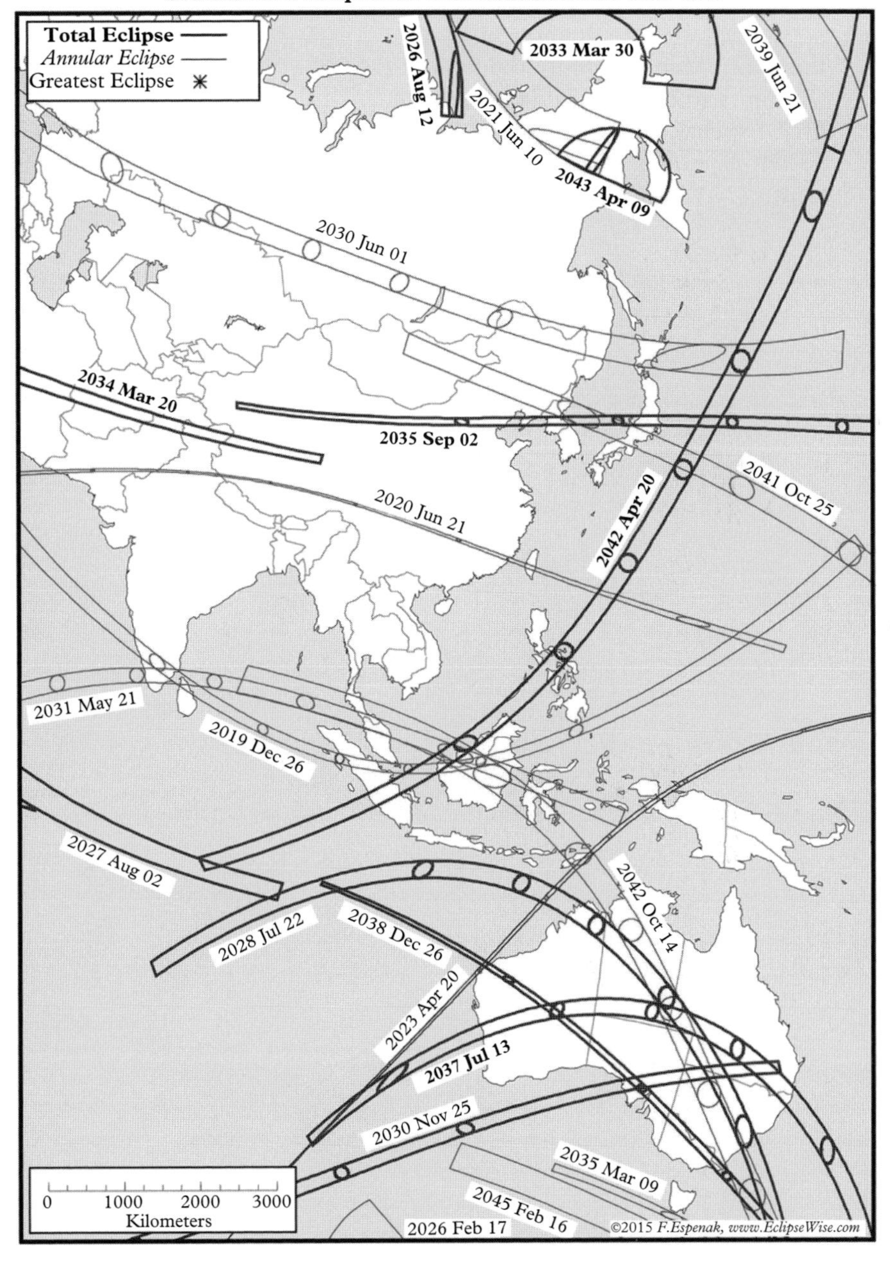

NOTES AND REFERENCES

1. Epigraph: Mabel Loomis Todd: *Total Eclipses of the Sun*, revised edition (Boston: Little, Brown, 1900), page 25.
2. Fred Espenak and Jean Meeus: *Five Millennium Canon of Solar Eclipses: −1999 to +3000* (Greenbelt, Maryland: NASA Goddard Space Flight Center [NASA/TP–2006–214,141], 2006.

Appendix B

Total, Annular, and Hybrid Eclipses: 2017–2070[1]

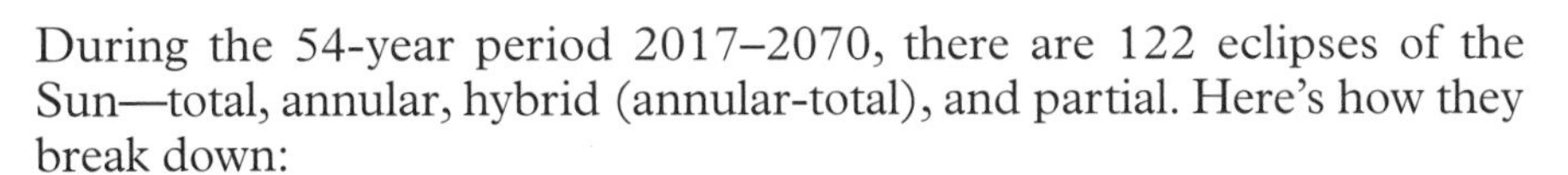

During the 54-year period 2017–2070, there are 122 eclipses of the Sun—total, annular, hybrid (annular-total), and partial. Here's how they break down:

Eclipse Type	2017–2070
Partial	44 = 36.1%
Annular	38 = 31.1%
Total	35 = 28.7%
Hybrid	5 = 4.1%

The table below gives the basic details for the 78 total, annular, and hybrid eclipses occurring during this period. When an eclipse path crosses the International Date Line, it can be seen on one of two dates depending on the observer's geographic position.

Date	Maximum Duration[2]	Eclipse Type	Geographic Region of Visibility
2017 Feb 26	*0m 44s*	*Annular*	*Pacific, Chile, Argentina, Atlantic Ocean, Angola, Zambia, DR Congo*
2017 Aug 21	**2m 40s**	**Total**	**Pacific Ocean, United States (Oregon, Idaho, Wyoming, Nebraska, Missouri, Illinois, Kentucky, Tennessee, North Carolina, Georgia, South Carolina), Atlantic Ocean**
2019 Jul 02	**4m 33s**	**Total**	**South Pacific Ocean, Chile, Argentina**
2019 Dec 26	*3m 39s*	*Annular*	*Saudi Arabia, Bahrain, Qatar, United Arab Emirates, Oman, southern India, Sri Lanka, Indonesia, Malaysia, Guam*

Date	Maximum Duration[2]	Eclipse Type	Geographic Region of Visibility
2020 Jun 21	*0m 39s*	*Annular*	*DR Congo, South Sudan, Ethiopia, Eritrea, Yemen, Saudi Arabia, Oman, Pakistan, northern India, China, Taiwan*
2020 Dec 14	**2m 10s**	**Total**	**Pacific Ocean, Chile, Argentina, south Atlantic Ocean**
2021 Jun 10	*3m 51s*	*Annular*	*Central Canada, Greenland, North Pole, Russia (Siberia)*
2021 Dec 04	**1m 54s**	**Total**	**Antarctica**
2023 Apr 20	***1m16s***	***Hybrid***	***South Indian Ocean, western Australia, Indonesia, Pacific Ocean (total except at beginning and end of path)***
2023 Oct 14	*5m 17s*	*Annular*	*Pacific Ocean, United States (Oregon, California, Nevada, Utah, Colorado, Arizona, New Mexico, Texas), Mexico (Yucatán Peninsula), Belize, Honduras, Nicaragua, Panama, Colombia, Brazil, Atlantic Ocean*
2024 Apr 8	**4m 28s**	**Total**	**Pacific Ocean, Mexico, United States (Texas, Oklahoma, Arkansas, Missouri, Kentucky, Illinois, Indiana, Ohio, Pennsylvania, New York, Vermont, New Hampshire, Maine), southeastern Canada, Atlantic Ocean**
2024 Oct 02	*7m 25s*	*Annular*	*South Pacific Ocean, Easter Island, Chile, Argentina, Atlantic Ocean*
2026 Feb 17	*2m 20s*	*Annular*	*Antarctica*
2026 Aug 12	**2m 18s**	**Total**	**Arctic Ocean, Greenland, Iceland, Atlantic Ocean, Spain**
2027 Feb 06	*7m 51s*	*Annular*	*Pacific Ocean, Chile, Argentina, Uruguay, Atlantic Ocean, Ivory Coast, Ghana, Togo, Benin, Nigeria*
2027 Aug 02	**6m 23s**	**Total**	**Atlantic Ocean, Morocco, Spain, Algeria, Tunisia, Libya, Egypt, Saudi Arabia, Yemen, Somalia, Indian Ocean**
2028 Jan 26	*10m 27s*	*Annular*	*Pacific Ocean, Galapagos, Ecuador, Peru, Brazil, French Guiana, Portugal, Spain*

Date	Maximum Duration[2]	Eclipse Type	Geographic Region of Visibility
2028 Jul 22	**5m 10s**	**Total**	**South Indian Ocean, Australia, New Zealand**
2030 Jun 01	*5m 21s*	*Annular*	*Algeria, Tunisia, Libya, Greece, Bulgaria, Turkey, Ukraine, Russia, Kazakhstan, northern China, Japan*
2030 Nov 25	**3m 44s**	**Total**	**Namibia, Botswana, South Africa, Lesotho, Indian Ocean, Australia**
2031 May 21	*5m 26s*	*Annular*	*Angola, Zambia, DR Congo, Malawi, Tanzania, southern India, Sri Lanka, Thailand, Malaysia, Indonesia*
2031 Nov 14/15	*1m 08s*	*Hybrid*	*Pacific Ocean (total), Panama (annular)*
2032 May 09	*0m 22s*	*Annular*	*South Atlantic Ocean*
2033 Mar 30	**2m 37s**	**Total**	**Russia (Siberia), United States (Alaska), Arctic Ocean**
2034 Mar 20	**4m 09s**	**Total**	**Atlantic Ocean, Nigeria, Cameroon, Chad, Sudan, Egypt, Saudi Arabia, Kuwait, Iran, Afghanistan, Pakistan, India, China**
2034 Sep 12	*2m 58s*	*Annular*	*Pacific Ocean, Chile, Bolivia, Argentina, Paraguay, Brazil, Atlantic*
2035 Mar 09/10	*0m 48s*	*Annular*	*New Zealand, South Pacific Ocean*
2035 Sep 02	**2m 54s**	**Total**	**China, North Korea, Japan, Pacific Ocean**
2037 Jul 13	**3m 58s**	**Total**	**Indian Ocean, Australia, New Zealand, South Pacific Ocean**
2038 Jan 05	*3m 19s*	*Annular*	*Cuba, Haiti, Dominican Republic, St. Lucia, St. Vincent, Barbados, Atlantic Ocean, Liberia, Ivory Coast, Ghana, Togo, Benin, Niger, Libya, Chad, Libya, Sudan, Egypt*
2038 Jul 02	*1m 00s*	*Annular*	*Pacific Ocean, Colombia, Venezuela, Atlantic Ocean, Western Sahara, Mauritania, Mali, Algeria, Niger, Chad, Sudan, outh Sudan, Ethiopia, Kenya*
2038 Dec 25/26	**2m 18s**	**Total**	**Indian Ocean, Australia, New Zealand, Pacific Ocean**

Date	Maximum Duration[2]	Eclipse Type	Geographic Region of Visibility
2039 Jun 21	*4m 05s*	*Annular*	*Pacific Ocean, United States (Alaska), Canada, Greenland, Arctic Ocean, Greenland, Atlantic, Norway, Sweden, Finland, Estonia, Latvia, Lithuania, Belarus, Russia*
2039 Dec 15	**1m 51s**	**Total**	**Antarctica**
2041 Apr 30	**1m 51s**	**Total**	**South Atlantic, Angola, DR Congo, Uganda, Kenya, Somalia, Indian Ocean**
2041 Oct 24/25	*6m 07s*	*Annular*	*Mongolia, China, North Korea, Japan, Pacific Ocean, Kiribati*
2042 Apr 19/20	**4m 51s**	**Total**	**Indonesia, Malaysia, Philippines, Pacific Ocean**
2042 Oct 14	*7m 44s*	*Annular*	*Bay of Bengal, Thailand, Malaysia, Indonesia, Australia, New Zealand, Pacific Ocean*
2043 Apr 09	**—**	**Total**	**Northeastern Russia/Siberia (shadow cone only grazes Earth)**
2043 Oct 03	*—*	*Annular*	*South Indian Ocean (shadow cone only grazes Earth)*
2044 Feb 28	*2m 27s*	*Annular*	*South Atlantic Ocean*
2044 Aug 23	**2m 04s**	**Total**	**Greenland, Canada, United States (Montana, North Dakota)**
2045 Feb 16/17	*7m 47s*	*Annular*	*New Zealand, Cook Islands, Pacific Ocean*
2045 Aug 12	**6m 06s**	**Total**	**Pacific Ocean, United States (California, Nevada, Utah, Colorado, Kansas, Oklahoma, Texas, Arkansas, Louisiana, Mississippi, Alabama, Georgia, Florida), Bahamas, Haiti, Dominican Republic, Trinidad and Tobago, Venezuela, Guyana, Suriname, French Guiana, Brazil, Atlantic Ocean**
2046 Feb 05/06	*9m 42s*	*Annular*	*Indonesia, Papua New Guinea, Pacific Ocean, Howland and Baker Islands, United States (Hawaii, California, Oregon, Nevada, Idaho)*
2046 Aug 02	**4m 51s**	**Total**	**Brazil, Atlantic Ocean, Angola, Namibia, Botswana, South Africa, Swaziland, Indian Ocean, Kerguelen**

Date	Maximum Duration[2]	Eclipse Type	Geographic Region of Visibility
2048 Jun 11	*4m 58s*	*Annular*	*United States (Colorado, Nebraska, Kansas, Iowa, Missouri, Minnesota, Wisconsin, Illinois, Michigan), Canada, Greenland, Iceland, Atlantic Ocean, Norway, Sweden, Estonia, Latvia, Lithuania, Belarus, Ukraine, Russia, Turkmenistan, Uzbekistan, Afghanistan*
2048 Dec 05	**3m 28s**	**Total**	**South Pacific Ocean, Chile, Argentina, Atlantic Ocean, Tristan da Cunha, Namibia, Botswana**
2049 May 31	*4m 45s*	*Annular*	*Pacific Ocean, Peru, Ecuador, Colombia, Brazil, Venezuela, Guyana, Suriname, Atlantic Ocean, Cape Verde, Senegal, The Gambia, Mali, Guinea, Burkina Faso, Ivory Coast, Ghana, Togo, Benin, Nigeria, Cameroon, Congo, DR Congo*
2049 Nov 25	***0m 38s***	***Hybrid***	***Total for part of Indian Ocean and parts of Indonesia; annular for Saudi Arabia, Yemen, part of Indian Ocean, eastern Indonesia, Pacific Ocean***
2050 May 20/21	***0m 21s***	***Hybrid***	***South Pacific Ocean***
2052 Mar 30	**04m 08s**	**Total**	**Pacific Ocean, Kiribati, Mexico, United States (Texas, Louisiana, Alabama, Georgia, Florida, South Carolina), Atlantic Ocean**
2052 Sep 22/23	*2m 51s*	*Annular*	*Indonesia, East Timor, Australia, Pacific Ocean*
2053 Mar 20	*0m 49s*	*Annular*	*Indian Ocean, Indonesia, Papua New Guinea*
2053 Sep 12	**3m 04s**	**Total**	**Atlantic Ocean, Spain, Morocco, Algeria, Tunisia, Libya, Egypt, Saudi Arabia, Yemen, Indian Ocean, Indonesia (Sumatra)**
2055 Jul 24	**3m 17s**	**Total**	**South Atlantic Ocean, South Africa, south Indian Ocean**
2056 Jan 16/17	*2m 52s*	*Annular*	*Pacific Ocean, Mexico, United States (Texas)*
2056 Jul 12	*1m 26s*	*Annular*	*Pacific Ocean, Colombia, Ecuador, Peru, Brazil*

Date	Maximum Duration[2]	Eclipse Type	Geographic Region of Visibility
2057 Jan 05	**2m 29s**	**Total**	**South Atlantic Ocean, south Indian Ocean (no land)**
2057 Jul 01/02	*4m 22s*	*Annular*	*China, Mongolia, Russia (Siberia), Canada, United States (Alaska, Minnesota, Michigan)*
2057 Dec 26	**1m 50s**	**Total**	**Antarctica**
2059 May 11	**2m 23s**	**Total**	**Pacific Ocean, Ecuador, Peru, Colombia, Brazil**
2059 Nov 05	*7m 00s*	*Annular*	*Atlantic Ocean, France, Spain, Mediterranean Sea, Sardinia, Sicily, Libya, Egypt, Sudan, Eritrea, Ethiopia, Somalia, Indian Ocean, Maldives, Indonesia (Sumatra)*
2060 Apr 30	**5m 15s**	**Total**	**Brazil, Atlantic Ocean, Ivory Coast, Ghana, Togo, Benin, Burkina Faso, Niger, Nigeria, Chad, Libya, Egypt, Cyprus, Turkey, Syria, Armenia, Iran, Azerbaijan, Turkmenistan, Kazakhstan, Uzbekistan, Turkmenistan, Kyrgyzstan, China**
2060 Oct 24	*8m 06s*	*Annular*	*Atlantic Ocean, Guinea, Sierra Leone, Liberia, Ivory Coast, Angola, Namibia, Botswana, South Africa, Pacific Ocean*
2061 Apr 20	**02m 37s**	**Total**	**Ukraine, Russia, Kazakhstan, Arctic Ocean, Svalbard (Norway)**
2061 Oct 13	*03m 41s*	*Annular*	*Chile, Argentina, Falklands, Antarctica*
2063 Feb 28	*07m 41s*	*Annular*	*Indian Ocean, Indonesia (Sumatra), Malaysia, Singapore, Brunei, Philippines*
2063 Aug 24	**05m 49s**	**Total**	**China, Mongolia, Russia (Siberia), Japan, Pacific Ocean**
2064 Feb 17	*08m 56s*	*Annular*	*Congo, Angola, DR Congo, Zambia, Tanzania, Seychelles, Indian Ocean, India, Nepal, Bangladesh, Bhutan, China*
2064 Aug 12	**04m 28s**	**Total**	**Pacific Ocean, Chile, Argentina, Atlantic Ocean**

Date	Maximum Duration[2]	Eclipse Type	Geographic Region of Visibility
2066 Jun 22	*04m 40s*	*Annular*	*Russia (Siberia), United States (Alaska), Canada, Atlantic Ocean*
2066 Dec 17	**03m 14s**	**Total**	**Indian Ocean, Australia, New Zealand, South Pacific**
2067 Jun 11	*04m 05s*	*Annular*	*Pacific Ocean, Kiribati, Ecuador, Peru*
2067 Dec 06	***00m 08s***	***Hybrid***	***Guatemala. Belize, Honduras, Nicaragua, Colombia, Venezuela, Guyana, Brazil, Atlantic Ocean, Nigeria, Cameroon, Chad, Sudan***
2068 May 31	**01m 06s**	**Total**	**Indian Ocean, Australia, New Zealand**
2070 Apr 11	**04m 04s**	**Total**	**Sri Lanka, Bay of Bengal, Myanmar, Thailand, Cambodia, Laos, Vietnam, Pacific Ocean**
2070 Oct 04	*02m 44s*	*Annular*	*Atlantic Ocean, Angola, Zambia, Zimbabwe, Mozambique, Madagascar, Indian Ocean*

NOTES AND REFERENCES

1. Based on Fred Espenak: *Thousand Year Canon of Solar Eclipses: 1501–2500* (Portal, Arizona, Astropixels Publishing, 2014; http://astropixels.com/pubs/).
2. Maximum duration of totality or annularity as seen from the central line in minutes and seconds.

Appendix C

Recent Total, Annular, and Hybrid Eclipses: 1970–2016[1]

The table below gives the basic details for the 67 total, annular, and hybrid eclipses occurring during the 47-year period 1970–2016. When an eclipse path crosses the International Date Line, it can be seen on one of two dates depending on the observer's geographic position.

Date	Maximum Duration[2]	Eclipse Type	Geographic Region of Visibility
1970 Mar 07	**3m 28s**	**Total**	**Pacific Ocean, Mexico, United States (Florida, Georgia, South Carolina, North Carolina, Virginia), eastern Canada, north Atlantic Ocean**
1970 Aug 31	*6m 48s*	*Annular*	*Papua New Guinea, south Pacific Ocean*
1972 Jan 16	*1m 53s*	*Annular*	*Antarctica*
1972 Jul 10	**2m 36s**	**Total**	**Russia, United States (Alaska), Canada, Atlantic Ocean**
1973 Jan 04	*7m 48s*	*Annular*	*South Pacific Ocean, Chile, Argentina, south Atlantic Ocean*
1973 Jun 30	**7m 04s**	**Total**	**Guyana, Suriname, French Guiana, Atlantic Ocean, Mauritania, Mali, Algeria, Niger, Chad, Central African Republic, Sudan, Uganda, Kenya, Somalia, Indian Ocean**

Date	Maximum Duration[2]	Eclipse Type	Geographic Region of Visibility
1973 Dec 24	*12m 03s*	*Annular*	*Costa Rica, Panama, Colombia, Venezuela, Brazil, Guyana, Atlantic Ocean, Mauritania, Mali, Algeria*
1974 Jun 20	**5m 09s**	**Total**	**South Indian Ocean, southwestern Australia**
1976 Apr 29	*6m 41s*	*Annular*	*Atlantic Ocean, Senegal, Mauritania, Mali, Algeria, Tunisia, Libya, Turkey, Iran, Turkmenistan, Afghanistan, India, China*
1976 Oct 23	**4m 46s**	**Total**	**Tanzania, south Indian Ocean, southern Australia**
1977 Apr 18	*7m 04s*	*Annular*	*South Atlantic Ocean, Namibia, Angola, Zambia, Congo, Tanzania, Indian Ocean*
1977 Oct 12	**2m 37s**	**Total**	**Pacific Ocean, Colombia, Venezuela**
1979 Feb 26	**2m 49s**	**Total**	**United States (Washington, Oregon, Idaho, Montana, North Dakota), Canada, Greenland**
1979 Aug 22	*6m 03s*	*Annular*	*South Pacific Ocean, Antarctica*
1980 Feb 16	**4m 08s**	**Total**	**Atlantic Ocean, Angola, Congo, Tanzania, Kenya, India, Bangladesh, Burma, China**
1980 Aug 10	*3m 23s*	*Annular*	*Pacific Ocean, Peru, Bolivia, Paraguay, Brazil*
1981 Feb 04	*0m 33s*	*Annular*	*Southern Australia, south Pacific Ocean*
1981 Jul 31	**2m 02s**	**Total**	**Russia, Pacific Ocean**
1983 Jun 11	**5m 11s**	**Total**	**Indian Ocean, Indonesia, Papua New Guinea, Pacific Ocean**
1983 Dec 04	*4m 01s*	*Annular*	*Atlantic Ocean, Gabon, Congo, Uganda, Kenya, Ethiopia, Somalia*

Date	Maximum Duration[2]	Eclipse Type	Geographic Region of Visibility
1984 May 30	*0m 12s*	*Annular*	*Pacific Ocean, Mexico, United States (Louisiana, Mississippi, Alabama, Georgia, South Carolina, North Carolina, Virginia, Maryland), Atlantic Ocean, Morocco, Algeria*
1984 Nov 22/23	2m 00s	Total	Indonesia, Papua New Guinea, south Pacific Ocean
1985 Nov 12	1m 59s	Total	South Pacific Ocean
1986 Oct 03	*0m 00.2s*	*Hybrid*	*North Atlantic Ocean*
1987 Mar 29	*0m 08s*	*Hybrid*	*annular in Argentina; total for part of south Atlantic Ocean, Gabon, Cameroon, and part of Central African Republic; then annular in Sudan, Ethiopia, Somalia*
1987 Sep 23	*3m 49s*	*Annular*	*Kazakstan, China, Mongolia, Pacific Ocean*
1988 Mar 18	3m 47s	Total	Indonesia, Philippines, Pacific Ocean
1988 Sep 11	*6m 57s*	*Annular*	*South Pacific Ocean*
1990 Jan 26	*2m 03s*	*Annular*	*South Atlantic Ocean, Antarctica*
1990 Jul 22	2m 33s	Total	Finland, Russia, north Pacific Ocean
1991 Jan 15	*7m 53s*	*Annular*	*Southern Australia, New Zealand, south Pacific Ocean*
1991 Jul 11	6m 53s	Total	United States (Hawaii), Mexico, Guatemala, El Salvador, Honduras, Nicaragua, Costa Rica, Panama, Colombia, Brazil
1992 Jan 04	*11m 41s*	*Annular*	*Pacific Ocean, United States (California)*
1992 Jun 30	5m 21s	Total	Uruguay, south Atlantic Ocean
1994 May 10	*6m 14s*	*Annular*	*Mexico, United States (diagonally from Arizona through Missouri, New York, and Maine), south-eastern Canada, Atlantic Ocean, Morocco*

Date	Maximum Duration[2]	Eclipse Type	Geographic Region of Visibility
1994 Nov 03	**4m 23s**	**Total**	**Peru, Chile, Bolivia, Paraguay, Brazil, south Atlantic Ocean**
1995 Apr 29	*6m 37s*	*Annular*	*South Pacific Ocean, Peru, Ecuador, Colombia, Brazil*
1995 Oct 24	**2m 10s**	**Total**	**Iran, Afghanistan, Pakistan, India, Bangladesh, Burma, Cambodia, Vietnam, Indonesia, Pacific Ocean**
1997 Mar 09	**2m 51s**	**Total**	**Mongolia, Russia, China, Arctic Ocean**
1998 Feb 26	**4m 09s**	**Total**	**Pacific Ocean, Colombia, Panama, Venezuela, southern Caribbean, Atlantic Ocean**
1998 Aug 22	*3m 14s*	*Annular*	*Indonesia, Malaysia, south Pacific islands*
1999 Feb 16	*0m 40s*	*Annular*	*South Indian Ocean, Australia*
1999 Aug 11	**2m 23s**	**Total**	**Atlantic Ocean, England, France, Luxembourg, Germany, Austria, Hungary, Romania, Bulgaria, Turkey, Iraq, Iran, Pakistan, India**
2001 Jun 21	**4m 57s**	**Total**	**Atlantic Ocean, Angola, Zambia, Zimbabwe, Mozambique, Madagascar**
2001 Dec 14	*3m 53s*	*Annular*	*Pacific Ocean, Costa Rica, Nicaragua, Caribbean Sea*
2002 Jun 10/11	*0m 23s*	*Annular*	*Pacific Ocean, touching the islands of Sangihe, Talaud, Rota, and Tinian*
2002 Dec 04	**2m 04s**	**Total**	**Angola, Zambia, Botswana, Zimbabwe, South Africa, Mozambique, south Indian Ocean, southern Australia**
2003 May 31	*3m 37s*	*Annular*	*Scotland, Iceland, Greenland (the Moon's shadow passes over the North Pole of the sunward-inclined northern hemisphere, so the path of the eclipse in this unusual case moves east to west)*

Date	Maximum Duration[2]	Eclipse Type	Geographic Region of Visibility
2003 Nov 23	**1m 57s**	**Total**	**Antarctica**
2005 Apr 08	*0m 42s*	*Hybrid*	*Starts annular in western Pacific Ocean, becomes total in eastern Pacific, then becomes annular again for Costa Rica, Panama, Colombia, and Venezuela*
2005 Oct 03	*4m 31s*	*Annular*	*Atlantic Ocean, Portugal, Spain, Algeria, Tunisia, Libya, Chad, Sudan, Ethiopia, Kenya, Somalia, Indian Ocean*
2006 Mar 29	**4m 07s**	**Total**	**Eastern Brazil, Atlantic Ocean, Ghana, Togo, Benin, Nigeria, Niger, Chad, Libya, Egypt, Turkey, Russia**
2006 Sep 22	*7m 09s*	*Annular*	*Guyana, Suriname, French Guiana, Brazil, south Atlantic Ocean*
2008 Feb 07	*2m 12s*	*Annular*	*South Pacific Ocean, Antarctica*
2008 Aug 01	**2m 27s**	**Total**	**Northern Canada, Greenland, Arctic Ocean, Russia, Mongolia, China**
2009 Jan 26	*7m 54s*	*Annular*	*South Atlantic Ocean, Indian Ocean, Indonesia (Sumatra, Java, Borneo)*
2009 Jul 22	**6m 39s**	**Total**	**India, Nepal, Bhutan, Myanmar, China, Pacific Ocean**
2010 Jan 15	*11m 08s*	*Annular*	*Central Africa Republic, Congo, Uganda, Kenya, Somalia, Indian Ocean, Maldives, southernmost India, Sri Lanka, Bangladesh, Myanmar, China*
2010 Jul 10/11	**5m 20s**	**Total**	**South Pacific Ocean, Easter Island, Chile, Argentina**
2012 May 20/21	*5m 46s*	*Annular*	*China, Taiwan, Japan, Pacific Ocean, United States (Oregon, California, Nevada, Utah, Arizona, Colorado, New Mexico, Texas)*
2012 Nov 13/14	**4m 02s**	**Total**	**Northern Australia, south Pacific Ocean**

Date	Maximum Duration[2]	Eclipse Type	Geographic Region of Visibility
2013 May 9/10	*6m 03s*	*Annular*	*Australia, Solomon Islands, Nauru, Pacific Ocean*
2013 Nov 03	***1m 40s***	***Hybrid***	***Atlantic Ocean, Gabon, Congo, Democratic Republic of the Congo, Uganda, Kenya, Ethiopia (annular only at beginning of path)***
2014 Apr 29	—	*Annular*	*Antarctica (extension of shadow cone only grazes Earth)*
2015 Mar 20	**2m 47s**	**Total**	**North Atlantic Ocean, Faroe Islands (Denmark), Arctic Ocean, Svalbard (Norway)**
2016 Mar 09	**4m 09s**	**Total**	**Indonesia (Sumatra, Borneo, Sulawesi, Halmahera), Pacific Ocean**
2016 Sep 01	*3m 06s*	*Annular*	*Atlantic Ocean, Gabon, Congo, Tanzania, Mozambique, Madagascar, Indian Ocean*

NOTES AND REFERENCES

1. Based on Fred Espenak: *ThousandYear Canon of Solar Eclipses: 1501–2500* (Portal, Arizona, Astropixels Publishing, 2014; http://astropixels.com/pubs/).
2. Maximum duration of totality or annularity as seen from the central line in minutes and seconds.

Appendix D

Total Eclipses in the United States: 1492–2100[1]

Listing based on current boundaries of the lower 48 states

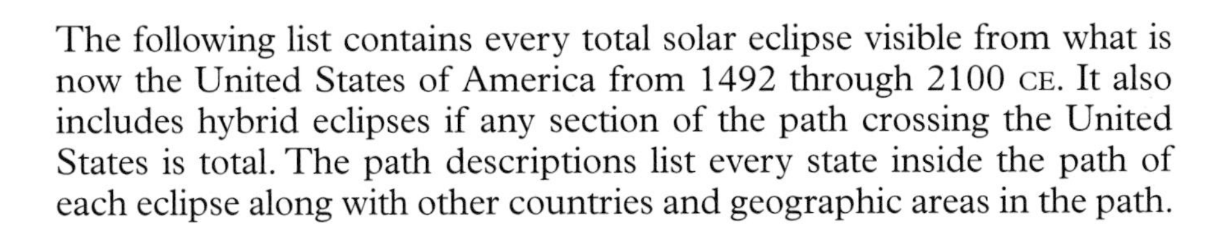

The following list contains every total solar eclipse visible from what is now the United States of America from 1492 through 2100 CE. It also includes hybrid eclipses if any section of the path crossing the United States is total. The path descriptions list every state inside the path of each eclipse along with other countries and geographic areas in the path.

Date	Maximum Duration	Locations in the Path of Totality
1496 August 8	5m 30s	Pacific Ocean, California, Mexico, Central America, Venezuela
1503 March 27	5m 04s	Pacific Ocean, Oregon, Idaho, Montana, North Dakota, South Dakota, Minnesota, Wisconsin, Michigan
1506 July 20	5m 08s	Texas, Arkansas, Louisiana, Mississippi, Tennessee, Georgia, North Carolina, Virginia, Maryland, Atlantic Ocean, Africa
1531 March 18	0m 21s	Pacific Ocean, Mexico, New Mexico to New Jersey (hybrid eclipse: total in New Mexico, Texas, and Oklahoma)
1557 April 28	5m 42s	Pacific Ocean, Hawaii, California, Nevada, Arizona, New Mexico, Texas, Arkansas, Louisiana, Mississippi, Alabama, Georgia, Florida

Date	Maximum Duration	Locations in the Path of Totality
1558 April 18	0m 50s	Kansas, Missouri, Illinois, Michigan, Canada, Greenland, Arctic Ocean
1562 February 3	0m 41s	Oregon, Washington, Idaho, Montana, Canada
1569 September 10	2m 55s	Arctic Ocean, Canada, New York, Vermont, New Hampshire, Maine, Nova Scotia, Atlantic Ocean
1576 April 28	0m 55s	Alaska, Arctic Ocean, Greenland
1578 September 1	3m 17s	Alaska (Aleutian Islands), Oregon, California, Nevada, Arizona, Mexico, Colombia, Brazil
1585 April 29	0m 03s	Pacific Ocean, California, Oregon, Idaho, Montana, North Dakota, Canada, Atlantic Ocean (hybrid: total in Idaho, Montana, and North Dakota)
1600 July 10	2m 08s	Florida, Atlantic Ocean, Portugal, Spain, Africa
1618 July 21	2m 13s	Pacific Ocean, Washington, Oregon, Idaho, Wyoming, Colorado, Kansas, Oklahoma, Texas, Louisiana, Cuba
1623 October 23	2m 31s	Siberia, Pacific Ocean, California, Arizona, New Mexico, Texas, Arkansas, Louisiana, Mississippi, Alabama, Georgia, South Carolina, North Carolina
1625 March 8	3m 50s	Pacific Ocean, Mexico, Florida, Atlantic Ocean
1632 October 13	1m 55s	Alaska, Pacific Ocean, California, Mexico, South America
1648 June 21	0m 49s	China, Siberia, Pacific Ocean, Alaska, Oregon, Nevada, Utah, New Mexico (hybrid eclipse: total in Oregon and Nevada)
1659 November 14	1m 56s	Canada, Maine, Atlantic Ocean, Africa
1670 April 19	3m 15s	Pacific Ocean, Hawaii, Washington, Canada, Greenland
1672 August 22	2m 15s	Alaska, Canada, New Hampshire, Maine, Atlantic Ocean

Date	Maximum Duration	Locations in the Path of Totality
1677 November 24	1m 36s	Pacific Ocean, California, Nevada, Utah
1679 April 10	4m 17s	Pacific Ocean, Hawaii, California, Utah, Arizona, Wyoming, Colorado, Nebraska, South Dakota, North Dakota, Minnesota, Canada, Atlantic Ocean, Ireland
1684 July 12	0m 24s	Mexico, Texas to Massachusetts, Atlantic Ocean, Portugal, Spain, Africa (hybrid eclipse: total in Massachusetts)
1713 December 17	0m 56s	Utah, New Mexico, Texas, Louisiana, Florida (Keys), Atlantic (hybrid eclipse: total in Texas and Florida [Keys])
1717 October 4	0m 56s	Alaska, Pacific Ocean, Canada, Montana, North Dakota, South Dakota, Iowa, Missouri, Illinois, Indiana, Kentucky, Tennessee, North Carolina, South Carolina, Atlantic Ocean (hybrid eclipse: total everywhere except Alaska)
1724 May 22	4m 33s	Pacific Ocean, California, Nevada, Arizona, Utah, Wyoming, Colorado, North Dakota, South Dakota, Minnesota, Canada, Ireland, England, France, Switzerland, Italy
1742 June 3	5m 00s	Southeast Asia, China, Japan, Pacific Ocean, Alaska, Oregon, Idaho, Nevada, Utah, Colorado, New Mexico
1752 May 13	5m 42s	Pacific Ocean, Mexico, Florida, Atlantic Ocean
1768 January 19	0m 13s	Pacific Ocean, Mexico, Texas to New York, Canada (hybrid eclipse: total in Mexico, Texas, Louisiana, and Mississippi)
1778 June 24	5m 52s	Pacific Ocean, Mexico, Texas, Louisiana, Mississippi, Alabama, Florida, Georgia, South Carolina, North Carolina, Virginia, Maryland, Atlantic Ocean, Africa
1780 October 27	2m 00s	Canada, Maine, Atlantic Ocean

Date	Maximum Duration	Locations in the Path of Totality
1803 February 21	4m 09s	Pacific Ocean, Mexico, Florida
1806 June 16	4m 55s	Pacific Ocean, Arizona, Colorado, Oklahoma, Texas, Kansas, Missouri, Iowa, Illinois, Indiana, Michigan, Ohio, Pennsylvania, New York, Connecticut, Vermont, New Hampshire, Rhode Island, Massachusetts, Atlantic Ocean, Africa
1825 December 9	1m 34s	Pacific Ocean, Mexico, Florida, Atlantic Ocean (hybrid eclipse: total in Mexico and Florida)
1834 November 30	2m 02s	Canada, Montana, Wyoming, Colorado, Nebraska, Kansas, Oklahoma, Arkansas, Mississippi, Alabama, Georgia, South Carolina, Atlantic Ocean
1860 July 18	3m 39s	Washington, Canada, Atlantic Ocean, Spain, Africa
1869 August 7	3m 48s	Russia, Pacific Ocean, Alaska, Canada, Montana, North Dakota, South Dakota, Minnesota, Nebraska, Iowa, Missouri, Illinois, Indiana, Kentucky, Tennessee, Virginia, North Carolina, South Carolina, Atlantic Ocean
1878 July 29	3m 11s	Russia, Alaska, Canada, Washington, Idaho, Montana, Wyoming, Colorado, Kansas, Oklahoma, New Mexico, Texas, Louisiana, Cuba
1880 January 11	2m 07s	Pacific Ocean, California, Nevada, Utah, Wyoming
1889 January 1	2m 17s	Alaska (Aleutians), Pacific Ocean, California, Nevada, Idaho, Wyoming, Montana, North Dakota, Canada
1900 May 28	2m 10s	Pacific Ocean, Mexico, Texas, Louisiana, Mississippi, Alabama, Georgia, South Carolina, North Carolina, Virginia, Atlantic Ocean, Portugal, Spain, Africa
1918 June 8	2m 23s	Pacific Ocean, Washington, Oregon, Idaho, Utah, Wyoming, Colorado, Kansas, Oklahoma, Arkansas, Louisiana, Mississippi, Alabama, Florida, Atlantic Ocean

Date	Maximum Duration	Locations in the Path of Totality
1923 September 10	3m 37s	Pacific Ocean, California, Mexico, Belize, Caribbean Sea
1925 January 24	2m 32s	Minnesota, Wisconsin, Michigan, Canada, Pennsylvania, New York, Connecticut, Rhode Island, Atlantic Ocean
1930 April 28	0m 01s	Pacific Ocean, California, Oregon, Nevada, Idaho, Montana, Canada, Atlantic Ocean (hybrid eclipse: total in California, Oregon, Nevada, and Idaho)
1932 August 31	1m 45s	Arctic Ocean, Canada, Vermont, New Hampshire, Maine, Massachusetts, Atlantic Ocean
1945 July 9	1m 15s	Idaho, Montana, Canada, Greenland, Atlantic Ocean, Norway, Sweden, Finland, Russia, Kazakhstan
1954 June 30	2m 35s	Nebraska, South Dakota, Iowa, Minnesota, Canada, Atlantic Ocean, Greenland, Norway, Sweden, Lithuania, Belarus, Ukraine, Russia, Iran, Afghanistan, Pakistan, India
1959 October 2	3m 02s	New Hampshire, Massachusetts, Atlantic Ocean, Africa
1963 July 20	1m 40s	Pacific Ocean, Alaska, Canada, Maine, Atlantic Ocean
1970 March 7	3m 28s	Pacific Ocean, Mexico, Florida, Georgia, South Carolina, North Carolina, Virginia, Massachusetts, Canada, Atlantic Ocean
1979 February 26	2m 49s	Pacific Ocean, Washington, Oregon, Idaho, Montana, North Dakota, Canada, Greenland
1991 July 11	6m 53s	Pacific Ocean, Hawaii, Mexico, Central America, South America
2017 August 21	2m 40s	Pacific Ocean, Oregon, Idaho, Wyoming, Nebraska, Kansas, Missouri, Illinois, Kentucky, Tennessee, Georgia, North Carolina, South Carolina, Atlantic Ocean

Date	Maximum Duration	Locations in the Path of Totality
2024 April 8	4m 28s	Pacific Ocean, Mexico, Texas, Oklahoma, Arkansas, Missouri, Illinois, Indiana, Kentucky, Ohio, Pennsylvania, New York, Vermont, New Hampshire, Maine, Canada, Atlantic Ocean
2044 August 23	2m 04s	Greenland, Canada, Montana, North Dakota
2045 August 12	6m 06s	Pacific Ocean, California, Nevada, Utah, Colorado, Kansas, Oklahoma, Texas, Arkansas, Mississippi, Louisiana, Alabama, Georgia, Florida, Caribbean, South America
2052 March 30	4m 08s	Pacific Ocean, Mexico, Texas, Louisiana, Alabama, Georgia, Florida, South Carolina, Atlantic Ocean
2078 May 11	5m 40s	Pacific Ocean, Mexico, Texas, Louisiana, Alabama, Georgia, Florida, South Carolina, North Carolina, Virginia, Atlantic Ocean
2079 May 1	2m 55s	Pennsylvania, New Jersey, New York, Connecticut, Rhode Island, Vermont, New Hampshire, Maine, Canada, Greenland, Arctic Ocean
2099 September 14	5m 18s	Pacific Ocean, Montana, North Dakota, Minnesota, Wisconsin, Michigan, Indiana, Ohio, West Virginia, Pennsylvania, Virginia, North Carolina, Atlantic Ocean

NOTE AND REFERENCE

1. Based on Fred Espenak: "Total Eclipses in the USA: 1001–3000," http://www.eclipsewise.com/solar/SEcountry/SEinUSA-T.html

Glossary

annular eclipse A central eclipse of the Sun in which the angular diameter of the Moon is too small to completely cover the disk of the Sun and a thin ring (*annulus*) of the Sun's bright apparent surface surrounds the dark disk of the Moon. Thus, an annular eclipse is actually a special kind of partial eclipse of the Sun. There are more annular eclipses than total eclipses.

annular-total eclipse A solar eclipse that begins as an annular eclipse, changes to a total eclipse along its path, and then returns to annular before the end of the eclipse path. It is also called a **hybrid eclipse**.

anomalistic month The time it takes (27.55 days) for the Moon to orbit the Earth as measured from its closest point to Earth (perigee) to its farthest point (apogee) and back to its closest point again.

anomalistic year The time it takes (365.26 days) for the Earth to orbit the Sun as measured from its closest point to the Sun (perihelion) to its farthest point (aphelion) and back to its closest point again.

antumbra The extension of the Moon's shadow cone so that it forms a mirror image of itself. A region experiencing an annular eclipse lies in the antumbra of the Moon.

aphelion The point for any object orbiting the Sun where it is farthest from the Sun.

apogee The point for any object orbiting the Earth where it is farthest from the Earth.

arc minute An angular measurement: 1 minute of arc is 60 seconds of arc and 1/60 of a degree of arc.

arc second An angular measurement: 1 second of arc is 1/60 of a minute of arc and 1/3600 of a degree of arc.

ascending node (of the Moon) The point on the Moon's orbit where it crosses the ecliptic (orbit of the Earth) going north.

Baily's beads An effect seen just before and after the total phase of a solar eclipse in which the Moon hides all the light from the Sun's disk except for a few bright points of sunlight passing through valleys at the rim of the Moon.

central eclipse (of the Sun) An eclipse in which the axis of the Moon's shadow touches the Earth. A central eclipse can be either total or annular.

chromosphere The reddish lower atmosphere of the Sun just above the photosphere. The chromosphere is only about 600 miles (1,000 kilometers) thick and the temperature is about 7,100°F (4,200°C).

contact (in a solar eclipse) Special numbered stages of a solar eclipse. For all eclipses (total, annular, and partial), **first contact** occurs when the leading edge of the Moon appears tangent to the western rim of the Sun, initiating the eclipse. In a *total eclipse*, **second contact** occurs when the Moon's leading edge appears tangent to the eastern rim of the Sun, initiating the total phase of the eclipse. In a *total eclipse*, **third contact** occurs when the Moon's trailing edge appears tangent to the western rim of the Sun, concluding the total phase of the eclipse. In an *annular eclipse*, **second contact** occurs when the Moon's trailing edge appears tangent to the Sun's western rim, initiating the ring of sunlight completely around the Moon. In an *annular eclipse*, **third contact** occurs when the Moon's leading edge appears tangent to the Sun's eastern rim, ending the ring of sunlight. **Fourth contact** (for all eclipses) occurs when the trailing edge of the Moon appears tangent to the eastern rim of the Sun, concluding the eclipse. Note that a partial eclipse has only first and fourth contacts (no second or third contacts). These same four contact points are used to describe lunar eclipses, planet transits across the face of the Sun, satellite transits across the face of a planet, and binary star transits across the face of one another.

core (of the Sun) The central regions of the Sun where it produces its energy in a nuclear fusion reaction by converting hydrogen into helium at a temperature of 27 million °F (15 million °C).

corona The rarefied upper atmosphere of the Sun that appears as a white halo around the totally eclipsed Sun. Speeds of atomic particles in the corona give it a temperature sometimes exceeding 2 million °F (1.1 million °C).

coronagraph A special telescope that produces an artificial solar eclipse by masking the Sun's apparent surface with an opaque disk; invented by Bernard Lyot in 1930.

coronal hole A region of the corona low in brightness and density. It is from coronal holes that solar particles escape most easily into space to become the solar wind.

coronal mass ejections Vast bubbles of gas ejected from the corona into space. The bubbles can expand to a size larger than the Sun. Coronal mass ejections seem to be the principal cause of aurorae and a significant hazard to electric power grids and spacecraft electronics when these high-velocity particles and ropes of magnetic fields hit the Earth.

degree of obscuration The fraction of the area of the Sun's disk obscured by the Moon at eclipse maximum, usually expressed as a percentage. (Degree of obscuration is not the same as the magnitude of an eclipse, which is the fraction of the Sun's diameter that is covered).

descending node (of the Moon) The point on the Moon's orbit where it crosses the ecliptic (orbit of the Earth) going south.

diamond ring effect The stage of a solar eclipse when only a tiny sliver of the Sun's photosphere shines along the edge of the Moon as the corona appears. The diamond ring effect occurs in the seconds before and after totality.

draconic month The time it takes (27.21 days) for the Moon to orbit the Earth as measured from ascending node through descending node and back to ascending node again. Eclipses of the Sun and Moon can only take place near a node.

eclipse limit (for the Sun) The maximum angular distance that the Sun can be from a node of the Moon and still be involved in an eclipse seen from Earth. For partial eclipses, the limit ranges from 15°21′ to 18°31′ according to the varying angular sizes of the Moon and Sun due to the elliptical orbits of the Moon and Earth. For a central eclipse, the maximum and minimum limits are 11°50′ and 9°55′.

eclipse magnitude The fraction of the Sun's *diameter* occulted by the Moon. The *eclipse magnitude* is less than 1.0 for *partial* and *annular* eclipses. The *eclipse magnitude* is equal to or greater than 1.0 for *total* eclipses.

eclipse obscuration The fraction of the Sun's *area* occulted by the Moon. The *eclipse obscuration* is less than 1.0 for *partial* and *annular eclipses*. The *eclipse obscuration* is equal to or greater than 1.0 for *total eclipses*.

eclipse season The period of time in which the apparent motion of the Sun places it close enough to a node of the Moon so that an eclipse is possible. The Sun crosses the ascending and descending nodes of the Moon in a period of 346.62 days, so eclipse seasons occur about 173.3 days apart. Depending on where the Sun is on the ecliptic (how fast the Earth is moving), a solar eclipse season may last from 31 to 37 days.

eclipse year The time (346.62 days) it takes for the apparent motion of the Sun to carry it from ascending node to the descending node and back to the ascending node of the Moon.

ecliptic The apparent annual path of the Sun around the star field as seen from the Earth as the Earth orbits the Sun in the course of a year. (Thus, the ecliptic is the plane of the Earth's orbit around the Sun.) The Sun's apparent path is called the ecliptic because all eclipses of the Sun and Moon occur on or very close to this track in the sky.

exeligmos An eclipse repetition cycle of 54 years 34 days, equal to 3 saros cycles and often called the triple saros. After one exeligmos cycle, a solar eclipse returns to almost the same longitude, but occurs about 600 miles (1,000 kilometers) north or south of its predecessor.

filament A dark threadlike feature seen against the face of the Sun. A filament is a **prominence** seen from the top rather than at the edge of the Sun.

flare Intense brightening in the upper atmosphere of the Sun, which erupts vast amounts of charged particles into space. Flares can reach temperatures of 36 million °F (20 million °C).

greatest eclipse The instant and location on the Earth's surface in a solar eclipse when the axis of the Moon's shadow cone passes closest to the Earth's center. This geometry is important in the computation of eclipses and is used as a standard in comparing eclipses. At this point along the path of a total eclipse, totality should last longest—and it does, to within a second or two. *Greatest eclipse* doesn't quite correspond to the *greatest duration* of totality because the calculation of *greatest eclipse* is made with a round Moon, ignoring mountains and valleys at the Moon's limb and because it ignores that the Earth isn't quite spherical.

greatest duration The instant and location on the Earth's surface in a solar eclipse when the length of the total (or annular) phase of the eclipse is longest. *Greatest duration* does not occur at the point of *greatest eclipse* because of factors such as the relative motion of the Moon's shadow across the curvature of the Earth's surface. *Greatest duration* typically differs from *greatest eclipse* by a second or two in time and in geographic location by as much as 60 miles (100 kilometers) or so. *Greatest duration* can be further refined to include the effects of the Moon's irregular profile. This *corrected greatest duration* can differ from *greatest duration* by a couple of seconds.

hybrid eclipse A solar eclipse that begins as an annular eclipse, changes to a total eclipse along its path, and then returns to annular before the end of the eclipse path. Also called an **annular-total eclipse**.

inex A period of 10,571.95 days (29 years less 20.1 days) after which another eclipse of the Sun or Moon will occur (although not of the same type, such as total). This period equals 358 synodic months and 388.5 draconic months.

lunation The time it takes (29.53 days) for the Moon to complete a phasing cycle (also called a synodic period).

magnitude (of a solar eclipse) The fraction of the apparent diameter of the solar disk covered by the Moon at eclipse maximum. Eclipse magnitude is usually expressed as a decimal fraction: below 1.000 is a partial eclipse; 1.000 or above is a total eclipse. (The magnitude of a solar eclipse is not the same as the degree of obscuration, which is the percentage of the area of the Sun's disk that is covered.)

mid-eclipse The instant in a central solar eclipse halfway between second and third contacts.

New Moon The phase of the Moon when it is most nearly in conjunction with the Sun (also called dark of the Moon). Solar eclipses can occur only at New Moon. (In ancient times, New Moon had a different meaning: the crescent Moon when it became visible after dark of the Moon.)

nodes The two points at which the orbit of a celestial body crosses a reference plane. The Moon crosses the orbital plane of the Earth (the ecliptic) going northward at the ascending node and going southward at the descending node.

partial eclipse (of the Sun) An eclipse in which a portion of the Sun's disk is not covered by the Moon.

penumbra The portion of a shadow from which only part of the light source is occulted by an opaque body. Seen from outside the shadow region, the penumbra is a fuzzy fringe to the dark umbra that declines in darkness outward from the umbra. A region experiencing a partial solar eclipse lies in the penumbra of the Moon.

perigee The point for any object orbiting the Earth where it is closest to the Earth.

perihelion The point for any object orbiting the Sun where it is closest to the Sun.

photosphere The apparent "surface" of the Sun. It is actually a layer of hot gases only about 300 miles (500 kilometers) thick where the Sun's atmosphere changes from opaque to transparent, and visible light escapes from the Sun. The temperature of the photosphere is about 10,000°F (5,500°C).

prominence An arch or filament of denser gas in the Sun's corona, shaped by the magnetic field of the Sun. Some prominences rise but most are descending, as if raining.

regression of the nodes The westward shift of the Moon's nodes along the ecliptic due to tidal forces on the Moon's orbit exerted by the Sun and Earth. The regression of the nodes is responsible for the eclipse year being 18.62 days shorter than the seasonal (tropical) year. The nodes complete a westward regression entirely around the ecliptic in 18.6 years.

saros An eclipse cycle of 6,585.32 days (18 years 11⅓ days or 18 years 10⅓ days if five leap years occur in the interval) in which an eclipse will occur that is very similar to the one that preceded it. The saros results from the near equivalence of 223 synodic months, 19 eclipse years, and 239 anomalistic months.

shadow bands Faint flickers or ripples of light sometimes seen on the ground or buildings shortly before or after the total phase of a solar eclipse. Shadow bands are caused by light from the thin crescent of the Sun passing through parcels of rising and falling air that have different densities and hence act as lenses to bend the light continuously in varying amounts.

solar constant The amount of power from the Sun falling on an average square meter of the Earth's surface (1.35 kilowatts).

solar wind A stream of changed particles (mostly protons, electrons, and helium nuclei) ejected from the Sun that flows by the Earth at 720,000 to 1.8 million miles per hour (320 to 800 kilometers per second). When enhanced by flares, the particles collide with molecules in the Earth's upper atmosphere so intensely that they cause the upper atmosphere to glow by fluorescence in displays of the aurora (the northern and southern lights).

spectroheliosc ope A solar spectroscope that blocks unwanted colors so that an observer can view the Sun in the light of one spectral line at a time; invented by Jules Janssen in 1868.

spectroscope A device (usually employing a prism or a diffraction grating) to spread out a beam of light into its component wavelengths for study. Spectroscopy can reveal the composition, temperature, radial velocity, rotation, magnetic fields, and other features of a light source.

spicule A jet-like spike of upward-moving gas in the chromosphere of the Sun. Viewed near the edge of the Sun, spicules resemble a forest. Each spicule lasts 10 minutes or so and ejects material into the corona at speeds of 12–19 miles per second (20–30 kilometers per second).

sunspot A darker area in the Sun's photosphere where magnetic fields are very strong. The temperatures of sunspots are 2,500 to 3,600°F (1,400 to 2,400°C) cooler than their surroundings, making them appear darker. Sunspots can last for a day up to several months.

sunspot cycle A period averaging 11.1 years in which the number of sunspots increases, decreases, and then begins to increase again.

synodic month The period of time (29.53 days) required for the Moon to orbit the Earth and catch up with the Sun again. Because the Moon's position with respect to the Sun determines the phase of the Moon, the synodic period is the time required for a complete set of phases by the Moon.

transition region The thin, irregular layer that separates the chromosphere from the corona. In this layer, the temperature rises suddenly from about 7,200°F (4,000°C) to about 1.8 million °F (1 million °C). The transition region is of variable thickness, sometimes no more than tens of miles.

tritos A period of 3,986.6295 days (11 years less 31 days) after which another eclipse will occur (although not the same type, such as total). This period, less accurate than the saros or inex, equals 135 synodic months, 146.5 draconic months, and roughly 144.5 anomalistic months.

total eclipse (of the Sun) An eclipse in which the angular size of the Moon is sufficient to totally cover the disk of the Sun. In a total eclipse, the umbral shadow of the Moon touches the surface of the Earth.

umbra The central, completely dark portion of a shadow from which all of the light source is occulted by an opaque body. A region experiencing a total solar eclipse lies in the umbra of the Moon.

Bibliography

Alexander, Hartley Burr. *Latin-American Mythology*, volume 11 of *The Mythology of All Races*. Boston: Marshall Jones, 1920.

Allen, David; and Carol Allen. *Eclipse*. Sydney; Boston: Allen & Unwin, 1987.

Ananikian, Mardiros. H. *Armenian Mythology*, volume 7 of *The Mythology of All Races* (Boston: Marshall Jones,1925),

Andrews, Tamra. *Legends of the Earth, Sea, and Sky: An Encyclopedia of Nature Myths*. Santa Barbara, California: ABC-CLIO, 1998.

Arago, François. *Popular Astronomy*. 2 volumes. Translated by W. H. Smyth and Robert Grant. London: Longman, Brown, Green, Longmans, and Roberts, 1858.

Ashbrook, Joseph. *The Astronomical Scrapbook*. Edited by Leif J. Robinson. Cambridge: Cambridge University Press; Cambridge, Massachusetts: Sky Publishing, 1984.

Associated Press (No author given). "Antarctica Concerns Grow as Tourism Numbers Rise." March 16, 2013 (online).

Aveni, Anthony F. *Skywatchers of Ancient Mexico*. Austin: University of Texas Press, 1980.

Aveni, Anthony. *Stairways to the Stars: Skywatching in Three Great Ancient Cultures*. New York: John Wiley & Sons, 1997.

Baily, Francis. "On a Remarkable Phenomenon that Occurs in Total and Annular Eclipses of the Sun." *Memoirs of the Royal Astronomical Society*. Volume 10, 1838, pages 1–40.

Baily, Francis. "Some Remarks on the Total Eclipse of the Sun, on July 8th, 1842." *Memoirs of the Royal Astronomical Society*. Volume 15, 1846, pages 1–8.

Bakich, Michael. "25 Facts You Should Know about the August 21, 2017 Total Solar Eclipse." Blog, August 6, 2014. Access: http://cs.astronomy.com/asy/b/astronomy/archive/2014/08/05/25-facts-you-should-know-about-the-august-21–2017-total-solar-eclipse.aspx

Bakich, Michael. "Two Dozen Tips for the August 21, 2017 Total Solar Eclipse." Blog, August 8, 2014. Access: http://cs.astronomy.com/asy/b/astronomy/archive/2014/08/08/25-tips-for-the-august-21– 2017-total-solar-eclipse.aspx

Berman, Bob. *The Sun's Heartbeat and Other Stories from the Life of the Star that Powers Our Planet*. New York: Little, Brown, 2011.
Bernardino de Sahagún. *Florentine Codex; General History of the Things of New Spain*. Book 7: *The Sun, Moon, and Stars, and the Binding of the Years*. Translated from the Aztec by Arthur J. O. Anderson and Charles E. Dibble. Santa Fe, New Mexico: School of American Research; and Salt Lake City: University of Utah, 1953).
Brewer, Bryan. *Eclipse*. Seattle: Earth View, 1978; 2nd edition 1991.
Bruce, Ian. *Eclipse: An Introduction to Total and Partial Eclipses of the Sun and Moon*. Harrogate, England: Take That, 1999.
Brunier, Serge; and Jean-Pierre Luminet. *Glorious Eclipses: Their Past, Present, and Future*. Translated by Storm Dunlop. Cambridge: Cambridge University Press, 2000.
Caesar, Ed. "Stonehenge: What Lies Beneath." *Smithsonian*. Volume 45, Number 5, September 2014, pages 30–41.
Calvin, William H. *How the Shaman Stole the Moon: In Search of Ancient Prophet-Scientists from Stonehenge to the Grand Canyon*. New York: Bantam Books, 1991.
Chambers, George F. *The Story of Eclipses*. Library of Valuable Knowledge. New York: D. Appleton, 1912.
Chang, Kenneth. "Life on Mars? Could Be, But How Will They Tell?" *New York Times*, March 29, 2005.
Clark, Ronald W. *Einstein, the Life and Times*. London: Hodder and Stoughton, 1973.
Clerke, Agnes M. *A Popular History of Astronomy during the Nineteenth Century*. 4th edition. London: A. and C. Black, 1902.
Codona, Johana L. "The Enigma of Shadow Bands," *Sky & Telescope*, volume 81 (May 1991), page 482.
Comte, Auguste. *The Essential Comte, Selected from Cours de philosophie positive*, translated by Margaret Clarke. London: Croom Helm, 1974.
Couderc, Paul. *Les éclipses*. (Que sais-je? series no. 940.) Paris: Presses Universitaires de France, 1961.
Covington, Michael. *Astrophotography for the Amateur*. Cambridge: Cambridge University Press, 1988.
Crommelin, Andrew C. D. "Results of the Total Solar Eclipse of May 29 and the Relativity Theory," *Nature*, volume 104, November 13, 1919, pages 280–281.
Crump, Thomas. *Solar Eclipse*. London: Constable, 1999.
De Ferrer, José Joaquín. "Observations of the Eclipse of the Sun, June 16th, 1806, Made at Kinderhook, in the State of New-York," *Transactions of the American Philosophical Society*, volume 6, 1809, pages 264–275.
Dillard, Annie. "Total Eclipse," in *An Annie Dillard Reader*. New York: Harper Perennial,1995.

Douglas, Allie Vibert. *The Life of Arthur Stanley Eddington*. London: T. Nelson, 1956.

Dyson, Frank W.; Andrew C. D. Crommelin; and Arthur S. Eddington. "Joint Eclipse Meeting of the Royal Society and the Royal Astronomical Society," *The Observatory*, volume 42, November 1919, pages 389–398.

Dyson, Frank W.; Arthur S. Eddington; and Charles R. Davidson. "A Determination of the Deflection of Light by the Sun's Gravitational Field, From Observations Made at the Total Eclipse of May 29, 1919," *Philosophical Transactions of the Royal Society of London*, series A, volume 220, 1920, pages 291–333.

Dyson, Frank; and Richard v.d. R. Woolley. *Eclipses of the Sun and Moon*. Oxford: At the Clarendon Press, 1937.

Espenak, Fred. *EclipseWise*. Website: http://eclipsewise.com/

Espenak, Fred. *Fifty Year Canon of Solar Eclipses: 1986–2035*. Washington, D.C.: NASA; Cambridge, Massachusetts: Sky Publishing, 1987. NASA Reference Publication 1178 Revised.

Espenak, Fred. *Road Atlas for the Total Solar Eclipse of 2017*. Portal, Arizona: Astropixels, 2015.

Espenak, Fred; and Jay Anderson. *Eclipse Bulletin: Total Solar Eclipse of 2017 August 21*. Portal, Arizona: Astropixels, 2015.

Espenak, Fred; and Jean Meeus. *Five Millennium Canon of Solar Eclipses: −1999 to +3000 (2000* BCE *to 3000* CE*)*. NASA Technical Publication 2006–214141. Greenbelt, Maryland: NASA Goddard Space Flight Center, 2006.

Farber, Jim. "Can't Get to Mars? Try Chile's Atacama Desert, the Driest Place on Earth and the Most Otherworldly." *New York Daily News*, January 16, 2012.

Ferguson, John C.: *Chinese Mythology*, volume 8 of *The Mythology of All Races*. Boston: Marshall Jones, 1928.

Fiala, Alan D.; James A. DeYoung; and Marie R. Lukac. *Solar Eclipses, 1991–2000*. (US Naval Observatory circular 170.) Washington, D.C.: US Naval Observatory, 1986.

Flammarion, Camille. *The Flammarion Book of Astronomy*. Edited by Gabrielle Camille Flammarion and André Danjon. Translated by Annabel and Bernard Pagel. New York: Simon and Schuster, 1964.

Fomalont, Edward B.; and Richard A. Sramek. "A Confirmation of Einstein's General Theory of Relativity by Measuring the Bending of Microwave Radiation in the Gravitational Field of the Sun," *Astrophysical Journal*, volume 199, August 1, 1975, pages 749–755.

Fowler, Alfred. "Sir Norman Lockyer, K. C. B., 1836–1920," *Proceedings of the Royal Society of London*, series A, volume 104, December 1, 1923, pages i–xiv.

Francillon, Gérard; and Patrick Menget, editors. *Soleil est mort: l'éclipse totale de soleil du 30 Juin 1973*. Nanterre: Laboratoire d'ethnologie et de sociologie comparative (Récherches thématiques, 1), 1979.

Frazer, James George. *Balder the Beautiful*, part 1, in *The Golden Bough*, volume 10. London: Macmillan, 1930,

Friedman, Herbert. *Sun and Earth*. New York: W. H. Freeman (Scientific American Library), 1986.

Guillermier, Pierre; and Serge Koutchmy. *Total Eclipses: Science, Observations, Myths and Legends*. Translated by Bob Mizon. Berlin: Springer-Verlag and Chichester, United Kingdom: Praxis, 1999.

Golub, Leon; and Jay M. Pasachoff. *The Solar Corona*. Cambridge: Cambridge University Press, 1997.

Hadingham, Evan. *Early Man and the Cosmos*. New York: Walker, 1984.

Harrington, Philip S. *Eclipse! The What, Where, When, Why, and How Guide to Watching Solar and Lunar Eclipses*. New York: John Wiley & Sons, 1997.

Harris, Joel; and Richard Talcott. *Chasing the Shadow*. Waukesha, Wisconsin: Kalmbach, 1994.

Held, Wolfgang. *Eclipses: 2005–2017*. Translated by Christian von Arnim. Edinburgh: Floris, 2005.

Herodotus. *The History*, volume 1, translated by George Rawlinson. Everyman's Library, volume 405. London: J.M. Dent, 1910.

Hetherington, Barry. *A Chronicle of Pre-Telescopic Astronomy*. Chichester, United Kingdom: John Wiley & Sons, 1996.

Hoffleit, Dorrit. *Some Firsts in Astronomical Photography*. Cambridge, Massachusetts: Harvard College Observatory, 1950.

Hoffmann, Banesh; with the collaboration of Helen Dukas. *Albert Einstein, Creator and Rebel*. New York: Viking Press, 1972.

Holloway, Richard P. *Solar Eclipse 1999*. London: Calculus International, 1999.

Ion, Victoria. *Indian Mythology*. New York: Peter Bedrick Books, 1984.

Johnson, Samuel J. *Eclipses, Past and Future; with General Hints for Observing the Heavens*. Oxford: J. Parker, 1874.

Joslin, Rebecca R. *Chasing Eclipses: The Total Solar Eclipses of 1905, 1914, 1925*. Boston: Walton Advertising and Printing, 1929.

Keith, Arthur Berriedale. *Indian Mythology*, volume 6 of *The Mythology of All Races*. Boston: Marshall Jones, 1917.

Kippenhahn, Rudolph. *Discovering the Secrets of the Sun*. New York: John Wiley & Sons, 1994.

Knappert, Jan. *Indian Mythology: An Encyclopedia of Myth and Legend*. London: HarperCollins, 1991.

Koestler, Arthur. *The Sleepwalkers*. New York: Grosset & Dunlap, 1963.

Krupp, E[dwin] C. *Beyond the Blue Horizon: Myths and Legends of the Sun, Moon, Stars, and Planets*. New York: HarperCollins, 1991.

Kudlek, Manfred; and Erich H. Mickler. *Solar and Lunar Eclipses of the Ancient Near East from 3000* BC *to 0 with Maps*. Neukirchen-Vluyn, Germany: Butzon & Bercker Kevelaer, 1971.

Lang, Kenneth R. *Sun, Earth, and Sky*. New York: Springer, 1995.

Lang, Kenneth R; and Owen Gingerich, editors. *A Source Book in Astronomy and Astrophysics, 1900–1975*. Cambridge, Massachusetts: Harvard University Press, 1979.

Le Bovier de Fontenelle, Bernard. *A Plurality of Worlds*, translated by John Glanvill. London: Nonesuch Press, 1929.

Legge, James (editor and translator). *The Chinese Classics*, volume 3, *The Shoo King* [*Shu Ching*], Hong Kong: Hong Kong University Press, 1960.

Levy, Dawn. "An Eclipsed Vacation," *Los Altos* [California] *Town Crier*, August 7, 1991, pages 22–23.

Lewis, Isabel M. *A Handbook of Solar Eclipses*. New York: Duffield, 1924.

Lewis, Isabel M. "The Maximum Duration of a Total Solar Eclipse." *Publications of the American Astronomical Society*. Volume 6, 1931, pages 265–266.

Little, Robert T. *Astrophotography: A Step-by-Step Approach*. New York: Macmillan, 1986.

Littmann, Mark; Fred Espenak, and Ken Willcox. *Totality: Eclipses of the Sun*. 3rd edition updated. New York: Oxford University Press, 2009.

Littmann, Mark; Fred Espenak, and Ken Willcox. *Totality: Eclipses of the Sun*. 3rd edition. New York: Oxford University Press, 2008.

Littmann, Mark; Ken Willcox, and Fred Espenak. *Totality: Eclipses of the Sun*. 2nd edition. New York: Oxford University Press, 1999.

Littmann, Mark; and Ken Willcox. *Totality: Eclipses of the Sun*. Honolulu: University of Hawaii Press, 1991.

Lockyer, William J. S. "The Total Eclipse of the Sun, April 1911, as Observed at Vavau, Tonga Islands," in Bernard Lovell, editor: *Astronomy*, volume 2, The Royal Institution Library of Science (Barking, Essex: Elsevier Publishing, 1970), pages 190–191.

Lovell, Bernard, editor. *Astronomy*. 2 volumes. The Royal Institution Library of Science. Barking, Essex; New York: Elsevier Publishing, 1970.

Lynn, William Thynne. *Remarkable Eclipses: A Sketch of the Most Interesting Circumstances Connected with the Observation of Solar and Lunar Eclipses, Both in Ancient and Modern Times*. London: Edward Stanford, 1896.

Marriott, Alice; and Carol K. Rachlin. *Plains Indian Mythology*. New York: Thomas Y. Crowell, 1975.

Marschall, Laurence A. "A Tale of Two Eclipses." *Sky & Telescope*. Volume 57, February 1979, pages 116–118.

Maunder, Michael. "Eclipse Chasing." (On eclipse photography.) Pages 139–157 in Patrick Moore, editor. *Yearbook of Astronomy*. New York: W. W. Norton, 1989.

Maunder, Michael; and Patrick Moore. *The Sun in Eclipse*. (Patrick Moore's Practical Astronomy Series). London: Springer-Verlag London, 1997.

McEvoy, J. P. *Eclipse: The Science and History of Nature's Most Spectacular Phenomenon*. London: Fourth Estate, 1999.

McPherson, Florence Andsager; as told to Julie Andsager. *Florence's Memories—as Remembered at Age 89–90*. Self-published family history, 1998.

Meadows, A. J. *Early Solar Physics*. Oxford: Pergamon Press, 1970.

Meadows, A. J. Science and Controversy: A Biography of Sir Norman Lockyer. London: Macmillan, 1972.

Meeus, Jean. *Astronomical Algorithms*. Richmond, Virginia: Willmann-Bell, 1991.

Meeus, Jean. *Elements of Solar Eclipses: 1951–2200*. Richmond, Virginia: Willmann-Bell, 1989.

Meeus, Jean. "The Frequency of Total and Annular Solar Eclipses at a Given Place," *Journal of the British Astronomical Association*, volume 92, April 1982, pages 124–126.

Meeus, Jean. *Mathematical Astronomy Morsels*. Richmond, Virginia: Willmann-Bell, 1997.

Meeus, Jean. "The Maximum Possible Duration of a Total Solar Eclipse," *Journal of the British Astronomical Association*, volume 113, number 6, 2003; pages 343–348

Meeus, Jean; Carl C. Grosjean; and Willy Vanderleen. *Canon of Solar Eclipses*. (Solar eclipses from 1898 to 2510 AD) Oxford: Pergamon Press, 1966.

Menzel, Donald H.; and Jay M. Pasachoff. *A Field Guide to the Stars and Planets*. 2nd edition. Boston: Houghton Mifflin, 1983.

Mitchell, Samuel A. *Eclipses of the Sun*. 5th edition. New York: Columbia University Press, 1951.

Mobberley, Martin. *Total Solar Eclipses and How to Observe Them*. Astronomers' Observing Guides. New York: Springer-Verlag New York, 2007.

Montelle, Clemency. *Chasing Shadows: Mathematics, Astronomy, and the Early History of Eclipse Reckoning*. Baltimore: Johns Hopkins University Press, 2011.

Moskvitch, Katia. "Rover Explores Chile Desert to Aid Mars Life Hunt." Space.com, June 24, 2013.

Mucke, Hermann; and Jean Meeus. *Canon of Solar Eclipses: −2003 to +2526.* Vienna: Astronomisches Büro, 1983.

Nakayama, Shigeru. *A History of Japanese Astronomy—Chinese Background and Western Impact.* Cambridge, Massachusetts: Harvard University Press, 1969.

Needham, Joseph; and Wang Ling. *Science and Civilisation in China.* Volume 3: *Mathematics and the Sciences of the Heavens and the Earth.* Cambridge: At the University Press, 1959.

Needham, Joseph; and Wang Ling. Colin A. Ronan, editor. *The Shorter Science and Civilisation in China.* Volume 2. Cambridge: Cambridge University Press, 1981.

Neugebauer, Otto. *The Exact Sciences in Antiquity.* 2nd edition. Providence: Brown University Press, 1957.

Newton, Robert R. *Ancient Astronomical Observations and the Accelerations of the Earth and Moon.* Baltimore: Johns Hopkins Press, 1970.

O'Baugh, Stephen. "Eclipse Had Top Rating," *The Age* (Melbourne, Australia), October 24, 1976.

Olcott, William Tyler. *Sun Lore of All Ages.* New York: G. P. Putnam's Sons, 1914.

Oppolzer, Theodor von. *Canon of Eclipses.* (Solar and lunar eclipses from 1207 BC to 2161 AD) Translated by Owen Gingerich. New York: Dover, 1962.

Osterbrock, Donald E.; John R. Gustafson; and W. J. Shiloh Unruh. *Eye on the Sky: Lick Observatory's First Century.* Berkeley: University of California Press, 1988.

Ottewell, Guy. *Astronomical Calendar.* Greenville, North Carolina: Universal Workshop, annually.

Ottewell, Guy. *The Astronomical Companion.* 2nd edition. Greenville, North Carolina: Universal Workshops, 2010.

Ottewell, Guy. *The Under-Standing of Eclipses.* 3rd edition. Greenville, North Carolina: Universal Workshop, 2004.

Pang, Alex SooJung-Kim. *Empire and the Sun: Victorian Solar Eclipse Expeditions.* Stanford, California: Stanford University Press, 2002.

Pankenier, David W. *Astrology and Cosmology in Early China: Conforming Earth to Heaven.* Cambridge: Cambridge University Press, 2013.

Pannekoek, Anton. *A History of Astronomy.* London: G. Allen & Unwin, 1961.

Parker Pearson, Mike; and the Stonehenge Riverside Project. *Stonehenge—A New Understanding: Solving the Mysteries of the Greatest Stone Age Monument.* New York: The Experiment, 2014.

Parrinder, Geoffrey. *Africa Mythology.* New York: Peter Bedrick Books, 1986.

Pasachoff, J[ay] M. "Halley as an Eclipse Pioneer: His Maps and Observations of the Total Solar Eclipses of 1715 and 1724," *Journal of Astronomical History and Heritage*, volume 2, no. 1, 1999, pages 39–54.

Pasachoff, Jay M.; and Michael A. Covington. *The Cambridge Eclipse Photography Guide*. Cambridge: Cambridge University Press, 1993.

Pepin, R. O.; J. A. Eddy; and R. B. Merrill, editors. *The Ancient Sun: Fossil Record in the Earth, Moon and Meteorites*. Proceedings of the Conference on the Ancient Sun; Boulder, Colorado; October 16–19, 1979. New York: Pergamon Press, 1980.

Pitts, Michael W. "Stones, Pits and Stonehenge," *Nature*, volume 290, March 5, 1981, pages 46–47.

Plutarch. *The Rise and Fall of Athens: Nine Greek Lives*, translated by Ian Scott-Kilvert. Baltimore: Penguin Books,1960.

Pound, Robert V.; and Glen A. Rebka, Jr. "Resonant Absorption of the 14.4-kev Gamma Ray from 0.10-microsecond Fe^{57}," *Physical Review Letters*, volume 3, December 15, 1959, pages 554–556.

Rao, Joe. *Your Guide to the Great Solar Eclipse of 1991*. Cambridge, Massachusetts: Sky Publishing, 1989.

Reynolds, Michael D.; and Richard A. Sweetsir. *Observe: Eclipses*. Washington, D.C.: Astronomical League, 1995.

Rogers, Michael. "Totality—A Report," *Rolling Stone*, October 11, 1973. Reprinted in Gannon, Robert, editor: *Best Science Writing: Reading and Insights*. Phoenix, Arizona: Oryx Press, 1991.

Ruggles, Clive. *Ancient Astronomy: An Encyclopedia of Cosmologies and Myth*. Santa Barbara, California: ABC Clio, 2005.

Ruggles, Clive. *Astronomy in Prehistoric Britain and Ireland*. New Haven, Connecticut: Yale University Press, 1999.

Ruggles, Clive. "Stonehenge for the 1990s," *Nature*, volume 381, May 23, 1996, pages 278–279.

Russo, Kate. *Total Addiction: The Life of an Eclipse Chaser*. Heidelberg, Germany, Springer, 2012.

Sands, Charles P. *Chasing the Shadow: The Dynamics of Eclipses*. Frederick, Maryland: PublishAmerica, 2005.

Scheub, Harold. *A Dictionary of African Mythology: The Mythmaker as Storyteller*. New York: Oxford University Press, 2000.

Schwegman, John E. "A Prehistoric Cultural and Religious Center in Southern Illinois." http://www.southernmostillinoishistory.net/kincaid-mounds.html.

Sébillot, Paul Y. *Le folk-lore de France*. Volume 1: *Le ciel et la terre*. Paris: Librairie orientale & américaine, 1904.

Shapiro, Irwin I. "New Method for the Detection of Light Deflection by Solar Gravity," *Science*, volume 157, August 18, 1967, pages 806–807.

Silverman, Sam; and Gary Mullen. "Eclipses: A Literature of Misadventures." *Natural History*. Volume 81, June–July 1972, pages 48–51, 82.

Steel, Duncan. *Eclipse: The Celestial Phenomenon Which Has Changed the Course of History*. London: Headline, 1999, and Washington, D.C.: National Academies Press, 2001.

Stegemann, Viktor. "Finsternisse." In Hanns Bächtold-Stäubli, editor. *Handwörterbuch des Deutschen Aberglaubens*. Volume 2. Berlin: W. de Gruyter, 1930. Columns 1509–1526.

Stephenson, F. Richard. *Historical Eclipses and Earth's Rotation*. Cambridge: Cambridge University Press, 1997.

Stephenson, F. Richard; and David H. Clark. *Applications of Early Astronomical Records.* Monographs on Astronomical Subjects, 4. New York: Oxford University Press, 1978.

Thompson, J. Eric S. *A Commentary on the Dresden Codex: A Maya Hieroglyphic Book*. Philadelphia: American Philosophical Society, 1972.

Todd, Mabel Loomis. *Total Eclipses of the Sun*. Revised. Boston: Little, Brown, 1900.

Van den Bergh, George. *Periodicity and Variation of Solar (and Lunar) Eclipses*. Haarlem: H. D. Tjeenk Willink, 1955

Vaquero, J. M.; and M. Vázquez. *The Sun Recorded Through History*. Heidelberg, Germany: Springer, 2009.

Wentzel, Donat G. *The Restless Sun*. Washington, D.C.: Smithsonian Institution Press, 1989.

Werner, E. T. C. *A Dictionary of Chinese Mythology*. New York: Julian Press, 1961.

Williams, Sheridan. *2012 & 2013 Solar Eclipses with the Transit of Venus*. London: Bradt Travel Guides, 2012.

Zahn, Jean-Paul; and Magda Stavinschi, editors. *Advances in Solar Research at Eclipses from Ground and from Space*. Dordrecht, Netherlands: Kluwer, 1999.

Zirker, Jack B. *Total Eclipses of the Sun*. 2nd edition. Princeton: Princeton University Press, 1995.

Index

Note
References to the text are given in roman type.
References to figures are given in *italic* type.
References to tables are given in **bold** type.